Douglas B-18 Bolo

Douglas
B-18 Bolo

The Ultimate Look:

From Drawing Board to U-boat Hunter

William Wolf

Schiffer Military History
Atglen, PA

Book Design by Ian Robertson.

Library of Congress Control Number: 2006932417

Printed in China.
ISBN: 978-0-7643-2581-6

We are interested in hearing from authors with book ideas on related topics.

Published by Schiffer Publishing Ltd.
4880 Lower Valley Road
Atglen, PA 19310
Phone: (610) 593-1777
FAX: (610) 593-2002
E-mail: Info@schifferbooks.com.
Visit our web site at: www.schifferbooks.com
Please write for a free catalog.
This book may be purchased from the publisher.
Please include $3.95 postage.
Try your bookstore first.

In Europe, Schiffer books are distributed by:
Bushwood Books
6 Marksbury Avenue
Kew Gardens
Surrey TW9 4JF, England
Phone: 44 (0) 20 8392-8585
FAX: 44 (0) 20 8392-9876
E-mail: Info@bushwoodbooks.co.uk.
Visit our website at: www.bushwoodbooks.co.uk
Free postage in the UK. Europe: air mail at cost.
Try your bookstore first.

Contents

Acknowledgments

My lifelong hobby has been WWII aerial combat, and over the past 35 years I have collected over 15,000 books and magazines, along with hundreds of reels of microfilm on the subject. I probably have nearly every book written on WWII aviation, and a complete collection of every aviation magazine published since 1939. Also included in my collection are many hundreds of aviation unit and pilot's histories, crew manuals, and aircraft technical, structural, and maintenance manuals. My microfilm collection includes vintage intelligence reports, USAF, USN, and USMC group and squadron histories, complete Japanese Monograph series, and U.S. Strategic Bombing Surveys, as well as USAF Historical Studies. Over the years I have been fortunate to meet many fighter aces, other pilots, and fellow aviation buffs who have shared stories, material, and photographs with me (I have over 5,000 photos of fighter aces alone). I have made many multi-day expeditions to various military libraries, museums, and photo depositories with my copy machine and camera, accumulating literally reams of information and 1,000s of photographs. I also had a photo darkroom where I developed 1,000s of rare photos from microfilm negatives.

I have always intended to write a book on the B-18, and for many years I have collected material and photographs for this project. The author wishes that every person who contributed over the past quarter century could be specifically mentioned. Over the years the origin of many of the 1,000s of photos I have been lent to copy, or have copied and collected myself have become obscured. Most are from military and government sources, but many are from private individuals, and I apologize in advance if some of the photos are miscredited. Also, some of the photos are not of the best quality because of their age and sources, especially those copied from microfilm, but were used because of their importance to the book.

A particular thank you goes to Scott Marchand, curator of the Pima Air & Space Museum, and Kate DeMeester, archivist of that museum, who were of invaluable help in aiding in photographing the museum's B-18 and B-23 restorations, and gathering and copying the B-18 and B-23 aircraft and crew manuals and photograph collection. Through them I was able to make this book into the detailed, ultimate book on the B-18. A belated thank you goes to Judy Endicott of the Albert F. Simpson Historical Research Center, Maxwell AFB, Alabama. Ms. Endicott was of great help during my ten-day expedition to that facility in the mid-1980s to collect material on fighter aces and pilots (see my books *Victory Roll* and *USAAF Jabos in the MTO & ETO*, also by Schiffer) and the B-29 and B-32 (see my other "Ultimate Look" books on the Boeing B-29 and Consolidated B-32 by Schiffer). Thanks also go to the personnel at the Air Force Museum Archives at Wright-Patterson, Dayton, Ohio, and those at the Ferndale Photographic facility, Washington, DC, who aided me during my visits there.

Again, thanks go to my persevering wife, Nancy, who allows me to spend many hours researching and writing, and patiently (mostly) waits while I browse bookstores and visit air museums, in search of new material and photos. Also, I thank her because her car sits out in the hot Arizona sun as my WWII library luxuriates in the remodeled, air-conditioned three-car garage.

Foreword

The Douglas B-18 Bolo in our Pima Air & Space Museum collection often evokes puzzled looks and comments from visitors as they quickly walk by to move towards the more well known North American B-25 Mitchell, Consolidated B-24 Liberator, and the Boeing B-17 Flying Fortress. It is a shame that they don't linger, as our Bolo has been meticulously restored by devoted volunteers who spent thousands of hours to make it one of five remaining examples in the world.

Of all the American bombers of World War II, the B-18 Bolo probably is the least known and the most maligned. In its day—the Depression years of the mid-1930s—the bomber was welcomed by its crews as a modern innovative bomber. However, the B-18 design was nothing more than a DC-3 airliner configured as a bomber. The design left it little room for development when the superior four engine B-17s and B-24s appeared, while as a twin engine bomber contender the B-25s and B-26s would also prove superior. The B-18 was ordered and manufactured in relatively large numbers, and was the most numerous American bomber at the time of Pearl Harbor. By default it became a jack of all trades and a master of none, but finally found a niche as America's first anti-submarine aircraft, and did yeoman's duty in this undertaking, as Dr. Wolf describes.

Since I was a young boy I have been interested in the B-18 after listening to stories from my uncle about the B-18s that were surreptitiously moved across the US-Canadian border before America entered the war to enter the RCAF as anti-submarine bombers operating out of Nova Scotia. Bill Wolf has written a fitting tribute to the bomber that has been unsung, unjustly denigrated, and misunderstood. A book such as this has been long awaited. The author has taken on a very daunting task, as there has been virtually nothing written on the subject, but using his large personal collection of literature and microfilm in his library; he has produced a well-written and well-organized text that has, for the first time, presented anyone with an interest in the B-18 a concise and definitive reference. As with his other "Ultimate Look" books on the Boeing B-29 and the Consolidated B-32, this book presents a meticulously researched text and an outstanding photo collection—many published here for the first time—that pay fitting tribute to all associated with the Bolo. The author has performed a very necessary and worthwhile task in having presented the history of the B-18 so thoroughly and effectively before living memory is lost.

Scott Marchand, M.Phil.
Director of Collections & Aircraft Restoration
Pima Air & Space Museum

Preface

Of all the American bombers of World War II, the Douglas B-18 Bolo probably is the least known. Over the years the story of the airliner turned bomber has languished in obscurity, and the few articles on the subject in popular aviation magazines have emphasized its faults and maligned it as a budget bomber that had few virtues. In the Depression years of the mid-1930s an economy-minded Congress decided to fund the production of the twin engine B-18 over the more promising but more expensive B-17 prototype that had the misfortune to crash during testing. In 1937, when Air Corps crews received their speedy and comfortable B-18s, the Douglas bombers seemed to be cutting edge when compared to the Martin B-10, which itself had been the innovative bomber of the early 1930s.

The B-18 was built using contemporary but dead end design specifications, and it soon showed itself to be what it was, an "airliner-bomber" that allowed little latitude for future development to enable it to attack distant targets with 5,000 pound bomb loads or more at speeds and altitudes that would enable it to defend itself against more contemporary fighter aircraft. Thus, the bomber would not measure up to the matchless four engine B-17s and B-24s or the twin engine B-25 or B-26s, and its mass production delayed the development and production of these superior, more modern bombers which could have made a difference in the grim days after Pearl Harbor.

But at the time of Pearl Harbor the B-18 was the most numerous American bomber to be based overseas, and then proved itself to be useful as a stopgap bomber and jack of all trades until the more modern bombers whose development it interrupted could become operational. However, as a stopgap the Bolo did leave an important legacy. At a time when the B-17 and B-24 could not be spared it was available in numbers, and played a significant role in America's early anti-submarine operations, first off the American coast, and later in the Caribbean, where the U-Boats were ravaging merchantmen carrying vital oil and bauxite cargoes. For the first time this is the complete story of the Douglas B-18 Bolo: the good, the bad, and the ugly.

1

Donald Douglas, the DC Airliners, and the B-18

Introduction

Donald Douglas, along with contemporaries Glenn Martin, Bill Boeing, and Reuben Fleet, were creative and business geniuses who led the American aircraft industry in the 1920s and 1930s. By the onset of World War II Douglas and Martin would have tried and failed to enter the bomber field, while the Boeing Company (without the retired Bill Boeing) and Fleet's Consolidated Aircraft Company would both have great success with their B-17 (and later the B-29) and B-24, respectively. At the time of his entry into the bomber competition Douglas had established his company and the DC-1/2/3 series of aircraft as the standard in the airline and air transport field. The reason Douglas would succeed with his DC series and fail with the B-18, and then the B-23 bomber spin off will be described in tracing the story of Donald Douglas and the Douglas Company.

Donald Douglas: The Early Years

Donald Wills Douglas was born the second son of William, an assistant bank teller, and Dorothy Douglas in Brooklyn, New York, on 6 April 1892. As a youngster Douglas developed an interest in aviation, and avidly followed the careers of the Wright Brothers. On 30 July 1909 he witnessed Orville Wright flying the *Wright Flyer* during its acceptance trials by the U.S. Signal Corps at nearby Fort Myer, and this experience made a lasting impression on Douglas that shaped his future. Douglas graduated from Trinity Prep School that year, and followed his brother Harold by attending the U.S. Naval Academy. After three years Douglas confidently resigned from the Academy and enrolled at the prestigious Massachusetts Institute of Technology, feeling that it was a better opportunity for a future in aviation. He majored in mechanical engineering, and graduated from the four year course in just two years. After graduation he was appointed to the prestigious position as the Assistant Aeronautical Engineer to Cmdr. Jerome Hunsaker in the M.I.T. Department of Naval Architecture. In this post he was to assist in the design of the first advanced American wind tunnel, and also helped to set up the school's first courses in aerodynamics and aeronautical engineering; all for the salary of $500 per year.

After teaching and working with Hunsaker, Douglas decided that he preferred a career in aeronautical design and production. In early 1915, Hunsaker recommended Douglas for a consulting position with the Connecticut Aircraft Company of New Haven that had been contracted to build the D-1, the Navy's first non-rigid dirigible. After a short time the dirigible was completed, and Douglas was ready to move onto developing heaver-than-air machines. At the time Hunsaker received a letter from Glenn L. Martin, whose company was building Army training aircraft in Los Angeles. Martin asked for a recommendation for an aeronautical engineer to become the company's chief engineer, and Hunsaker tendered Douglas' name. During their first meeting in a L.A. hotel lobby, Martin at first completely ignored the very youthful looking 23-year old when he was approached, but after trying again Douglas reasserted that he truly was Donald W. Douglas, Martin's new chief engineer. In those pioneering days detailed drawings and stress analysis were unknown, but soon Douglas convinced Martin that he could save the company time and money, and proved his point by exceeding performance specs in his first design, the Model S seaplane manufactured for Holland. During his Martin sojourn Douglas met visiting Indianan Charlotte Ogg during a blind date that was so successful that the couple soon married. Working for Martin, Douglas was confined to the West Coast, and he felt isolated from the East Coast, which he considered was more of a hub of aeronautical knowledge and activity, particularly with the Great War in Europe escalating.

In November 1916 Douglas resigned from Martin, and left for a position as chief civilian engineer with the Army Signal Corps, Aviation Section. His first assignment was to tour all eastern aircraft factories to determine the status of American aviation. At the time there were many British and French aircraft types available in America, and Douglas gathered the latest technology in aircraft design and fabrication from these examples. Douglas and the Army's first aeronautical engineer, the virtuoso Grover Loening, worked hard to advance the status of the Aviation Section, and stressed the importance of airpower to the infantry-minded Army generals. At the time both the Navy, and more so the Army, lacked personnel and aircraft. Once the $649 million Congressional appropriation

Donald Douglas' first job was with Glenn L. Martin on the West Coast. After resigning, he rejoined Martin in Cleveland, where his first assignment was to design the MB-1 biplane, which was to be the largest American bomber ever built. Here he (left) and Martin (center) examine the fuselage of the MB-1. (Douglas)

was approved, both Loening and Douglas understood that the only way that aircraft could be produced in huge numbers (20,000 combat aircraft and 9,000 trainers) was to build them on an assembly line much like the fledgling automobile industry was doing at the time. Douglas' primary responsibility during this time was to redesign British aircraft, but all of his designs were rejected or ran into red tape. Douglas became disgruntled and resigned his position when automobile industry executives, knowing nothing about aircraft, were imported to administer the new aircraft production program.

Douglas' next stop was to join Glenn Martin again after the company relocated in Cleveland. His first assignment was to design the MB-1 biplane, which was to be the largest American bomber ever built. It was crewed by three men, measured 46 feet, 10 inches long, had a 71 foot, 5 inch wingspan, was 14 feet, 7 inches high, and weighed 10,225 pounds. It was powered by two, 12-cylinder, 400hp Liberty engines that drove the bomber to a top speed of 118mph with a full 1,500 pound bomb load. The giant bomber first flew on 17 August 1918, and was delivered to the Army just before the Armistice in November. On 20 July 1921 these bombers were employed by Gen. Billy Mitchell in his famous demonstration of airpower, when the heavily armored German battleship *Ostrfieisland* was bombed and sunk off the Virginia Capes. After the war there was a glut of aircraft, as the Army was selling its surplus aircraft at deep discounts, and Martin decided not to pursue the further development of the MB-1, despite the fact that the versatile aircraft could be converted to a commercial transport version carrying 12 passengers and two crewmen over a distance of 600 miles.

Douglas Forms an Aircraft Company

The Ohio winter of 1919-20 was so harsh that Douglas sent his wife and two babies to sunny California. Douglas stayed behind as Martin paid him the, then, princely salary of $10,000 per year, and assured him of job security in the weak post war aviation job market. However, Douglas wanted to start his own company, and on a visit to his family in California he made the momentous decision to resign from Martin in March 1920. Douglas made the rounds of the banking community to borrow start up money, but with so many failures of aeronautical companies he was turned down. He was introduced to David Davis, a millionaire sportsman and aviation enthusiast, who wanted to build an airplane to fly non-stop, coast-to-coast. With $40,000 from Davis and $2,000 of his wife's money Douglas incorporated the Davis-Douglas Airplane Company in South Dakota, and lured five friends from the Martin Company to Los Angeles. Douglas rented the second story of a former mill in downtown Los Angeles and borrowed tools, and began work on the *Cloudster* that was to be the first aircraft that was able to lift a payload equal to its own weight. The large biplane had a length of 35 feet, a height of 13 feet, and a wingspan of 56 feet. A 400hp Liberty engine gave it a range of 2,800 miles at a cruising speed of 85mph. As each section was built, it was lowered into a truck and transported to the former Goodyear hangar in East Los Angeles for assembly and flight testing. The aircraft ground looped during its first attempt at flight, as the runway wasn't long enough, and the initial flight was postponed to February 24, 1921, when it was successful. After breaking the Pacific Coast altitude record of 19,160 feet on 29 March, the first American cross-country flight was attempted by Davis and former Martin chief test pilot, Eric Springer. The flight began on 27 June 1921, but was cut short when the Liberty engine quit over El Paso. Before a second attempt could be made two Army pilots, Lieutenants Oakley Kelly and John

Douglas (left) returned to the West Coast, and with millionaire sportsman and aviation enthusiast David Davis, incorporated the Davis-Douglas Airplane Company in the rented second story of a former mill in downtown Los Angeles. (Douglas)

The *Cloudster* was the first aircraft able to lift a payload equal to its own weight. The large biplane had a length of 35 feet, a height of 13 feet, and a wingspan of 56 feet. A 400hp Liberty engine gave the aircraft a range of 2,800 miles at a cruising speed of 85mph. (Douglas)

Macready flew a Fokker T-2 monoplane from Roosevelt Field, Long Island, to Rockwell Field, San Diego: 2,500 miles in 28 hours, 50 minutes. With the record gone, Davis sold the *Cloudster* that was ultimately sold to T. Claude Ryan, who had established his aircraft company and airline in San Diego. Ryan converted the aircraft to carry 12 passengers in its luxurious cabin, and this aircraft was considered America's first commercial airliner.

Davis soon lost interest in the company, and sold his 2,500 shares in exchange for a promissory note backed by Douglas' father William. Douglas renamed the company appropriately enough the "Douglas Company," and reincorporated it in California. Douglas developed a design for a torpedo bomber that was partially based on the *Cloudster*, and was able to sell three test examples to the Navy as the DT-1 (Douglas Torpedo #1). The biplane was powered by the reliable 400hp Liberty engine and operated as a seaplane, with folding wings and twin pontoons that could be replaced with wheels. Tests during 1922 demonstrated that the DT-1 was the best design submitted, and the Navy awarded a contract as the DT-2, which was similar to the DT-1, except for the redesigned tail and movement of the radiator from the side of the cowling to the front. The Navy made only an initial partial payment on the $120,000 contract, and Douglas needed $15,000 to remain viable, and after making the rounds of local banks was able to secure a loan. Douglas began to build the torpedo bombers in the Goodyear dirigible hangar, but the hangar was unsuited for large scale production, and Douglas purchased a bankrupt Hermann movie studio lot in Wilshire to set up his first true production line to complete the initial order for 45 DT-2 aircraft. Douglas design for the DT-2 aircraft was so successful that it was licensed to the Naval Aircraft Factory, and the Dayton Aircraft Company to complete an additional Navy contract for 55 aircraft. The DT series was important to Douglas, as it gave him footing as a player in the aircraft industry, and the injection of fresh capital into the fledgling company. Soon the Post Office Department and Army ordered several DTs for use as mail planes (M-1) and observation aircraft (O-2).

Douglas' next design challenge came when the Army Air Service purchased five DT-2s from the Navy production run as part of a secret scheme to modify the aircraft to be capable of flying around

Douglas developed a design for a torpedo bomber that was partially based on the *Cloudster*, and was able to sell three to the Navy as the DT-1 (Douglas Torpedo #1), followed by 45 DT-2s. The biplane was powered by the reliable 400hp Liberty engine, and operated as a seaplane, with folding wings, and twin pontoons that could be replaced with wheels. (USN)

The DWC "Douglas World Cruiser" was Douglas' next design challenge. The Army Air Service purchased five DT-2s from the Navy production run as part of a secret scheme to revamp the aircraft to be capable of flying around the world. The Douglas design was able to set down on land or water, and had folding wings for easy storage. (USAF)

the world, with the purpose of validating aerial communications and the feasibility of air transport. The Douglas design, known as the DWC "Douglas World Cruiser," was able to set down on land or water, and had folding wings for easy storage. After testing at Langley Field, VA, beginning in December 1923 four DWCs (the fifth was a spare)—the *Seattle*, *Boston*, *Chicago*, and *New Orleans*—began their quest, taking off from Clover Field adjoining the Douglas factory on 17 March 1924. After stopping at Seattle and flying into a fierce storm over the Aleutian Islands off Alaska the *Seattle* was lost, but the remaining three aircraft continued. Two thirds into the flight the *Boston* was lost over the North Atlantic off Iceland, and was replaced by the spare fifth DWC named the *Boston II*. The three DWCs arrived at Clover Field on 23 September, and were welcomed by a huge crowd of 200,000. On the 28th the aircraft left for Seattle via San Francisco, completing the 27,550 mile, 175 day journey involving 36 stops requiring 371 hours of flying time to firmly establish Douglas as the leading designer and manufacturer of long range aircraft.

Orders rolled in, as the Army ordered 27 six to seven passenger transport versions of the DWC designated as the C-1, and four observation seaplanes (DOS) that were given the Army designation of O-6. Along with the DWC versions, the O-2 (DT-2) was converted into 25 O-7/8/9 variants, and 59 were converted to the BT-1; one each were re-engined as the O-32 and O-34 variants. The O-2 was to remain in production for ten years, as it was to become the standard Army observation aircraft. These orders caused Douglas to increase his payroll to 112 employees at the beginning of 1925. The Douglas Engineering Department, already including the talented Jack Northrop, Jerry Vultee, and Edward Burton, hired recent M.I.T. graduate Arthur Raymond, and a high school drop out named Ed Heinemann. To lead the 20-man Engineering Department Douglas hired a new Chief Engineer, the redoubtable James "Dutch" Kindleberger who had previously replaced Douglas at Glenn Martin Company.

During this period Douglas and his father William were the sole owners of the company, and from 1921 to 1928 had made a profit of $1.2 million. The rapid growth of the company necessitated reorganization, and on 30 November 1928 the new Douglas Aircraft Company was incorporated in Delaware with an authorized one million shares, of which 300,000 were issued; 200,000 to Douglas, and 100,000 sold to the public for $10 per share to raise $1,000,000 in capital. Douglas invested $500,000 to build a new 7.75 acre factory in Santa Monica adjacent to Clover Field, and he banked the remaining $500,000.

After building the DT-2s for the Navy and the O-2s and variants for the Army, Douglas' next project was to develop a small commercial flying boat in 1929 to sell for $45,000. The design, initially named the *Sinbad* as the prototype, was powered by two Pratt & Whitney radial engines that carried the aircraft at 153mph (at sea level) over a range of 770 miles, carrying six to eight passengers in a spacious cabin containing a lavatory, and a 30 cubic foot rear baggage compartment. The onset of the Depression de-

The rapid growth of his company necessitated Douglas to reorganize the new Douglas Aircraft Company in November 1928. With the proceeds from a public stock offering, Douglas invested $500,000 to build a new 7.75 acre factory in Santa Monica adjacent to Clover Field, and he banked the remaining $500,000. (Douglas)

After building DT-2s for the Navy and O-2s and variants for the Army, Douglas' next project was to develop a small commercial flying boat in 1929 to sell for $45,000. The design, initially named *Sinbad* as the prototype, was powered by two Pratt & Whitney radial engines that carried the aircraft at 153mph (at sea level) over a range of 770 miles, carrying six to eight passengers in a spacious cabin containing a lavatory and a 30 cubic foot rear baggage compartment. (Douglas)

flated the American commercial aircraft market, but the military bought all but 47 of the 58 production aircraft by then renamed the *Dolphin*. The Navy bought 12 RD *Dolphins* (including two for the Marine Corps as the RD-3), and the Coast Guard bought ten as the RD-4, while the Army purchased 24 (eight *Dolphin Is* as C-21s, 14 *Dolphin IIIs* as C-26s, and two nine passenger *Dolphin IIIs* as C-29s). Among the 11 remaining *Dolphins*, one was sold to the Argentine Navy, three to the Wilmington Catalina Airline Company, and one to Pan American Airways. The remaining *Dolphins* were sold for private use: two to the Vanderbilt family, one to Standard Oil, one to Crosley Radio, one to a French millionaire and, interestingly, one to Boeing Aircraft founder William Boeing, who had it outfitted as his personal executive aircraft named *Rover*.

The Civilian Aircraft Market: 1926-1931

America in the early 1920s saw the beginnings of the air transport business, and the use of ex-military aircraft that were minimally improvised into "airliners," some of which sometimes exposed their passengers to the weather. Even after passenger cabins were introduced, tradition mandated that the pilot's cockpit remain open. Also, after the war the military closed many airfields, so there was a lack of satisfactory fields close to major cities, while the country's excellent railway system was more convenient, more comfortable, less expensive, and thus more utilized.

Soon the nascent airline industry realized that safety, comfort, and speed were necessary to attract prospective passengers. Initially, the image of the airline industry suffered from an abnormal number of accidents, some due to operating conditions, such as maintaining schedules in poor weather, but most were due to unreliable, underpowered engines that could not cope with normal circumstances that could have been easily overcome by more reliable and powerful engines. Beyond the safety factor, contemporary aircraft were not economical because they were underpowered, and could only carry a limited cargo and a small number of passengers. Over the post war decade aircraft engine design, especially using air-cooled radial engines, increased reliability and power, and improved the airlines' safety record. Anthony Fokker designed his aircraft with airframes that could accommodate future engine developments, a major factor that potential buyers considered. Fokker's tri-motor designs were the prime examples of this engine adaptability, and Ford's Tri-motor AT-4 and AT-5 and Junker's F 13 would follow this example.

By the mid to late 1920s American passenger air transport lagged behind that in Europe. The reasons included: U.S. government apathy toward air passenger transport as opposed to its subsidy of mail transport; lack of suitable aircraft that were fast and comfortable; lack of airfields in a country with far-reaching geography; competition from the excellent railway system; and poor public relations, as the industry focused on the transport of mail over courting passenger travel. After World War I European airlines had relied on substantial subsidies from their governments, but in the U.S., commercial aviation had to compete with the glut of surplus aircraft that the American military sold below cost to any interested buyer who wished to start an airline. In 1924, the British government merged the country's four major airlines to form Imperial Airways, and created economical global and transcontinental routes to service the far-flung British Empire.

Thus, in the America of the 1920s, the development of passenger air transport was a secondary consideration to the transport of mail. From 1918 to 1927 the U.S. Post Office Air Mail Service monopolized air commerce. Future President, and then Secretary of Commerce Herbert Hoover, introduced legislation in 1922 to aid the U.S. commercial aviation industry, but he was not successful until 1925, when the Kelly Act was passed by Congress, authorizing the Postmaster General to solicit bids for airmail service on eight designated Contract Air Mail (CAM) routes from private contractors capable of meeting government requirements. The inducement was that 80% of the revenues could be retained by the mail hauler if they *could* also carry passengers *if* its aircraft had that capacity. "Could" and "If" were the operative words, as the airlines had no real inducement to carry passengers when they were being subsidized on a pound of mail per mile basis. The bidding drew the 26 year old future Pan American Airlines entrepreneur Juan Trippe of Colonial Air Transport, who was awarded the Boston-New York route using two six passenger Fokker Tri-motors. The Chicago-San Francisco route was awarded to William Boeing and Edward Hubbard of Boeing Air Transport, who placed 25 two passenger Boeing 40s into service. National Air Transport (NAT) won the

Chicago-Dallas/Fort Worth, and later the New York-Chicago route. In May 1926 Congress passed the Air Commerce Act, which was the foundation of American commercial aviation. The Act also authorized the Bureau of Aeronautics (BA) within the Department of Commerce, which was sanctioned to license all U.S. planes and pilots, establish and enforce air traffic rules and regulations, investigate accidents, and test new aircraft and engines for safety.

The Kelly Act of 1925 had contracted 15 airlines to carry the mail as their main source of revenue through the end of the decade and into the early 1930s. Only Western Air Express and Ryan Airlines had offered dedicated passenger-only service in 1926, and other "airmail" airlines carried passengers as bonus income, and then only when approved by the Post Office. Passengers on these airmail flights often were seated on folding chairs among the postal sacks in the mail compartments of the Ryan M-1 or the Douglas M-1 or M-2. However, in 1927 the Lindbergh Trans-Atlantic flight focused the public's attention on the possibilities of air transport, and the introduction of Anthony Fokker's and Ford's "Tin Goose" tri-motor aircraft made high speed air travel possible. The airline industry, while trying to convince the public that the less efficient tri-motor design was safer, also convinced itself. From inception speed had been the main selling point of airline advertising to their passengers. Airlines had to be faster than land transportation, particularly railways, which were more comfortable than the noisy, vibrating aircraft of the day. So by the beginning of the 1930s the airlines had finally made the idea of air travel attractive to the public, but had not cut operating costs. The aircraft of the time that were able to carry a worthwhile payload over an acceptable range were slow and obsolete, while the faster aircraft could only carry a smaller load over a shorter range. From 1925 through 1931 a large number of aircraft designs using different engine configurations were introduced that were capable of primarily carrying passengers. The first purpose built airliner was the eight passenger Stout 2-AT tri-motor that went into service in 1925. Between 1926 and 1931 the Stout was followed by a number of tri-motors, led by the eight passenger Fokker F-VIII and 14 passenger Ford 4-AT in 1926, and the 20 passenger Boeing 80, 12-14 passenger Fokker F-19, and 16 passenger Ford 5-AT in 1928. In 1930 Ruben Fleet developed his twin engine 6-8 passenger Consolidated *Fleetser* and 18-22 passenger Commodore amphibian, while Glenn Curtiss introduced his 18 passenger Model 18 *Condor*. The first four engine airliners were developed: the 32 passenger Fokker F-32 in 1930, and the 24-40 passenger Sikorsky S-40/41 amphibian in 1931.

By the end of the decade, the Stock Market crash of 1929 had decimated the smaller air carriers, and only a handful of major carriers survived. President Hoover's Postmaster General, Walter F. Brown, began an investigation in March 1929, believing that the government subsidies given to commercial air carriers encouraged them not to expand passenger service as long as they found it more profitable to carry subsidized airmail. The powerful Brown, who had control of airmail contract awards, and had established an integrated air route system across America, pressured Congress to change the Kelly Act of 1925 to eliminate the pound-per-mile payment method, and pay on the basis of cargo space available. The result was the McNary-Waters Act of 1930 that was to encourage air transport operators to purchase larger capacity aircraft that would now have to be filled with people to earn revenues. The Act was intended to increase competition among commercial carriers so that the airline industry could become self sufficient. This expectation was not to be met, as the equipment financing under the Act caused many smaller carriers to go out of business in the deepening Depression, and there were only three large carriers in 1930. Western Air and Transcontinental Air Transport merged into Transcontinental and Western Air (TWA). United, which flew between the west coast and Chicago, bought out east coast based National Air Transport to become a transcontinental carrier. In the southern U.S., American Airways bought a number of smaller regional carriers to become dominate in that area. As a result TWA and United survived as the nation's major airlines, with American becoming a contender in third place. United had routes that carried freight and passengers from New York across the northern tier of states, while TWA flew the central route form New York to the West Coast. Brown was so disgruntled by the emergence of giants TWA and United that he awarded American the southern mail and passenger route, where it would become the most profitable airline of the period. Officials of all three of the airlines recognized that the company that purchased the new and better aircraft would have the advantage on their two rivals.

Air accidents were common and well reported, as they are today. On 31 March 1931 beloved Notre Dame football coach Knute Rockne was flying from Kansas City to Wichita in a TWA wooden Fokker F-10A tri-motor that suffered structural failure. All aboard died in the crash, and the Press and public were outraged and demanded change. The accident resulted in the first government grounding of a commercial aircraft, and the downfall of Fokker aircraft in the U.S. until the late 1960s (F-27 Fellowship). The Bureau of Air Commerce dictated that all aircraft with wooden spars and ribs have periodic thorough inspections; a procedure which became so time consuming and expensive that it caused the phasing out of wooden aircraft. Even the few brave passengers who had patronized the fledgling airline industry quit flying, and the airlines realized that they needed a new airliner design that was safer and bigger, and that goal could only be accomplished by utilizing an all-metal design. More than ever, it was crucial for the aircraft industry to develop a revolutionary airplane that combined the requirements of comfort, speed, and safety with optimum payload and range potential.

Boeing and the Civilian Transport Market: 1930s and the Model 247

Soon Boeing announced its plans to build the Model 247 commercial transport, utilizing the experience it had gained with the development of its Model 80 airliners and B-9 bombers. In mid-1932 the United Aircraft & Transport group decided to consolidate Boeing Air Transport, Pacific Air Transport, National Air Transport, and Varney Air Lines into the 120 plane United Air Lines and Transport Company, with Philip Johnson as Chairman and Frederick Rentschler as President. Johnson would retain his position as the Boeing Airplane Company President, but would delegate much of his duties to the young 40 year old Claire Egtvedt as his vice president. At this time the bi-plane versus monoplane controversy continued in airliner design. Johnson wanted to equip his newly formed United Airlines Company with the best aircraft available, and he and Egtvedt felt their successful experience with the *Monomail* and B-9 made the monoplane design the wave of the future. They assigned Robert Minshall to lead the program, with Frank Canney as his project manager. The result was the development of the world's first true airliner, the superlative twin engine, all-metal, low-wing monoplane, the Model 247, which flew its maiden flight on 8 February 1933. The Model 247s were powered by either the 525hp Pratt & Whitney *Wasps* or Twin Wasp *Juniors*. The aircraft could carry ten passengers at 180mph (more than 60mph faster than the tri-motors of the day) over 500 miles, at altitudes of up to 20,000 feet. United Airlines placed a $3 million, 60 aircraft order to be delivered at the end of 1932. Despite 24-hour shifts to meet Johnson's target delivery date, the Model 247 did not make its first flight until 8 February 1933. The 247 revolutionized the airline industry, and Boeing Chairman William Boeing was awarded the Guggenheim Medal for "successful pioneering and achievement in aircraft manufacture and air transportation." Bill Boeing was only 51 years old but wished to retire, and in August 1933 he relinquished the Boeing Airplane Company chairmanship to Johnson, and Egtvedt became the Boeing president. At this juncture the Boeing Company committed a critical business error by refusing to sell the Model 247 to other airlines, particularly to arch rival TWA, but Boeing actually had no choice, as it first needed to furnish in-house United Airlines its allocation of the outstanding airliner that would give the operator the definite competitive edge.

At the time TWA was flying Ford and Fokker tri-motors on its transcontinental routes, and company President Richard Robbins realized that Boeing's 247 would give United an advantage over his company. In response to the Boeing 247 rebuff, TWA convened a board meeting headed by Robbins and attended by TWA chief technical advisor Charles Lindbergh and Vice President of Operations Jack Frye. After a discussion of basic requirements Frye was instructed to have his engineering staff draw up performance specifications for an airliner that was to be superior to the Boeing 247,

Jack Frye, TWA Vice President of Operations, instructed his engineering staff to draw up performance specifications for an airliner that was to be superior to the Boeing 247, as he realized that Boeing's 247 would give United an advantage over TWA's Ford and Fokker tri-motors. (TWA)

The Boeing Model 247 was the world's first true airliner, a superlative all-metal, low-wing monoplane, that flew its maiden flight on 8 February 1933. The aircraft could carry ten passengers at 180mph over 500 miles at altitudes of up to 20,000 feet. (Boeing)

and the recently introduced Junkers Ju-52 tri-motor, which was the flagship of the German national Lufthansa Airlines. In the safety conscious climate of the time the aircraft was to carry the latest radio and communications equipment, and be able to fly on one engine out, so three engines continued to be deemed safer than two for a large aircraft. However, there was no question about the advantage of the monoplane design. The specifications called for an all-metal tri-motor monoplane transport that could carry at least 12 passengers in a roomy cabin at 145mph over a range of 1,080 miles at a minimum service ceiling of 21,000 feet. The gross weight of the aircraft was to be 14,200 pounds, with a 2,300 pound payload. On 2 August 1932 Frye sent out a two paragraph bid invitation letter to Douglas, Martin, Ford, Curtiss-Wright, and Consolidated to design and manufacture ten aircraft that would meet the enclosed specifications.

Douglas Enters the Civilian Transport Market: The DC-1 and DC-2

Although the TWA bid invitation did not mention a price per plane, the 38 year-old Douglas was confident that his company was ready for the commercial transport market. While the rest of America and the aviation industry were suffering the onset of the Great Depression, Douglas was more than solvent after selling 430 aircraft in the 1920s, and still had orders to fill for 53 aircraft from the Army, 18 from the Navy, and ten from China. Nonetheless, while earning $2 million in military sales in 1932, the year still was to be the worst in company history, and Douglas realized that the company needed commercial contracts to turn to, as Congress was cutting back on military contracts during the Depression. North American Aviation President Harold Talbot, whose company owned 89,000 shares of Douglas stock, and who also was a TWA director, urged Douglas to enter the TWA airliner derby, as TWA urgently needed an aircraft to compete with the forthcoming Boeing 247. Talbot mentioned TWA's high regard for Douglas and his design team, and called attention to Henry Ford's expected departure from aviation. The Ford Motor Company was the major factor in American aviation, but during the late 1920s the commercial transport market became so saturated that no company was making a profit, much less breaking even. Ford produced over 200 of its immortal 4-AT and 5-AT tri-motor air transports at a cost beginning at $55,000, and reaching $68,000; supplying 95 airlines worldwide. However, it is estimated that Ford still lost over $3 million (Ford never released any figures), and by the beginning of 1932 Ford was ready to leave the aviation business. Talbot also revealed that several companies were planning on entering the TWA design derby. General Aviation Aircraft Corporation, which had recently acquired Fokker Aircraft in the U.S., was planning on entering a tri-motor proposal, and that there also was talk of Sikorsky working on a 40 passenger design (the future S-40). Curtiss-Wright, which was already supplying its 18 passenger Model 18 *Condor* to American Airways, would also be working on a TWA contender. With Talbot's reinforcement, Douglas directed his engineering department to initiate studies on the TWA specifications using a twin engine hypothesis on a low wing monoplane design.

Donald Douglas was not interested in the oversold-on-safety, extremely noisy, and inefficient tri-motor concept, and had not entered the commercial aviation market earlier for that reason, stating that his company's previous twin engine designs easily flew safely on one engine in emergency situations. Engine makers Pratt & Whitney and Wright Aeronautical were developing new engines, two of which furnished more horsepower than the three engines powering existing tri-motor aircraft. These engines were mounted in the newly developed—by the National Advisory Committee for Aeronautics (NACA)—cowlings that produced a streamlining that reduced air resistance by half.

The Douglas engineering team was led by Chief Engineer Dutch Kindleberger and Deputy Chief Engineer Arthur Raymond, and had a superlative staff of Senior Engineers, including Ed Burton, Fred Stineman, Fred Herman, and Lee Atwood. Also present was Donald Douglas' right hand man since 1922, Harry Wetzel, who was company Vice President and General Manager, and who had the ability to take the engineers' designs and correlate them into cost and manufacturing requisites. The team worked for a week, not only studying the TWA specification, but also Boeing's 247 design to make the Douglas proposal a superior airplane to the Boeing aircraft. The Douglas focus was on passenger comfort, with more head room, a clear central aisle, and better soundproofing. The main wing spar in the Boeing design cut through the passenger cabin, dividing it in half, and its fuselage was too narrow in cross section, and made the cabin feel constricted. During their brainstorming sessions the Dou-

The Douglas DC-1 engineering team was led by Chief Engineer Dutch Kindleberger (shown with Atwood) and Deputy Chief Engineer Arthur Raymond, and had a superlative staff of Senior Engineers, including Ed Burton, Fred Stineman, Fred Herman, and Lee Atwood. Kindleberger and Atwood would later leave Douglas to form North American Aviation. (Douglas)

glas engineers decided to use a modified version of the tapered wing developed by Jack Northrop that would give added lift without adding too much area. Douglas had experience with the wing, and had jigs and machine tools available for its fabrication. The structure of the wing was of multi-cellular aluminum construction with an internally braced rib and spar. The wing was to be built in three sections, with the stub center section integrated with the fuselage so that it was strong enough to support the engines, and eliminated the main wing spar running through the cabin as it did in the 247; dividing the cabin into two. Using this wing design, the engine mounts could extend out of the wings, giving more lift as the thrust of the propellers was directed over the airfoil, giving the extra lift. Since the Boeing aircraft had retractable landing gear, it was essential that the Douglas project have it also, as retracting the gear into the nacelle reduced drag by 20%. The TWA specs required a landing speed of 65mph, and a flap had to be developed to increase wing area for slower landings, and to give more lift on take off. The hydraulic flaps operated by the pilot from the cockpit were split trailing edge flaps built into the lower side of the wing to increase lift for take off and drag for slow landings.

After ten days the plans for the Douglas Commercial #1 (the DC-1) were ready, and an appointment was made to meet with TWA's Robbins and Lindbergh in New York City. Wetzel and Raymond left by train from California, and on the way they finalized the design specs, and organized, wrote, and rewrote their presentation. Once they arrived in New York they were scheduled for the first meeting the next morning, with Robbins and Lindbergh, and TWA senior pilots D.W. "Tommy" Tomlinson and Paul Richter. Also present was Donald Douglas' good friend Harold Talbot. The meeting continued for three weeks, and the design was discussed and analyzed at length, with problems and questions being resolved. The last major hurdle was Lindbergh's insistence that the aircraft be able to take off with a full load from any TWA field on *one* engine, and then climb and fly over any mountain range along the route. Douglas asked Kindleberger for the answer, and his reply was that this had never been done before with such a large aircraft, and the only way to find out was to build it and find out. On 20 September 1932 a contract was signed for a service test aircraft to cost no more than $125,000 (with Douglas to be accountable for the cost of any contract over runs), and for some unspecified reason Douglas insisted it was to be paid for in gold bullion. The contract also stipulated that TWA had the right to buy all, or part, of the 60 aircraft in batches of 10, 15, or 20 aircraft at $58,000 each, with the engines to be supplied by TWA. Douglas and Kindleberger both knew that the $125,000 would not come close to covering the cost of the test aircraft, but the announcement of the contract caused Douglas stock to jump from $7.12 to $16 per share overnight.

On his way back home Raymond flew coast-to-coast for the first time, and personally sampled the 1932 air travel experience. He flew in a TWA Ford Tri-motor, and immediately had to stuff cotton in his ears to block out the noise. The cotton blocked the noise and any attempt at conversation, but had no effect on the vibration. Once over the Rocky Mountains the cabin became so cold that his feet became numb, as did his behind from the narrow leather-covered wicker seats. He found the passenger cabin small, and the rear lavatory even smaller. As a climax to the uncomfortable journey, when the airliner landed on the wet field that was covered with puddles, the muddy water was sucked through the cabin air vents and sprayed everyone. Then and there Raymond knew that the design specs were not only about numbers and performance, but also about comfort, and declared, "We've got to build comfort and put wings on it." Douglas engineers were urged to think about passenger comfort: better seats and leg room, a more spacious passenger cabin and lavatory, and better cabin heating and sound proofing.

In the fall of 1932, the eight acre Santa Monica Douglas facility covered about 350,000 square feet of floor space, employed 900 personnel, and lay adjacent to Clover Field, which had all paved taxiways and runways. At this time the Douglas factory was occupied with military orders for Army observation planes, Navy torpedo bombers, and a new twin-engine amphibian, so room had to be found for the new DC-1 project. The process for developing and building a new aircraft followed a series of steps. After the design was approved and blueprints drafted; these ideas and paper plans were transformed into a reduced 1/11th scale model for wind tunnel testing at the California Institute of Technology. CIT, under the leadership of the brilliant Dr. Robert Millikan, had one of the world's most modern aerodynamics departments, and an excellent wind tunnel. Douglas recruited recent CIT aerodynamics graduate Dr. W. Bailey Oswald to determine the ideal aerodynamics of the scale model, which was shown to have initial instability and center of

Dr. W. Bailey Oswald, a recent Cal Tech aerodynamics graduate, was recruited by Douglas to determine the ideal aerodynamics of the scale DC-1 model, which was shown to have initial instability and center of gravity problems. Oswald would become an important member of the Douglas team. (Douglas)

gravity problems. After the scale model passed its wind tunnel tests, mock ups were made of each system, such as the fuel, brake/hydraulic, and oil, and then tested. A full-sized mock up was constructed to determine the optimal location and installation of instruments, systems, and equipment. Next a hand-fabricated flying prototype was constructed, but due to the pioneering nature of the new multi-cellular wing, the all-metal fuselage framework, and the aircraft's stressed skin, the construction was essentially done by trial and error.

On 8 February 1933, the Boeing 247 made its first flight, and put the pressure on Douglas engineers and the DC-1 project. The first major problem came when TWA, which was to supply the power plant, had not made a decision between the air-cooled radial Pratt & Whitney *Hornet* and the Wright *Cyclone* engines. Douglas engineers had designed the aircraft to be capable of using either engine, so there was intense competition between the two engine companies for the contract, and they sent their field engineers and technicians to Santa Monica to work with the Douglas airframe and engine departments. Each company set up work sites on either side of the hangar, and the competition became so intense that an actual chalk line delineating a "no man's land" had been painted to separate the two factions, and high screens erected to hide each company's "secrets." Finally, because of Wright's new cooling fin and cylinder design the nine cylinder, air-cooled, 600hp radial SGR-1820 *Cyclone* was chosen by TWA. When TWA approved the comfortable passenger cabin mock up on 15 March 1933, it was found that the added passenger cabin amenities ballooned the DC-1 design from 14,000 to 17,000 pounds, making it 1,000 pounds over the maximum weight. There was an anxiety at Douglas that a fully loaded DC-1 would be unable to get off the ground with two engines, much less one. However, at the time the Hamilton-Standard Company was developing their revolutionary adjustable pitch propeller and new gearing that would allow more efficient operation, not only during take off, but also while cruising at high altitudes. The adjustable pitch propeller permitted the pilot to automatically

The roomy DC-1 pilot's compartment had controls for a pilot and co-pilot, and a center console with twin throttles, prop, and mixture controls. (Douglas)

adjust the angle of the propeller blades so that they could take larger bites of air at take off, supplying increased thrust and providing more lift. For cruising speeds the blade angle could be decreased to conserve gasoline and decrease engine speed and wear. The final propeller configuration was a three-blade, two-position, hydro-controllable Hamilton-Standard unit. There were two 180 gallon main fuel tanks and two auxiliary 75 gallon tanks (510 gallons total) to give the DC-1 a range of 1,200 miles. While some weight paring would still have to be done, the Hamilton-Standard propeller was the answer to keeping the DC-1 as comfortable as possible.

On 22 June 1933, only ten months from Jack Frye's bid invitation, Douglas Commercial Model-1 rolled out of the Douglas hangar onto the Clover Field ramp carrying the experimental license X223Y, and the Douglas serial number 1137. The large 16,000 pound, sleek aluminum DC-1 was truly impressive, with its tapered cylindrical fuselage that, at 60 feet long and having a wing span of

The DC-1 (Douglas Commercial Model-1) rolled out of the Douglas hangar on 22 June 1933, only ten months from Jack Frye's bid invitation. The large 16,000 pound, sleek aluminum DC-1 was truly impressive, with its tapered cylindrical fuselage that, at 60 feet long and having a wing span of 85 feet, was the largest American twin-engine monoplane land plane ever built. (Douglas)

85 feet, made it the largest American twin-engine monoplane land plane ever built. It sat on a conventional landing gear, with the wheels on shock-absorbing struts, and a small tail wheel. The passenger compartment was still bare bones, and would be completed later. The passenger "salon" was to be 26 feet, 4 inches long; 5 feet, 6 inches wide; and 6 feet, 3 inches high, allowing a tall passenger to walk upright the entire length of the cabin. Two rows of seven seats were separated by a 16 inch aisle, and each seat had its own window. The comfortably upholstered lounge seats were 40 inches wide and fully adjustable for reclining, and could be reversed to face the passenger behind.

The first flight test of the DC-1 was scheduled for 1 July at noontime, so that Douglas factory employees could watch. On this first flight, Project Manager Fred Herman accompanied Carl Cover, Douglas Chief Test Pilot and Vice President of Sales. Cover had his early training in 1917 at the Officer's Training School at the University of California, Berkeley, and was commissioned into the Army Air Corps in 1918. After World War I he was the Air Corps representative in California and Hawaii, and joined the Douglas staff in 1930 as Chief Test Pilot and manager of the military sales sector; over the next 13 years he would fly every new Douglas model except the XB-19. With the Douglas workers waiting to see the results of their handiwork, Cover taxied the aircraft to the end of the runway into the wind. After about 30 seconds of climbing after take off the left engine quit, but Cover was able to gain a few hundred feet of altitude when the right engine also quit. The nose dropped, but unexpectedly the engines started again, and Cover began to climb; only to lose power in both engines again. The unpowered aircraft once again begin to dive; Cover got the engines to start again and put the plane into a climb. Over the next ten minutes, with the engines stopping in a climb and starting in a dive; the skillful Cover was to get the plane to climb to about 1,500 feet, where he was able to bank and make a rough but successful landing. Immediately the concerned Douglas engineers pulled the engines for inspection. After running the engines on test blocks for hours without a problem the engineers finally tilted the test stand and discovered that the carburetor floats prevented gasoline from flowing upwards in a climb and shut off the fuel, which was again able to flow when the plane went into a dive.

Carl Cover, Douglas Chief Test Pilot, flew the first flight test of the DC-1 on 1 July at noontime, so that Douglas employees could watch. During the flight the engines continued to cut out, but Cover was able to save the day by successfully landing the valuable prototype. (Douglas)

Once the carburetor glitch was solved testing continued, and Cover found that the aircraft fishtailed. The engineers went to work, and added more rudder surface and changed the airfoil attitude. For landing trials the DC-1 was flown to Mines Field, where tests were performed by redoubtable test pilot and aeronautical genius Edmund "Eddie" Allen, and TWA's Tommy Tomlinson as the co-pilot, with Cal Tech's W. Bailey Oswald and chief flight test engineer Frank Collbohm aboard to record the action of the landing gear. To do so Oswald would lie on the floor and stick his head out the cabin door that had been removed and observe the gear action, while Collbohm would raise and lower the gear in the cockpit. Oswald and Collbohm switched positions without the pilot's noticing, and on the next touch and go the pilots did not tell Oswald to lower the gear. As Collbohm was watching, suddenly runway asphalt and gravel flew into the doorway, and he yelled up to Allen, who knew immediately the gear was still up, and the propeller tips were chewing up the runway. Allen frantically pulled up out of danger and circled to make a successful gear down landing. There was no structural damage, but the plane had to be returned to the factory, and was out of commission until mid-August. Once repaired and the passenger compartment completed the aircraft was taken up for daily testing. The superchargers were found to make the aircraft a superlative high altitude performer. The aircraft was so quiet that Tomlinson reported that he thought the engines sounded like they were not pulling full power. The aircraft was purposely overloaded and had no problems taking off well within limits. For the ultimate overloaded, one engine take off test the DC-1 was flown to Winslow, AZ, the highest airport on the TWA route at 4,500 feet. On the take off Allen nursed the plane for a long run before the plane lifted off the ground. He then climbed to 8,000 feet and flew to Albuquerque on one engine, beating the Ford Tri-motor chase plane that had taken off a quarter hour before. During testing Oswald felt that the pilots were not getting the maximum performance from the aircraft, and set about devising the first accurate performance charts that would transform the industry's thinking on predicting aircraft performance. After the aircraft was accepted for service the airlines were provided with precise performance charts that made operations more efficient and reduced costs.

The DC-2 Evolves into the Remarkable DC-3

After two months of the most rigorous flight testing of any aircraft to that time, Douglas had met all TWA's requirements, and on 4 September 1933 the airline ordered 20 DC-2s, which was an improved version of the DC-1, with its fuselage elongated by two feet to provide an extra row of seats. The new airliner had a take off weight of 18,000 pounds, including 14 passengers, 1,740 pounds of cargo, and 510 gallons of fuel that carried it over a range of 1,060 miles at 196mph. The first aircraft (NC13711 c/n 1237) was completed in May 1934, and made its first flight on 11 May. It was delivered to TWA on the 14th, and on the 17th it was flying on the New York-Los Angeles route as Ship 301, the *City of Chicago*, where it broke the New York to Chicago speed record four times in eight days. By June TWA ordered 55 more, with the official delivery ceremony occurring in December 1934 at the Los Angeles Municipal Airport, when Donald Douglas accepted a $125,000 check. The initial DC-2s required 59,000 man hours per aircraft to build, and Douglas engineers determined that this figure needed to be reduced to 38,000 as the break even point. On the first 25 aircraft delivered to TWA Douglas lost a little over $10,000 per aircraft. The answer was to purchase large hydraulic presses so that the parts could be stamped out, as was the procedure in the automobile industry, and the number of man hours dropped to 32,000 on each aircraft that was produced every two weeks and sent to TWA's Kansas City hub for acceptance. After eight months DC-2s were flying for 21 different airlines, with U.S. mainland operators and Pan American Airways in Latin America flying a combined 15 million miles, and other foreign operators and private owners flying another 5 million miles.

One month after the introduction of the DC-2, Congress passed the Air Mail Act of 1934 (the Black-McKellar Bill), which gave the Post Office Department control of awarding air mail contracts and determining routes and schedules; the Interstate Commerce Commission was to establish mail rates, and the Bureau of Air Commerce was to regulate the airways, and license pilots and aircraft. The Act made it illegal for aircraft manufacturers to own any segment of an airline operation, as Congress felt that large air holding companies and manufacturers were heading toward monopolizing the airline industry. The Act benefited Douglas the most as the company, unlike Boeing/United, had never been directly affiliated with an airline, and soon many airlines that had been freed from their parent companies were coming to Douglas to order the superior DC-2.

While United's Boeing 247s were beginning to have troubles competing with TWA DC-2s, American Airlines' aging 12 passenger bi-wing, fabric-covered Curtiss *Condors*—used as "sleeper" aircraft—were fast becoming obsolete, and American President Cyrus Smith considered the DC-2 as his airline's new flagship. In November 1934 Smith and his Chief Engineer, William Littlewood, met with Douglas, and presented their requirements for an improved wide-body, long range sleeper aircraft based on the DC-2. Without funding, Smith committed his company to a multi-million dollar verbal order for 20 of these aircraft. Smith went to Roosevelt's New Deal Reconstruction Finance Corporation, headed by his old friend, the powerful Jesse Jones, and convinced the RFC to loan him $4.5 million. Although it was difficult for Douglas to admit, especially having DC-2 orders that stretched his plant to capacity, he recognized that the DC-2 had its problems, as pointed out by Smith and Littlewood. Its handling characteristics, especially heavy ailerons and rudder, were likened to "flying a barn door." It was difficult to land, had directional instability, and insufficient vertical fin area continued to cause "fishtailing." Douglas assigned over 400 engineers and draftsmen to the redesign work that lasted into March 1935, with Ed Burton heading the layout team, Dr. W. Bailey Oswald the aerodynamics, and Lee Atwood the stress unit. Dutch Kindleberger left Douglas in July 1934 to become President of General Aviation, which was soon renamed North American Aviation of Inglewood, CA. To make his defection worse, Kindleberger took the Douglas Chief of Advanced Design, Lee Atwood, with him. Nonetheless, on 10 May 1935 Arthur Raymond released his Aircraft Report 1004 for the specifications for the DST (Douglas Sleeper Transport). The DST was virtually a whole new wider and longer aircraft, with more wing span and larger tail, a stronger landing gear, and more powerful engines. The airliner's passenger cabin was 19 feet, 5.5 inches long; 7 feet, 8 inches wide; and 6 feet, 4.5 inches in height. This was accomplished by splitting the fuselage lengthwise, and adding the 26 inch insert to allow two seats on one side of the aisle and one on the other. The fuselage was lengthened

In September 1933 TWA ordered 20 DC-2s, which was an improved version of the DC-1, with its fuselage elongated by two feet to provide an extra row of seats. Eight months after introduction DC-2s were flying for 21 different airlines, including Boeing controlled United. (Douglas)

The DC-3 design was so successful that after 191 DC-2s were built they were phased out, and Douglas production was devoted to the DC-3, which was dubbed as "aviation's Model T." 803 commercial DC-3s were eventually built, along with 10,123 military versions, mostly C-47s, on contracts that generated over an astonishing $1 billion in business for Douglas. The DC-3 truly lived up to the company's "Wings for the World" motto. Pictured is the American Airlines Flagship. (Douglas)

to accommodate eight rows of three seats, and three double rows of berths on each side of the aisle for the sleeper version (12 berths), plus two extra (or four seats) in a private "Sky Room" in the right of the front fuselage. Later, one row of seats was forfeited to give more forward baggage storage, reducing seating to 21. The "deeply upholstered" seats were 36 inches wide, and faced each other in the daytime, folding into a lower berth at night. A 30 inch upper berth folded into the ceiling when not in use. Both upper and lower berths were 6 feet, 5 inches long. There were separate dressing rooms provided for men and women located next to the rear lavatories. The engines were to be the more powerful 1,000hp Wright *Cyclone* 1820 G-series, which would enable the aircraft to carry a larger payload without increasing the per mile costs. American Airlines wanted their New York to Chicago route to be non-stop, and their transcontinental routes to have a minimum of four stops. The enlarged DST airliner could carry 16 sleeping berths during overnight flights, and 24 seats during the daytime in a "daylight" version dubbed the DC-3. Raymond quoted a price of $82,000, $81,000, or $79,000 each for multiple orders of 5, 6, or 10 aircraft, respectively. On 8 July Smith officially ordered ten aircraft, and the first DST X14988 (there were no prototypes) was completed and ready for its first flight on 17 December 1935 at Clover Field, Santa Monica. Carl Cover was the test pilot, with engineers Frank Collbohm and Ed Stineman, and aircraft mechanic Jack Grant on board for the first half hour flight, which was followed by two more flights that day. The test DST was then flown daily by American Airlines test pilots Elling Veblen and Dan Beard, and on 21 May 1936 it was certified. The aircraft were delivered to American Airlines in Phoenix, AZ, to avoid the California sales tax, and then immediately flown back to Santa Monica for the Department of Commerce's 50 hour airline proving flights. On 21 June 1936 American rolled out the second DST (NC16001), and flew it non-stop from Chicago Midway to Newark, NJ, round trip in 8:07. On 16 August 1936 the inaugural flight of the "daylight" DC-3 (NC16009), fitted with only seats, was made. The DST/DC-3 was not only faster than the DC-2; it was larger, and had a payload that was one third larger than that of any rival, and operated on a per mile cost equal to the Ford Tri-motor and Boeing 247. So successful was the DC-3 design that after 191 DC-2s were built they were phased out, and Douglas production was devoted to the DC-3, which was dubbed as "aviation's Model T." In November 1936 United Airlines President William Patterson was forced to concede that the DC-3 was a superior aircraft, placed an order for it, and the first DC-3 (NC16061) flew for United on 30 June 1937, and as the maxim goes, "The rest is history." 803 commercial DC-3s were eventually built, along with 10,123 military versions, mostly C-47s; contracts that generated over an astonishing $1 billion in business for Douglas. The DC-3 truly lived up to the company's "Wings for the World" motto.

2

Prewar American Bomber Development and Procurement Policies

Introduction

To understand the development of the prewar American bomber, the nation's convoluted procurement policy needs to be examined, as it had a profound effect on not only the quantity and quality of the bomber, but also in the conception of its tactical and strategic role as America was drawn closer to war.

Air Corps Act of 1926

The Air Corps Act of 1926 was intended to establish a statutory means to procure new aircraft. It provided for a design competition that would lead to the purchase of one or more prototypes, the issue of contracts for "experimental" aircraft by the Secretary of War at "his discretion without competition," and competition where aircraft could be procured on grounds other than provided by the first two provisions, with the Secretary able to "exercise discretion in determining the lowest responsible bidder." Another procurement possibility was for a negotiated purchase contract without the competition of a design "of sufficient interest to justify immediate procurement." The Act required the use of a design competition, with the designs to be submitted to and evaluated at Wright Field, and a winner was to be selected and awarded a contract to build one prototype for service testing. If the service tests were successful, a production order was then to be issued. However, in effect, design competition was impractical, as when bids went out Wright Field received a large number of design proposals that the designers claimed met or exceeded the specifications. Until a prototype was built, it could not be determined from the submitted design if specifications had been actually met and, if not, time passed and money was spent without result. Also, bidders were given inadequate time (several months) to design and submit their proposals, and once the winner was chosen, its design had to be detailed, and then it was often found that the original dollar bid was inadequate, and the manufacturer would lose money on the building of the prototype and production models. This ineffective design competition gave way to the negotiated purchase contract, but the manufacturer with the winning bid usually lost money on the prototype, as they intentionally low-balled the bid, expecting to recoup this loss on the quantity production order. But the Act made no provision for an automatic quantity order once the prototype had been accepted, and a new bid for a quantity order was to be issued, and another manufacturer could be contracted to build the winning design of another manufacturer, who built the prototype at a loss. In June 1929 the Navy announced that, despite the fact that Reuben Fleet's Consolidated Aircraft Company had invested an additional $500,000 more for the development of the *Admiral* Flying Boat than the original $150,000 Navy contract awarded, the Navy would accept bids from "qualified companies" (later Glenn Martin) to manufacture nine flying boats to Consolidated's design. Thus, manufacturers were reluctant to submit designs and bids, and Army Regulation 5-240 was resurrected from the mass of regulations to accommodate procurement. This Regulation stipulated that "competition might be avoided in certain special circumstances in which competition was impractical." By interpreting AR 5-240 to classify the manufacturer of an experimental aircraft purchased under the Air Corps Act of 1926 as the only source the manufacturer could be awarded the production contract. Between 1926 and 1934, under the terms of AR 5-240 awards of $16 million were issued for contracts under the "experimental" provision of the Act, and $22 million under the non-competitive provisions. There had been a public record and annual aircraft procurement report to Congress every year from 1926 to 1934, and each of the procurements was completely legal. Nonetheless, in January 1934 the *Washington Post* reported that the House of Representatives was about to investigate seven years of wrongful aircraft procurement by the War Department in violation of the Air Corps Act of 1926.

Delaney Committee

In late 1933 the ambitious Senator from Alabama, Hugo D. Black, was aggressively investigating the federal subsidies to private airmail contractors, which were U.S. airlines. The investigation was so vigorous that the newspapers soon led the public to believe that most of the nation's airlines were guilty of flagrant wrongdoing and excessive profits. Since many of America's aircraft manufacturers were associated with airlines (e.g. Boeing/United Airlines), they were also incriminated by the newspapers. When the Navy and Army came to Congress to present their appropriations bill, in

January and February 1934, respectively, both were attacked with accusations of allowing profiteering, and excess profits by the aircraft manufacturers at the expense of the taxpayers. The Chairman of the House Naval Affairs Committee, Georgia Representative Carl Vinson, appointed a subcommittee headed by New York Representative J.J. Delaney, to investigate the supposed widespread procurement corruption. After two months and hearing 800 pages of testimony, the Delaney Committee's final report found that the charges against the Navy's procurement policies were unfounded and were, in fact, "prudent and practicable," and fostered competition. The Committee found that the major airframe manufacturers made only 0.2% on cost profit on their sales to the military and commercial interests. The average profit earned by airframe and engine manufacturers between 1926 and 1934 was a not-so-excessive 9% on cost. Further figures showed that aircraft manufacturers lost an average of 50% on cost on "experimental" aircraft, and when combining these losses with production profits, if (a big IF) the aircraft went into production, the return was 11.5%, which was not considered excessive for the high risk involved. The Committee had vindicated the Navy, however, while the New York *Times* had headlined the Committee's appointment, and printed almost daily titillating accusation stories during the investigation; it only spent one day reporting the story exonerating the Navy, and that was carried on page 15! To exacerbate the situation, one member of the Committee charged publicly that the majority had "whitewashed" the investigation of the Navy, and then wrote a minority report that reached publication in the *Congressional Record*. Of course, the newspapers cited this minority report, and the mentioned "indication of new evidence of illegal procurement." Strangely, the majority members did not refute this report, or urge the divulgence of the supposed "indication of new evidence" that would never surface, but meanwhile the majority report languished in obscurity.

Rogers Committee

In the turmoil of the investigation of the Navy, the Air Corps also soon came under fire. South Carolina Representative John McSwain, Chairman of the Military Affairs Committee, put New Hampshire Representative W.N. Rogers in charge of the eight man committee that took his name. In closed hearings the Rogers Committee found the Chief of the Air Corps guilty of "gross misconduct," and "deliberate and willful and intentional violation of the law," and the Air Corps both "inefficient" and "expensive" using "various subterfuges" for "pernicious" and "unlawful" procurement. It made the recommendation that there be a return to "aggressive design competition for experimental aircraft," and "competition on all contracts for procurement in quantity." Perhaps due to the closed nature of the hearings to the press, Congress did nothing about the Air Corps Act of 1926, but did pass a law limiting manufacturer profits, and provided for the recapture of all earnings in excess of 10%, but provided nothing to put a floor under manufacturers' losses. This excess profits law raised questions, and a mandate for further revisions to it.

Secretary of War Woodring Circumvents the 1926 Air Corps Act

A new administrative procurement policy was fashioned by Assistant Secretary of War H.H. Woodring, which essentially supported competitive bids, and thus circumvented the Congress from amending the Air Corps Act by statute. During fiscal 1933 Congress appropriated $10 million for the AAC, but the new Roosevelt Administration, under the economic pressure of the Depression, impounded $7 million as an economic emergency measure. The AAC urgently needed more than 700 new aircraft to equip active units, and many more were needed to replace aircraft that were, or were soon going to be obsolete. At the end of the year the Roosevelt Administration transferred $7.5 million to the AAC from the Public Works Administration (PWA). To expedite the purchase of the best aircraft available, the AAC negotiated production contracts with the manufacturers of top quality aircraft using Army Regulation 5-240, maintaining that the manufacturer was the "sole source" of the required aircraft. Two companies complained about not receiving AR 5-240 negotiated contract awards. The Depression put the War Department under pressure to award contracts to a number of aircraft companies to keep them viable, and Woodring was forced to reconsider the contract awards. In order to award contracts equably throughout the aircraft industry, and also obtain aircraft of maximum performance at a minimum cost, Woodring found himself in a Catch-22 situation. Aircraft contracted on the basis of price competition would save the government money, but not insure purchasing aircraft with the best performance. On the other hand, aircraft contracted on the basis of having the best performance would cost more. Either way, the intention of spreading the wealth was not met, as whether contracts were awarded on cost or performance, most of the contracts tended to be awarded to a few efficient companies who had the best designs and production capacity. Woodring asked the Air Corps to develop a policy before Congress reconvened in January 1934. The AAC responded with the 1934 War Department Aircraft Procurement Policy, which had "competition" as its foundation. The AAC's solution was to allow each manufacturer to bid on its own specification, but required a minimum high speed, thus allowing competition as to performance, but disqualified all companies but those that fell within a narrow margin of specified performance. The competition was also limited to companies that had previously submitted similar aircraft for approval to Wright Field for evaluation, so that there was some assurance that the submitted aircraft proposal had some design and safety substance behind it. The 1934 Procurement Policy went on to state that if the aircraft with the highest performance was not the lowest bidder then the Secretary could award the contract "at his discretion to the best advantage of the Government." Each submitting company was required to supply a prototype for flight testing, eliminating "paper promises," and would provide a basis for an assessment of the aircraft for production contracts. The Secretary's timely submission of the 1934 Procurement Policy effectively undercut a Congressional Committee that was infuriated over the alleged profiteering by the aircraft industry, and bent on amending the Air Corps Act.

The 1934 Procurement Policy directed that the new competition procedure was to mail out a circular, which was an invitation for a design proposal containing "type specifications in terms of the minimum acceptable performance." The aircraft's maximum performance was then left to the talents of the manufacturer's design team, whose design performance was to be verified by flight tests of the prototype aircraft. The AAC needed to require a necessary "degree of uniformity and standardization" on the aircraft industry to prevent the "collection of heterogeneous aircraft and equipment," and to "insure a high degree of uniformity and interchangeability." So, along with their invitations to the aircraft industry for designs and bids, the AAC issued the *Handbook for Aircraft Designers*, and an "index of all pertinent Army, Navy, and federal specifications for materials and subassemblies." In addition, the industry was required to use Government Furnished Equipment (GFE), which included instruments, armament, communications, oxygen equipment, etc., and use mandatory engine and propeller installations. The *Handbook* and use of GFE decreased the number of variables to be incorporated in the prototype design, and thus limited the range of the competing designs, and made the competition more evenhanded among the qualified competitors. The aircraft manufacturers were then to submit designs based on the performance specifications issued by the AAC.

In June 1936, the Secretary of War reported to Congress that the new policy was a success, as it had increased the number of bidders, and the designs submitted were far advanced compared to contemporary aircraft. But the question remained as to equating price to performance. To which bidder should a contract be awarded when one manufacturer submitted a superior design at a higher cost, while another submitted a much lower bid on an inferior design? If performance was the main prerequisite, then the manufacturer with the superior design could ask a higher price for his design. During the design competition for an AAC transport aircraft, the larger, twin engine Douglas DC-2, already in successful service as a commercial airliner, was clearly the far superior design over the single engine transport designs submitted by Curtiss Wright (*Condor*) and Fairchild (C-8). But the Douglas bid was $49,500 per aircraft, as compared to the $29,500 for Curtiss Wright and $29,150 for Fairchild. The transport design proposal stipulated that the primary consideration would be performance, not price, and the Douglas DC-2 was given the production contract. But Fairchild protested, and the Comptroller General deferred payment to Douglas, pointing out that the performance of the Fairchild transport "was far in excess of the minimum performance required." However, the Comptroller General did not realize that in combat, having the minimum acceptable performance would not be sufficient against the superior aircraft being designed in Europe. The CG believed that the AAC's competition was illegal, as it did not provide any method of establishing a precise relationship between cost and performance, and left the choice only on performance. The Secretary of War held that no formula could evaluate price vs. performance, and that the Air Corps Act of 1926 gave him legal discretion to make decisions regarding the weight of price versus performance when evaluating bids "in order to serve the best interests of the air arm." Meanwhile Douglas delivered aircraft on the contract, but was not paid, as the disagreement between the CG and Secretary of War continued. The Attorney General was asked to intervene, and after four months ruled significantly in favor of the Secretary of War, and Douglas was finally paid. Even though the Attorney General had ruled in favor of the Secretary of War, AAC procurement officers realized that price would remain a problem with the Comptroller General and General Accounting Office. In order to expedite their procurement programs, and for their aircraft contractors to receive timely payments, the AAC agreed to include price as a factor for evaluation in all future competition. The AAC's evaluation proposal was to determine a "figure of merit" on the basis of performance that was to be divided by the dollar cost bid by the manufacturer. This "price factor" would favor the bidder with the lowest price and the highest performance. However, the War Department continued to be adamant that final selection would be the decision of the Secretary of War, and the figure of merit and price factor would serve as guidelines for the final selection.

With this procurement procedure in place the manufacturers found drawbacks lingered. The circulars (design proposals) had to be composed so that the manufacturers had enough design autonomy to incorporate innovations, but it also had to be specific as to the design requirements, so that the manufacturer knew what the circular required. To make the competition as fair as possible, the manufacturers could not consult Wright Field engineers which, in turn, prevented Wright from offering suggestions that could enhance the design. The manufacturers were also not allowed to submit mock up aircraft to Wright for evaluation that could discover design defects which could be more easily corrected in this mock up stage than in the more prototype phase. Changes in the prototype phase, once it reached Wright Field, could only be made by change orders to amend the issued contract, which was time and money intensive. The overriding factor in issuing a circular was getting a design into quantity production as soon as possible, and the AAC assumed that the manufacturers would submit a wholly developed prototype that would be ready to go into production. But to win the competition, the manufacturers had to design aircraft with innovations that were not yet combat proven, and thus the prototypes tended to be more experimental in nature, and would later require numerous contract change orders. Prototypes were very expensive to build, especially four engine bombers, whose airframe costs rose 300 to 400%, and the time for their fabrication increased from months to as much as two years during the 1930s. Also, the manufacturers always faced the very real possibility of not having their design and prototype accepted, and having to absorb the entire cost of the project, as there was a very slight chance that the rejected design could be sold to foreign air forces or used commercially. The manufacturers were between the proverbial rock and hard place, as they were forced into bidding simply because they needed the business in the poor economic environment of the Depression, and their failure to enter the competition would result in leaving the manufacturer behind its contemporaries in developing combat technology, but then it would save them the high expense of developing a design and building a prototype without assurance of a contract. In the late 1930s, the result of the 1934 Procurement Policy was a declining number of bidders for government contracts, as the economic conditions improved, and the large manufacturers were receiving contracts from commercial airlines and air transport companies. In 1938, manufacturers led by Consolidated Aircraft Corporation's Reuben Fleet suggested remedies to increase bidding. Fleet had once been a procurement officer for the Army in the early 1920s, and had been committed to the problems of aircraft procurement and legislation. Over the years Fleet had made several proposals that culminated in his suggesting that legislation be passed to authorize the War Department to procure aircraft in production quantities by negotiated contracts, rather than bids involving prototypes. Of course, the War Department could not endorse this idea without rankling Congress, and its desire for competition. When in a predicament, the government's solution was to convene yet another board to consider and revise yet another procurement recommendation. The AAC and various aircraft manufacturers testified, and the board recommended a solution that was a compromise between competition with prototypes and a simple design competition leading to a negotiated contract. Before issuing circular proposals for production aircraft, the AAC would invite aircraft manufacturers to submit designs for evaluation. One or more designs would then be selected, and be granted experimental contracts for the construction of one or more prototypes. There was also the proviso that quality ideas and data from losing designs could be purchased and integrated into the accepted design(s). Detailed type specifications would not be prepared until the winning design(s) passed the final mock up phase, allowing the AAC and manufacturers to consider and discuss changes. Once the design was finalized, the AAC was to issue its circular proposal for a prototype aircraft to be built by manufacturers interested in procuring a contract for the production aircraft. The War Department would subsidize the building of the prototype of the winning manufacturer(s), but other manufacturers who could afford to build an unsubsidized prototype to specifications could also enter the competition, and preclude any accusations that free competition was being thwarted. The design winner could expect to be, and mostly was, awarded a production contract with his prototype. An impartial evaluation of the design was now based on the performance of the prototype, and was to reduce the number of design changes, and get the aircraft into production sooner. It was not until October 1938 that the Chief of Staff approved the board's proposal and a workable procurement proposal finally seemed to have been realized, but the threat of an impending European war and the need for large quantities of aircraft soon made the proposal mute.

The Air Materiel Command and the Long-Range Bomber Requirements

In the early 1930s, the Army Air Corps realized they needed a bomber with both speed and range. However, range was contingent on aircraft size; the more fuel carried for longer range meant a bigger aircraft with a larger wing to carry the increased fuel supply. Large aircraft were dependent on the availability of a sufficient power plant(s). In 1933 the requirements of a long-range bomber were discussed at Wright Field by chief engineer James Howard, his assistant Al Lyons, aircraft branch chief James Taylor, Hugh Kneer of the field service section, and Leonard "Jake" Harman, the Air Materiel Command representative. The group decided to categorize the future bomber types that would be required:

Wingspan, 75 feet; gross weight 15,000 pounds (that was already contemporary in the B-9 and B-10)
100 feet span; 40,000 pounds gross weight
150 feet span; 60,000 pounds gross weight
200 feet span; 150,000 pounds gross weight
250 feet span; 200,000 pounds gross weight
Etc.

In the fall of 1933 it was decided to disregard the No.2 category, and category No.3 was considered as a feasible choice; it was to be identified as "Project A." Jake Harman drew up the requirement, and Gen. Conger Pratt of the Air Materiel Command authorized the money for engineering designs, and sent the request to Washington for approval. At the time strategic air power had few advocates, but long-range defensive patrol aircraft did have their proponents in Air Chief Benjamin Foulois and the Army General Staff. In 1933 Foulois directed his assistant, Brig.Gen. Oscar Westover, to fly missions testing the defense of the West Coast. Westover's report to the General Staff found the observation and patrol aircraft that flew these missions to be "woefully obsolete." But he suggested that modern bombers flying in formation with increased speed and range could frustrate any "known agency." Pratt was able to sell the idea of the 5,000 mile bomber, as it could protect Alaska and Hawaii, and this was a step in establishing a mobile GHQ Air Force.

Baker Board and the GHQ Air Force

In the early 1930s there was a dichotomy in the War Department. The conflict engaged the traditionalists, led by Army Chief of Staff, Gen. John J. Pershing, who believed that the infantry ruled the battlefield, while the Army Air Arm, led by the defrocked Gen. Billy Mitchell, espoused the future role of air power in warfare. The government organized a number of boards and committees to study air doctrine and its role in future warfare. The most influential of these groups was the Baker Board, and its report of July 1934 that recommended America's national defense policy was to be based on the defense of America and its overseas possessions, with the Army air forces fending off any invader until ground armed units and civilian forces could be mobilized. While seemingly to advocate air power, there was no mention of the use of the air force in any aggressive or strategic manner, and the air force was relegated to a supporting role to ground forces by the Baker Board:

"The idea that aviation can replace any of the other elements of our armed forces is found on analysis, to be erroneous. Since ground forces alone are capable of occupying territory, the Army with its own air forces remains the ultimate decisive factor in war."

Baker Board member Jimmy Doolittle voiced the minority dissent, supporting the separation of the Army and its air forces, and advocating the development of an air doctrine for its employment.

Although the Baker Board rejected an independent strategic air force, it did support the creation in March 1935 of a General Headquarters Air Force (GHQAF) that would be under the control of the Army commander in the field. However, there were conflicting opinions of the role of the GHQAF. One was that it would be allocated and attached to field armies, and utilized under their direct control. Another view had the GHQAF engaged as an integrated force acting to further the mission of the Army. The view held by most airmen was to have it act as a unified force, not only to support the Army, but also to act beyond the realm of the Army flying missions of its own. Maj.Gen. Frank Andrews was the CO of the GHQAF, and forcefully advocated the airmen's point of view. As the possibility of a war in Europe approached, the War Department was in a state of flux, and the GHQAF and its concepts were replaced by ideas formulated first by the Kilner Board in 1939, and then by the Joint Army-Navy Board, that culminated in the First Strategic Air Plan (AWPD-1) of 1941, which was formulated after the B-18 was already rendered obsolete by the B-17 and B-24, and the B-29 was far along in the planning stage.

Kilner Board Recommendation of 1939

In late 1938, Maj.Gen. Arnold called a secret meeting of AAC officers and aviation experts to discuss the future of the AAC and the bomber predicament. Charles Lindbergh had visited Hitler's Germany, and was impressed and alarmed over the emphasis that the *Luftwaffe* had put on bombardment aircraft. Arnold asked Lindbergh to serve on a board headed by Brig.Gen. Walter Kilner, and assisted by Col. Carl Spaatz, Col. Earl Naiden, and Maj. Alfred Lyon. The Kilner Board issued its report in June 1939, and members of the group thought that in a European war Germany would not only overrun the Continent, but also conquer Great Britain and then Africa, and perhaps establish bases in South America. If the Japanese began a war in the Pacific, it would have to be fought over long expanses of ocean, extending from Alaska to Hawaii to the Philippines, to as far as Australia. In this scenario the B-17 would surely be inadequate, and the super bomber notion was revived. The role of the super bomber was no longer thought to be as a strategic bomber, but was to be a defensive bomber able to fly long distances to strike an enemy advancing on the American Continents. The Board proposed a five-year research and development program for new engines and aircraft. Engine development was to be liquid-cooled, and range in horsepower from 1,500 to 2,400, and eventually to 3,000hp. A 2,000 mile four-engine heavy bomber was to be developed, along with another heavy bomber with a large fuselage that could carry bombs and/or enough fuel to give it a range of 5,000 miles. From the former specification the B-17 and B-24 were in the process of completing development, and to the latter specification the B-29 and ill-fated B-32 would eventually be developed "for hemispheric defense." The Board recommended that two-engine light and medium bombers be developed, with the B-18 being the only contemporary type available, and with the B-25 and B-26 available in the near future. Although the Kilner recommendation was accepted by the AAC, unfortunately it was at a time when there was an isolationist sentiment in America, and the AAC could do nothing until Congress appropriated funds.

AWPD-1 Renders Procurement Mute

On 9 July 1941, President Roosevelt sent a letter to the Secretaries of War and Navy, asking them to prepare "an estimate of overall production requirements required to defeat our potential enemies." To forestall the War Plans Division from preparing a plan that would relegate the Army Air Force into a supporting role to the Army, Arnold's Air Chief of Staff, Brig.Gen. Carl "Toohey" Spaatz, and a member of the 1939 Kilner Board, formed the Air War Plans Division (AWPD). The AWPD was to be headed by Lt.Col. Harold George, who was CO of the Second Bombardment Group, and the newly appointed Chief of the War Plans Division. Also appointed was Lt.Col. Kenneth Walker, Chief of the War Plans Group; Maj. Haywood Hansell, Chief of the European section of the War Plans Group; Lt.Colonels Lawrence Kuter, G-3 from the General Staff, and Orvil Anderson and Howard Craig. The responsibility of the AWPD was to formulate a general plan for the employment of air power to defeat the Axis that would utilize air power to its fullest capability. The AWPD was to determine the size of the air forces required to defeat Germany first, and then defeat Japan. Their plan, called AWPD-1, was submitted to the Army General Staff War Plans Division on 12 August 1941. Generally, it called for 24,500 combat aircraft, of which approximately 10,000 were to be four and six engine bombers. On 11 September 1941 Gen. Marshall accepted AWPD-1, and the aircraft production floodgates were opened; America became the arsenal of democracy, but without the Douglas B-18, which drowned in the flood.

3

Development of the B-18 DC Bomber

Introduction

On 14 August 1934, almost two years after the TWA Jack Frye bid invitation letter, Donald Douglas received another letter that also would have profound effect on this company's future. The letter was Circular 35-26 from the Army Air Corp's Air Materiels Division at Wright Field, delineating a specification for a new bomber. The Air Materiel Command had changed its thinking, and decided that a second bomber that was not as grand, as the "Project A" bomber could more easily be realized and put into production. Since both the Depression and Congress limited aircraft procurement funds, the new design would be more conservative, with a moderate performance envelop. The four or six man bomber was to have a minimum speed of 200mph and desired speed of 250mph, and was to carry a ton of bombs a minimum of 1,020 miles round trip, with the desired range set at 2,200 miles. The design for the potential order of 220 aircraft was to be tendered in a year. The Douglas proposal, designated the XB-18, was influenced by its DC-1/2/3 design, and would result in the largest order for a new bomber to date. The new bomber was to replace the then innovative Martin B-10, which went into service in 1934.

The Martin B-10, "Air Power Wonder of its Day"

After losing the backers in his Cleveland factory, the flamboyant Glenn Martin had spent the 1920s in California building the designs of other companies for the Navy, such as the Curtiss T3M and T4M, or underbidding rivals. Martin's 1924 low ball bid against Curtiss on the CS-1 was a move that saved Martin from bankruptcy, and allowed him enough profit to relocate his company to the East Coast in 1928. Inauspiciously, Martin moved into the new 300,000 square foot Baltimore plant with 250 employees only a month before the Black Friday stock market crash of 1929. The balance sheet of that year showed that Martin had lost $4 million on the sale of only two aircraft, his stock was virtually worthless, and the future was bleak, with less than a year's worth of order backlog even at reduced production rates. In December 1929 the Army's Air Materiel Command requested proposals for a revolutionary new bomber. Martin's engineering staff produced two unimaginative designs that were not even contemporary, much less avant-garde. The AAC offered to help Martin secure a contract if it would privately develop a creative new all-metal monocoque design at its own expense, and with no promise of a production contract. After six months of work

The all-metal B-10 monoplane bomber (Martin 139) looked sleek for its time, with its enclosed nose turret, and cockpit enclosures for its three man crew. The wing had a span of 70.5 feet, the fuselage length was 44.7 feet, and its height was 15.4 feet. It was a stable aircraft for bombing, and its two 675hp Wright R-1820 Cyclones made it faster at 207mph than any contemporary fighter. It was the first American bomber to have internal bomb stowage and retractable landing gear. (UASF)

Martin's design offering—the Martin 123, designated the XB-907 by the AAC—still fell short of expectations, having a fixed landing gear, open pilot's cockpit, and open dorsal and nose gunner's positions. Again Wright Field engineers stepped in and made suggestions. The wing and tail were to be redesigned, and a retractable landing gear added. In wind tunnel tests the aircraft showed even more problems, the most serious being poor directional stability, which required two months to correct. The problems were thought to be solved, but on its first flight on 26 February 1932, the prototype XB-907 (33-139) performed so poorly, and its engine vibrated so excessively, that the project was threatened with cancellation. Martin placed the blame on Wright, the engine manufacturer, and another two months passed before the matter was resolved. Gen. Frank Andrews of the Air Materiel Command was so exasperated that he decided he would request a $1,300 per aircraft refund from Martin, who put up a loud argument, and threatened to halt production. The AMC sent its refund request to Assistant Secretary of War Woodring, who added an extra $250 to the refund! Martin balked and stopped production, but finally relented, as his company was in the throes of potential bankruptcy, and at the time the Boeing B-9 was also in its testing phase and, while more expensive than the B-10, it was not encountering many problems.

At this point talented Wright Field engineers took over the project, virtually relegating Martin to a subcontractor status. The aircraft was thoroughly reevaluated, and sent back to the factory with numerous recommendations for changes. Without the B-10 contract the Martin Company would go into bankruptcy, and the company readily acquiesced to the seeming endless number of subsequent changes ordered by the Air Materiel Command.

The improved XB-907A aircraft was finally ready for retesting on 7 October 1932 at Wright Field, and was awarded the 1932 Collier Trophy for the number of innovations it displayed. The all-metal B-10 monoplane bomber (Martin 139) looked sleek for its time, with its enclosed nose turret and cockpit enclosures for its three man crew. The wing had a span of 70 feet, 6 inches, and an area of 678 square feet; the fuselage length was 44 feet, 8.6 inches, and its height was 15 feet 5 inches. It was a stable aircraft for bombing, and its two 675hp Wright R-1820 *Cyclones* made it faster at 207mph than any contemporary fighter. It was the first American bomber to have internal bomb stowage and retractable landing gear. After Martin spent $375,000 the AAC awarded the company a contract for 48 aircraft at a unit cost of $50,000, and reimbursed the company for the prototype. The AAC was so pleased with the B-10 that in 1934 it spent nearly all its yearly procurement funds on 88 B-10Bs, and then purchased 15 more in 1935. A total of 154 were built: 121 as B-10 and B-10Bs, and later 32 as B-12s and one XB-14—keeping Martin production lines busy for three years. The B-10B was the most produced type, and was powered by a 790hp Wright R-1820-33 engine. The performance and toughness of the B-10 was demonstrated on 19 July 1934, when ten aircraft led by Lt.Col. Henry Arnold flew a 7,360 mile round trip between Washington, DC, and Fairbanks, Alaska. One of the aircraft ditched in the Cook Inlet off Alaska, but due to its watertight wings and horizontal stabilizers was salvaged, and later flew the return trip. The bomber served with 28 squadrons over seven years of operational duty. After its introduction Hap Arnold described the B-10 as "the air power wonder of its day." The B-10 led to the B-12, which was to be overshadowed by the B-18, and then the redoubtable B-17.

Douglas Enters the Bomber Market with the DB-1/XB-18

Considering that they had one year until the 14 August 1935 deadline to meet the Air Materiel Command bomber specifications, Douglas and his engineers thought that they could easily convert the faster, further, higher performance with greater payloads DC airliner into a bomber. However, if the company did win the competition, it did not have the time or space to build another assembly line, as the DC-2 line was at full capacity, and orders were coming in for the new aircraft. Douglas knew that eventually the commercial air transport market would reach a saturation point, and he would need a government contract. Douglas consulted with Carl Cover, who was in charge of military sales, and asked his advice on producing a bomber. Douglas had previously considered a bomber version of the DC-1, and had sent Cover to Wright Field to explore this possibility, and to scout out any scuttlebutt on the competition. On his return Cover told his boss that in the competition Martin would offer an improved version of its B-10, and rumor was that Boeing had a new four engine design named the Model 299 on the drawing boards.

Douglas hardly hesitated in making his decision, and authorized his engineering department (Raymond, Kindleberger, Herman, Stineman, and Oswald) that had worked so long and successfully in developing the DC-1 and DC-2 to start the new bomber project. Carl Cover was of particular value due to his military experience and contacts. The primary design problem was to utilize the basic DC design as much as possible to avoid retooling, which would add expense and take valuable time. The DC-2 low wing configuration would have to be discarded, as it prevented the bomb bays from being located as near as possible to the aircraft's center of gravity for the best stability. To solve the wing problem, the new DC bomber would be a mid-wing design with the fuselage extending slightly below the wing, which would give the bomber its distinctive slightly potbellied look. Also, the wing of the new bomber needed to be lengthened. The original DC-1 wing had been lengthened in the DC-2, but to extend it even further for the new bomber, it was thought that a new joint would have to be developed, but Douglas engineers easily increased the wing span with the addition of rounded wing tips. With the new mid-wing configuration new engine nacelles had to be designed, which was always a challenging project. Also, the design team decided to use the basic, but a slightly larger DC-2 tail. To convert the forward fuselage of an airliner to that of a bomber caused another problem, as a bombardier's compartment and defensive armament had to be placed without upsetting the weight balance and aerodynamics of the aircraft. The bombardier's compartment was placed in the nose, forward and below the pilot's cabin, which gave the aircraft a distinctive tough, pug-nosed appearance. All the new holes cut into the fuselage for bomb bays and turrets required that the airframe be strengthened. The landing gears also had to be strengthened, due to the increase in weight and faster landing speeds. Once the design was completed, it was best described by Dutch Kindleberger as "looking like a DC-

The Douglas Company designated their design as the DB-1 (Douglas Bomber #1), while the AAC called it the XB-18. It was best described by Dutch Kindleberger as "looking like a DC-2 only a little bit pregnant." The aircraft was to have six crewmen: pilot, co-pilot, bombardier, and three gunners operating three .30 caliber machine guns. (USAF)

2 only a little bit pregnant." The aircraft was to have six crewmen: pilot, co-pilot, bombardier, and three gunners. Defensive armament consisted of three .30 caliber machine guns, one in each manually operated nose and dorsal turret, and one firing from a ventral hatch. The dorsal turret was located just forward of the vertical stabilizer and was fully retractable, and had a rectangular top, so that it could lay flush with the upper fuselage when retracted.

The Douglas Company designated their design as the DB-1 (Douglas Bomber #1), while the AAC called it the XB-18. In April 1935 the prototype was ready, and was subjected to a series of factory flight tests that showed its DC-2 pedigree, displaying the proven flight characteristics of the best designed and best built aircraft in the world. The DB-1 was ready by the July deadline, and was flown to Wright Field to meet its competitors.

The Competition

Martin Model 46/B-12

As production of the first 48 B-10s began, the Martin Company was forced to declare bankruptcy in the spring of 1934, as it lost assets in the failure of a Baltimore bank. Martin employed 2,000 workers, and because it had been awarded the B-10 contract, the government decided that the company was vital to the defense of America and required aid. The RFC extended a $1.5 million loan, with Glenn Martin using his facility as collateral. The B-10 had been operational for only a brief time when the Air Materiel Command circular of 14 August 1934 arrived. Because of its strained finances, and under the prevailing AAC procurement policy, Martin tried to meet its competition by upgrading its B-10 as the Martin Model 146 (Army B-12). The mid-wing aircraft measured 45 feet, 3 inches long, and had a longer wing span at 75 feet, fitted with Flower flaps. It was powered by two Wright 775hp engines that gave it a top speed of 236mph, and its increased fuel capacity increased its range. It carried a crew of four (pilot, co-pilot, bombardier, and rear gunner) in a long, narrow fuselage that seated the crew in tandem. The pilots and navigator sat in the forward position under the canopy, and the gunner just behind the trailing edge of the wing under another canopy of bulletproof glass. The aircraft's distinguishing feature was the bird cage nose turret that carried the bombardier/gunner. There were three .30 machine guns: one in the nose turret, and two in the rear gunner's position. The B-12's gross weight was 12,900 pounds, and was scheduled to carry 2,260 pounds of bombs over 2,000 miles. Sitting next to the DB-1 on the Wright Field tarmac, the B-12 was dwarfed. The Douglas entry had a wing span that was 14 feet longer, a length that was 11 feet longer; and a gross weight that, at 25,000 pounds, was almost double that the Martin bomber.

But all the Martin upgrades increased the B-12's weight, which downgraded its performance, and it never was a contender in the

Martin tried to meet its competition by upgrading its B-10 as the Martin Model 146 (B-12). But all the upgrades increased the B-12's weight, which downgraded its performance, and it never was a contender in the competition, although the AAC ordered 32. (USAF)

The Boeing 299 was a four-engine, low-wing monoplane that was essentially an upscaled version of Boeing's streamlined Model 247 twin-engine airliner. During AAC evaluation tests of the three entries at Wright Field the Model 299's better performance and reliability earned it the lead in gaining a contract. Then, on 30 October 1935, in the midst of testing, the Model 299 became uncontrollable and crashed just after take off, and was unable to complete the testing program, giving Douglas the contract by default. (Boeing via Pima)

competition, although the AAC ordered 32. Another cause for the elimination of the Martin B-12 from consideration can be blamed on the company, as the AAC had had it with Martin's reluctance and inability to solve its problems during the bomber's development as described previously. But by the time of the B-12's development, Gen. Andrews and the Air Materiel Command were thoroughly irritated with Martin and his aircraft, and there were two strikes against the B-12 before it entered the competition. With the lack of orders Martin was forced to enter the export business selling to threatened nations, particularly the Dutch, who were scrambling to purchase warplanes during the rise of Hitler and Mussolini in prewar Europe.

Boeing Model 299

While Martin and Douglas followed the AAC's modest design specifications, submitting rather unpretentious twin engine proposals based on contemporary technology, Boeing chose to consider the performance specs as guidelines only. The result was a four engine bomber that was completely different from its competitors, and would develop into the superlative B-17 Flying Fortress.

The AAC proposal was for "multi-engines," which to the AAC meant two, but to Boeing it meant four. To work on preliminary designs for the Model 299, Boeing President Claire Egtvedt assigned E.G. Emery as the project engineer, with Ed Wells as his assistant, and Frederick Laudan as the construction supervisor. The bomber was to be a low-wing monoplane that would have four engines to power it, and was essentially an upgraded version of Boeing's streamlined Model 247 twin-engine airliner that had been first flown in February 1934. Concurrently, Egtvedt ordered Emery to develop a four-engine airliner based on the Model 299 design that was to become the Model 307 Stratoliner. With Emery running the Model 307 project, Egtvedt placed 24-year-old Edward Wells as the project manager for the Model 299, which had been unofficially designated as the "XB-17" by Boeing before it had been approved by the AAC. Due to the importance of the two programs, Wells assigned an astounding 153 engineers to the project, which started the initial drawings in August 1934, and completed the final drawings only three weeks later. On 26 September 1934, the board of directors of the newly independent Boeing Airplane Company met for the first time after parent United Aircraft & Transport was finally dissolved. The company had $583,000 in cash from its share in United, but its payroll was down to 712 employees from a peak of 2,264 in 1933. The new Boeing board bravely voted $275,000 to design and build a bomber to meet the new AAC specification using the final blueprints that were completed in December. After the usual numerous design problems the Model 299 was rolled out after using over $400,000 in company reserves. It made its first flight on 28 July 1935, and journalist Richard Williams of the *Seattle Times* was so impressed by it size he was the first to dub it the "Flying Fortress."

It was not until 20 August that project test pilot Leslie "Les" Tower flew the 299 cross country 2,100 miles to Wright Field for evaluation, and landed with a flourish at an average speed of 232mph in a record time of nine hours and three minutes. If the DB-1 had dwarfed the B-12, then the 299 was a giant in comparison to the Douglas DB-1 bomber. The four engine bomber weighed almost 35,000 pounds vs. 25,000, was almost 12 feet longer, and had a wing span over 14 feet longer. But the aircraft not only dwarfed the Douglas physically, but also in performance: its four single row Pratt & Whitney 750hp engines propelled it to a top speed of 250mph over a maximum range of 3,000 miles at 30,000 feet, all while carrying an eight man crew (pilot, co-pilot, bombardier/gunner, radio operator/navigator, and four gunners), and carried a 5,000 pound bomb load. The 299 bristled with five manual guns mounted in turrets in the nose, belly, and on top of and on both sides of the fuselage.

From mid-August and into October 1935 the AAC evaluated the three entries at Wright Field, and the Model 299's better perfor-

The crash of the Boeing Model 299 certainly had a devastating effect on Congressional thinking and the ACC's procurement decision, which led to the contracting of 133 Douglas B-18s, and only 13 Boeing YB-17s. Pictured is a YB-17 flying near Mount Rainier. (USAF)

mance and reliability earned it the lead in gaining a contract. Then, on 30 October 1935, in the midst of testing, Boeing test pilot Les Tower, and AAC pilot Maj. P.P. "Pete" Hill, and the test crew took off in the Model 299. But once it was airborne the bomber became uncontrollable, and as it passed over the end of the runway it crashed in a nearby pasture. Hill was killed on impact, but Tower survived with horrible burns, and died 20 days later. Lt. Donald Putt was pulled from the burning wreckage and given little chance of survival, but Putt did survive, and four years later was to become the project manager of the B-29 program. Investigation showed that the ground crew had not unlocked the tail control surfaces, causing the crash that would put the Boeing program in jeopardy, as it was disqualified for not completing the test program. The crash protracted the testing of the DB-1 and B-12, and the final decision. There were strong arguments supporting the Boeing bomber based on its superior testing results before the crash, and it would not be until near the end of the year that the decision was made.

Why the B-18 and not the YB-17?

The crash of the Boeing Model 299 certainly had a devastating effect on Congressional thinking, and the ACC's procurement decision that led to the contracting of 133 Douglas B-18s, and only 13 Boeing YB-17s. The 13 aircraft order kept Boeing financially viable, as its losses on the Model 299 after insurance reimbursement was $500,000; a great amount of money for that time. YB-17 manufacture would at least keep Boeing employees busy on a project that the company firmly believed would be very successful.

With the AAC's decision the B-18 was to be put into immediate production with minor modifications, and then to be put into squadron service as soon as possible. Compared to the four engine YB-17 the B-18 was much smaller and lighter, and its performance was slower, its range shorter, ceiling lower, and it carried a smaller bomb load; it was considered to be a medium bomber. The XB-17 had more range, more speed, more bomb carrying capacity, more defensive firepower, and had the potential for far greater design modification. In June 1936, Congress authorized an increase of serviceable aircraft from the 1,800 funded by the 1926 Air Corps Act to the 2,320 recommended by the Baker Board. The mid-1930s was a time when American air policy emphasized hemispheric defense. The GHQ Air Force and the AAC wanted more four engine bombers to use to implement its defense objectives; long range aerial reconnaissance, and to reinforce and supply Hawaii, the Panama Canal, and Alaska in case of an emergency. However, the General Staff considered four engine bombers as offensive weapons, and it was the Navy's responsibility for the long range defense of coastal America. The medium bombers were considered more desirable, as they could be more easily adapted to the Army's intention to use them in support of the ground army. More medium bombers could be purchased for the same money as a lesser number of heavy bombers, and the goal of 2,320 aircraft could be fulfilled, and replacements would cost less. The B-18 became known as the "budget bomber," a decision that would have consequences in coming years.

Comparative Costs of the August 1935 Bomber Competitors

	Martin B-12	Douglas B-18	Boeing YB-17
Quantity Ordered	32	133	13
Total Cost per Aircraft	$72,000+	$105,000	$302,100++
Airframe Cost with Changes	$45,450	$75,243	$246,030
GFE* Cost	$13,650	$16,957	$23,261
Cost per Each Engine	$6,500	$8,200	$8,200
Engine Cost per aircraft	$13,000	$16,400	$32,800

+low cost per aircraft due to previous development as the B-10
++production cost of a B-17 was projected to be $176,000 per aircraft *GFE=Government Furnished Equipment

Army Air Corps Officers led by Gen. Frank Andrew's "Langley" faction realized that the B-18 did not fulfill their requirements, even though they were less expensive. Using the results of an invasion exercise by an "enemy" task force led by the Battleship *Utah*, approaching San Francisco, Gen. Andrews and his staff prepared tables that demonstrated that a comparable fleet of B-17s was superior to a fleet of B-18s on a performance and cost basis. This exercise will be described later in detail. Their figures from this exercise established that it would require 50 squadrons of B-18s to match the work of 31 squadrons of the Flying Fortresses. With B-17s priced at a production run $176,000 each, and the B-18 at $103,000, the cost of the two striking forces on an efficiency basis would about equal at $79 million for the B-17 vs. $77 million for the B-18. Using these bomber efficiency figures, it was determined with the total 2,320 aircraft authorized for the AAC, the required fleet of B-17s left 1,857 other aircraft available for pursuit, attack, observation, transport, and training. If the less expensive B-

18 was purchased this figure was reduced to 1,569 aircraft. This B-17 biased data did not persuade Gen. Malin Craig of the War Department, who directed that no more heavy bombers be purchased, except "for experimental purposes, or until the international situation required it." Craig contended that the B-18 "satisfied practical requirements at a reasonable cost." The AAC then ordered two more B-18s for a total of 133, that was followed by an order for 217 B-18Bs. The first B-18 was sent to Wright Field in the fall of 1936, and then to Langley for testing. By mid-1937 the Bolo was released for unit use, and personnel liked the aircraft better than the Martin B-10s and B-12s it replaced. However, the upper echelons of AAC favored the Model 299/YB-17, despite the devastating effect of the 30 October crash of the prototype. The Air Materiel Command felt that the problem laid in the need "to develop the most advanced types too rapidly" (e.g. the crash of the ground breaking Model 299), and it would be better to settle on a more conservative proposal that could be produced without fear of crashes (i.e. B-18), and not meeting specifications (i.e. B-10). With the conservative B-18 design, the AAC would be assured of equipping its squadrons with a dependable and less costly, albeit mediocre, bomber. It was ironic that, before the war was over, the Douglas Long Beach facility would assemble over 3,000 Boeing B-17s!

4

The B-18 Description

The B-18 is Ordered and Put into Production

On 17 January 1936 the Army General Staff decided to place an initial order for 82 B-18s, and then upped the order to 133 B-18s. The General Staff ordered only 13 Model 299s, which were now designated as the YB-17 (the "Y" meant that the aircraft was not experimental, but a "service test" model that could, or not, precede production aircraft). When Boeing put the YB-17 into production it was designated as the Y1B-17 (the 1 because they were purchased out of F-1 fiscal year supplementary funds). The first Y1B-17 rolled off the Seattle line on 2 December 1936, and the last of the 13 was finished on 5 August 1937. One Y1B-17 was sent to Wright for testing, and the other 12 were assigned to the Second Bombardment Group at Langley Field, VA, under Lt.Col. Robert Olds.

Early B-18 Variants

DB-1 (XB-18)

The DB-1 (AAC designation: XB-18) prototype was brought up to B-18 specifications and redelivered to the AAC as 37-51. The DB-1 prototype was later modified and used in feasibility tests for the use of large caliber cannon in aircraft, which will be discussed later.

DB-2

In late 1937 the last aircraft in the order, 37-34, was scheduled for delivery to Wright Field to be developed into the DB-2 prototype. However, the aircraft was delivered on 8 November, out of sequence on the original B-18 contract, and was actually the 36th B-18 when it was delivered to Wright Field (via Joe Baugher). It was fitted with a redesigned nose equipped with a powered nose turret, and a modified bombardier position with large surrounding Plexiglas windows. The revision was not successful, and the aircraft was rebuilt as a standard B-18 and sent to the 18th Recon Squadron at Mitchel Field, NY.

B-18

The DB-1 design was brought up to production standards mainly by redesigning its nose to contain more lateral windows, and a bomb-

B-18As awaiting delivery outside the Santa Monica plant sometime in late 1938. (Douglas via Pima)

Since the B-18 was a variant on the Douglas DC airliner series, much of the manufacturing process could easily be adapted to its production at the Douglas Santa Monica plant. Pictured are early B-18s awaiting the installation of their Wright R-1820-45 engines and outboard wing panels. (Douglas via Pima)

The DB-1 (AAC XB-18) prototype was brought up to B-18 specifications, and redelivered to the AAC as 37-51. (USAF)

In late 1937 a B-18 was delivered to Wright Field to be developed into the DB-2 prototype, which was fitted with a redesigned nose with a powered nose turret, and a modified bombardier position with large surrounding Plexiglas windows. (USAF)

The B-18 was the DB-1 design brought up to production standards, mainly by redesigning its nose to contain more lateral windows, and a bomb-aiming window in its lower forward section, reducing its fuselage length by seven inches, and adding military equipment. (USAF)

aiming window in its lower forward section, reducing its fuselage length by seven inches, and adding military equipment. The additional equipment and changes increased its weight by nearly 1,000 pounds, which caused the performance to fall off slightly. The B-18 was powered by a pair of Wright R-1820-45 radial engines accommodated in modified nacelles. The initial production order under contract AC 8307 was for 133 aircraft: AAC serial numbers 36-262 through 343; 36-431 through 446; 37-001 through -34; and 37-51 were added later.

Further Developments

B-18A

The B-18 had its drawbacks, particularly the nose configuration, which was cramped, and caused the gunner and bombardier to be uncomfortable and inefficient. This shortcoming led to the next and last production version of the B-18, the B-18A. Douglas altered the nose design by having the bombardier and gunner change positions. The top position was moved forward, with the nose being extended into what was deemed a "shark nose" to accommodate the bombardier and bomb sight. The flexible forward-firing nose gun was relocated, and was mounted in a spherical ball turret located further aft and below, which resulted in the very unusual arrangement of having the bombardier being located above the gunner. This nose configuration increased the aircraft length by 14 inches to 57 feet, 10 inches. The 930hp Wright R-1820-45 powering the B-18 was replaced by the 1,000hp R-1820-53, which featured fully feathering hydromatic propellers. A transparent domed cap was added over the top of the dorsal turret that caused it to be no longer flush when fully retracted. The weight of these changes increased the gross weight of the B-18A some 993 pounds to 22,123 pounds, and reduced the top speed to 215mph. So, after 134 production B-18s, the Douglas line converted to the B-18A. On 10 June 1937 177 B-18As were ordered, and an additional 78 were added to the contract on 30 June 1937. On 15 April 1938 the first B-18A was

The B-18 shows its Douglas DC airliner lineage in its wings and empennage configuration, but this ventral view shows its pot belly. (USAF)

The B-18 gunner and bombardier were cramped and inefficient, and this shortcoming led to the next and last production version of the B-18, the B-18A. Douglas altered the nose design by having the bombardier and gunner change positions. The top position was moved forward, with the nose being extended into a "shark nose" configuration to accommodate the bombardier and bomb sight. (USAF)

A dorsal view of the B-18A better shows its DC airliner ancestry. (USAF)

flown, and then was delivered to Wright Field a month and a half later. The AAC ordered 255 B-18As at a cost of $65,800 each on Contract 9977, but only 217 were delivered, with the last 38 on the contract being manufactured as the B-23. The last B-18A was delivered in January 1940. The serials of the B-18As were: 37-458 to 634, 38-585 to 609, and 39-12 to 26.

B-18B

Of the 217 B-18As delivered to the ACC, 122 were modified to B-18B specifications to be converted into anti-submarine warfare (ASW) aircraft to counter the German U-Boat threat off the U.S. Atlantic coast, and in the Caribbean in 1942. At the time the AAC needed an aircraft that could be fitted with ASW location/detection equipment, but couldn't draw on any of the new B-24s and B-17s coming off the production lines, as they were sorely needed elsewhere. Initially production B-18s were available, and scheduled to be converted to Bs for a new role as an anti-submarine patrol aircraft. However, tests showed that modifying the B-18 would be too extensive and time consuming, and instead, B-18As were chosen and ferried to San Antonio, TX, for the ASW conversion.

The shark nose was removed, and was replaced by a large bulbous radome that housed the SCR-517 ASV (Air to Surface Vessel) radar. The bombardier's position was moved below and behind the radome. The tail on some B-18Bs was modified by the addition of

Of the 217 B-18As delivered to the ACC, 122 were modified to B-18B specifications to be converted into anti-submarine warfare (ASW) aircraft to counter the German U-Boat threat off the U.S. Atlantic coast, and in the Caribbean in 1941-42. The shark nose was removed, and replaced by a large bulbous radome that housed the SCR-517 ASV (Air to Surface Vessel) radar. (USAF)

a Magnetic Anomaly Detector System (MAD) that was housed in a long tubular boom located behind and below the rudder. The B-18B also carried long range navigation equipment, and a number of aircraft had additional underwing bomb racks to carry an increased number of bombs and/or depth charges. The B-18B was credited with two U-Boat kills, but its ASW role was effectively 18 months. As soon as the Consolidated B-24 Liberator became available it replaced the B-18, and the last Bolo missions were flown in August 1943.

B-18AM

The B-18 series underwent two modifications when 22 B-18s and 18 B-18As were designated as B-18Ms and B-18AMs, respectively. On 1 April 1940 ACC T.O. 01-40E-87 directed that the Type D-3 and D-7 bomb shackles on all B-18s and B-18As based in the continental U.S. be removed (designating them as M models), and were to be installed in B-18s based in Panama, Puerto Rico, and Hawaii, so that these bombers (designated AMs) could carry their maximum bomb loads. The "M" designation was to be removed when the shackles were replaced, which they were in only a few instances. The "M" type were mostly utilized as bomber crew trainers for transition to the B-26, but once the crews flew the B-26 they were in for a surprise, as the docile and forgiving B-18 was nothing like the "hot" Marauder.

C-58

There were at least three B-18As that were stripped of all offensive and defensive armament, and went back to their roots as transport aircraft designated as the C-58. B-18s numbered 37-508 and 37-563 were converted and sent to Waller Field, Trinidad, and assigned to the 39TRS on 31 December 1943. B-18 (38-604) was flown from Trinidad at the end of July 1943 to be converted in the U.S.

Digby Mk1

In 1938 Douglas was able to interest the Royal Canadian Air Force (RCAF) in acquiring 20 B-18A bombers that they designated as the Digby Mk1. The RCAF Digby only differed from the B-18 by utilizing the .303 caliber machine gun and British and Canadian equipment. The Digby entered Canadian service mainly with the 10(BR) Squadron, assigned to anti-submarine patrol off the Canadian North Atlantic coast, where it was credited with one U-Boat kill. The RCAF serial numbers were 738 through 757. The Digby will be discussed in detail later.

XB-22

The XB-22 was a proposed B-18 airframe mounting 1,600hp, Wright R-2600-1 *Cyclone* engines that were equipped with a single stage, single speed supercharger. No B-22s were built.

"Bolo"

The name Bolo was an unofficial designation, and was used mainly by the B-18Bs in the Caribbean. In Latin America the bolo is a large machete-like knife.

B-18 Description and Dimensions

B-18 vs. the B-10

When the AAC ordered the B-18, it was confident that it would be successful, considering its Douglas DC pedigree, as much of its design was identical to its famous parent, particularly the wings and tail assembly. The sleek, all-metal Martin B-10 had been the first step out of the wire, fabric, and wood era of aircraft design, and while the B-18 featured and refined the B-10 innovations, it brought nothing new to bomber design. In reality, the B-18 was not much better than the B-10 it replaced, and the best that could be said about the design was that, while it imitated its parent, the DC

The sleek, all-metal Martin B-10 had been the first step out of the wire, fabric, and wood era of aircraft design, and while the B-18 featured and refined these B-10 innovations, it brought nothing new to bomber design. In reality, the B-18 was not much better than the B-10 it replaced, and the best that could be said about the design was that while it imitated its parent (the DC-1) and its predecessor (the B-10), it did validate the all-metal design for a bomber. (USAF)

airliners and the B-10; it did validate the all-metal design for a bomber. The bomber conversion denigrated the DC-1's streamlined appearance, flattening the nose into a pug shape, and the addition of a bomb bay gave the bomber a pot belly. Both bombers were well built and rugged, but the B-18 was more roomy, comfortable, and sophisticated than the Martin. But for a new bomber design its defensive armament was inadequate, and the same as the B-10, mounting three .30 caliber hand-held machine guns, one each in a nose turret, dorsal turret, and in the rear fuselage floor. Each turret rotated, and was manually operated. The dorsal turret was retractable and "popped up" when needed, but when raised it increased drag nearly as much as lowering the landing gear. Although the B-18 possessed over 500hp more than the B-10 (two 675hp vs. 930hp Wright *Cyclones*), it had to propel a bomber weighing nearly four tons more than the B-10. The B-18's top speed of 225mph was only 10mph faster, it cruised at nearly identical speed (167mph), climbed only 100 feet per minute faster (1,355 vs. 1,450), and its service ceiling was only 2,000 feet higher. The B-18 was 12 feet longer, and had a wing span 19 feet longer than the B-10, and wing area 281 square feet larger than the Martin. Although larger, the B-18 was only slightly more advanced than the B-10, and afterward each succeeding generation of American bomber design would leap frog the next in design and technology, and the YB-17 would be the forerunner of this trend.

Douglas manufacturer's plate for B-18A completed on 10/17/39. (Pima)

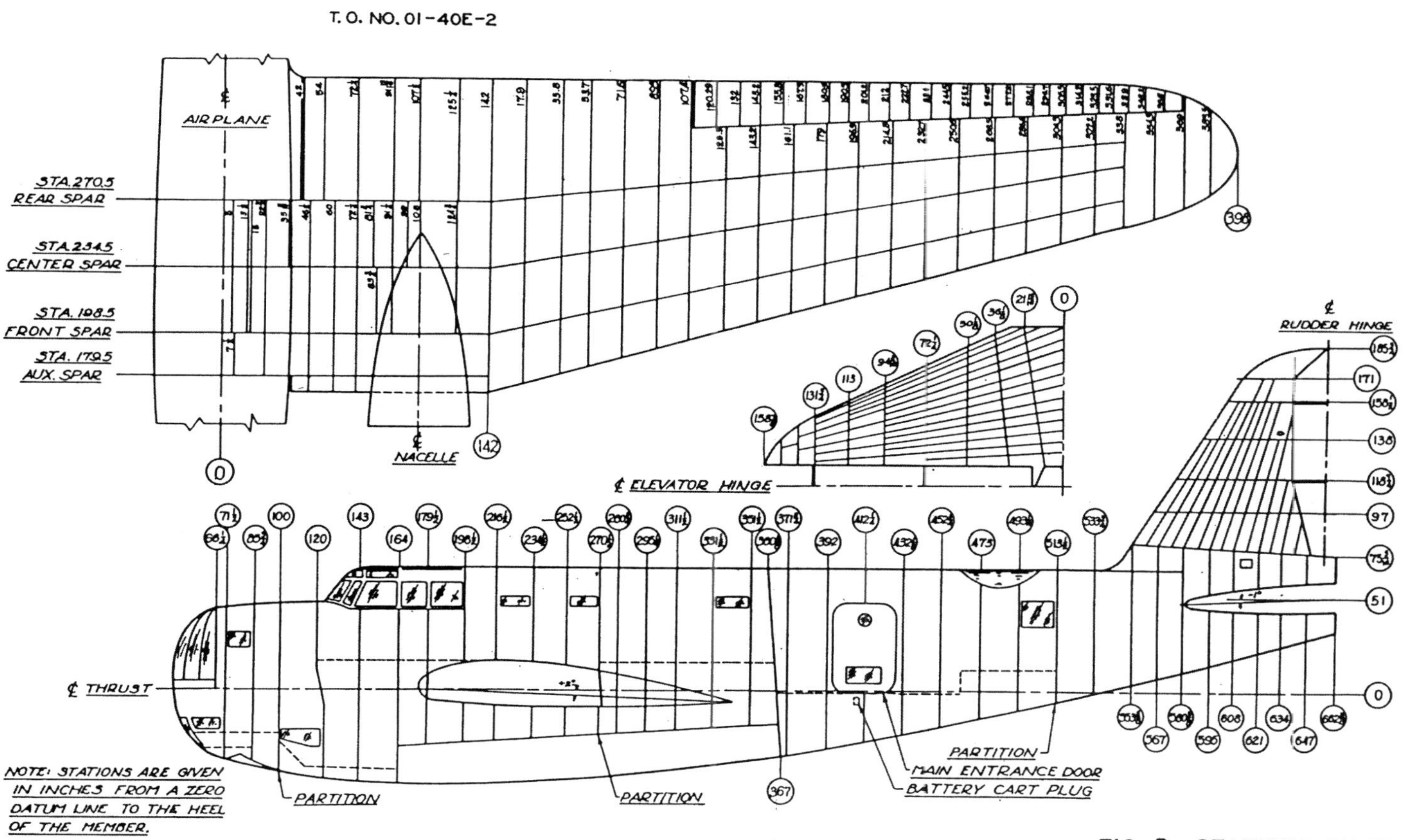

FIG. 2 - STATIONS DIAGRAM

(USAF)

B-18 General Dimensions

Overall Length	56ft./9.2in.
Overall Height (at rest)	13ft./4in.
Height at Propeller Hub (at rest)	8ft./6in.
Wing Span	90ft.
Wing Area (including ailerons)	965 sq.ft.
Wing Flap Area	64.2 sq.ft.
Aileron Area (each)	48 sq.ft.
Horizontal Stabilizer Overall Span	26ft./8in.
Horizontal Stabilizer Total Area	95.8 sq.ft.
Elevators Overall Span	26ft./8in.
Elevators Total Area (each)	41.7 sq.ft.
Vertical Stabilizer Height	13ft./4in.
Vertical Stabilizer Area	37.9 sq.ft.
Rudder Area	46.6 sq.ft.
Landing Gear Width	18 ft.

Fuselage

The semi-monocoque fuselage was divided into six compartments running from Station 0 to 632 (in inches):

Front Gunner's Compartment: Located in the nose of the aircraft and entered through Station 100.

Bombardier's Compartment: Was referred to as the Bomber's Compartment in the T.O. It was located immediately aft of the front gunner's compartment and below the pilot's compartment. A floor mat was provided, along with a seat cushion that was attached to the bombardier's back rest. The side walls of the compartment were lined with soundproofing material.

Pilot's Compartment: Was located forward and above the wing, and was entered through a door from cabin. The floor was covered with a dark green rug, and the side walls were lined with soundproofing material.

Cabin

Bomb Compartment: Located in the lower section of the fuselage between bulkhead Stations 164 and 368.

Tail Compartment: Located immediately aft of the cabin, and was accessible through a door in the rear cabin wall. The aft end of the fuselage terminated in a tail cone.

The fuselage was constructed so that the section below the wing could be removed as an integral part of the wing.

B-18 B Compartment Color Gallery

Co-pilot's position. (Author/Pima)

Looking into pilot's compartment from the cabin. (Author/ Pima)

Pilot's position. (Author/Pima)

Entrance to nose (radar equipment and gunner's) compartment. (Author/Pima)

Nose compartment, with .30 caliber machine gun on floor, and two 250 round ammo boxes. The B-18B had an ASV radar set located above the gun (the ASV radar unit is pictured in Chapter 9). (Author/Pima)

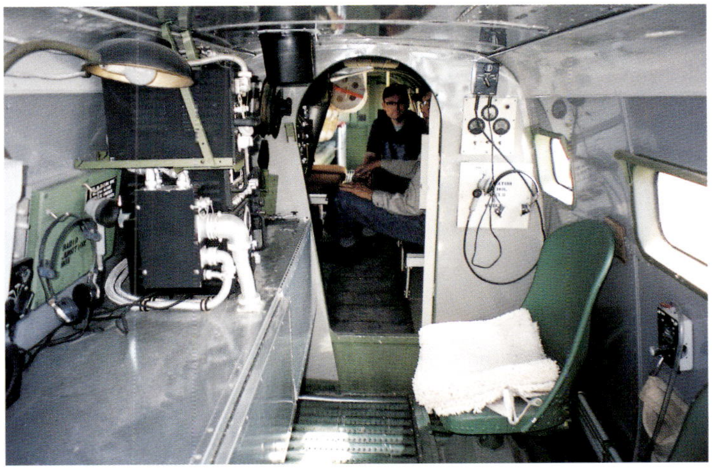

Looking aft from the pilot's position, showing the radio and radar equipment. (Author/Pima)

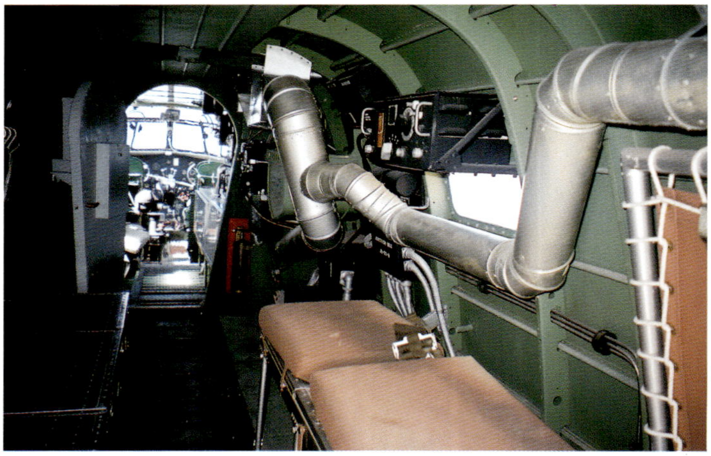

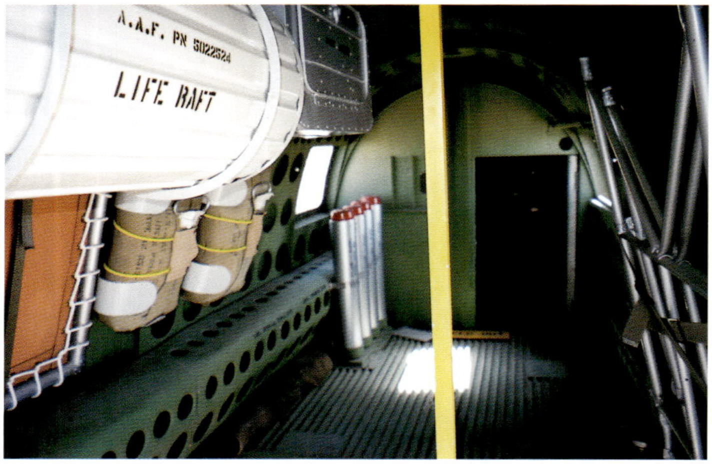

Above: Cabin at the main door looking forward. (Author/Pima)
Above left: Mid-cabin looking forward, with bunk/seats and ducting for heating apparatus. (Author/Pima)
Second left: Cabin at the main door looking aft toward tail compartment. (Author/Pima)
Lower left: Looking into tail compartment at tail landing gear. (Author/Pima)

Forward fuselage and bomb compartment (at arrow). (Author/Pima)

B-18 bombardier's/gunner's nose windows, pilot's windows, and escape hatch, located in the floor in the right hand side of the front gunner's compartment. (USAF)

B-18A "Shark Nose" window configuration and escape hatch. (USAF)

Doors

Main Entrance Door

The main entrance door was located between Stations 392 and 432, and entered into the cabin on the port side; it could be used as an emergency exit by removing the hinge pins that were attached to a cable located in the forward door. A removable V-shaped ladder was constructed of aluminum alloy, and was attached to the fuselage by two brackets. A strip of sponge rubber protected the fuselage from the point of contact of the step.

Emergency Exit Doors

The escape hatch was located in the floor in the right hand side of the front gunner's compartment, and was covered with a corrugated metal guard. It could be opened from the inside or outside, and when lowered it could be used for entrance and exit.

An emergency escape hatch for use by the pilot and co-pilot was located above and slightly aft of the pilot's seat. The forward end was hinged on removable pins. To release the hatch, the lock on the aft end was released, and the pins removed from the forward end. The hatch then was to be pushed upward on the forward end to allow the slipstream to carry it away.

Fuselage entrance door removable step. (Author/Pima)

Fuselage entrance door. (Author/Pima)

Pilot's windows. (USAF)

Pilot's window looking out. (Author/Pima)

Windows

Pilot's Compartment Windows

The pilot's windshield was "V-shaped," with each side of the "V" being made up of one stationary glass panel and one movable glass panel. The glass was 1/4 inch laminated shatterproof glass. The movable panel could be opened by loosening four thumb screws that held it in place, and sliding the section toward the point of the "V." The thumb screws pressed the glass against a formed rubber molding, making it watertight. However, when the molding aged it became less watertight, especially when the Bolo flew in the tropical weather of the Caribbean.

Pilot's side windows were made up of six panels, three on each side. The forward panels were movable, sliding the section forward to close and aft to open. The stationary panels were made of 1/8 inch laminated shatterproof glass.

The pilot's ceiling windows were made up of six panels: two upper front sections; two upper side sections, and two top sections of 1/4 inch laminated shatterproof glass. There were two adjustable sun visors for the pilot and co-pilot.

Front Gunner's Windows

There were six windows in the compartment to give the gunner full vision.

Bombardier's Windows

In the front of the bombardier's floor was an adjustable viewing section composed of two sections; one of metal and one of glass. A hand crank on the right hand side moved the section inward, giving forward vision when raised and downward vision when extended. Two windows, one on each side, provided sideways vision.

Fuselage Equipment

Seats and Tables

Pilot's Seat: The pilot's seat was fabricated from a formed aluminum alloy sheet, and was installed side-by-side with the co-pilot's seat in the forward area of the compartment. The pilot's seat was

Nose compartment looking out. (Author/Pima)

Pilot's seat. (Author/Pima)

mounted on two vertical tubes punched with holes so that a spring loaded plunger could be inserted to hold the seat in any desired position, with a plunger release lever located on the left hand side of the seat. A length of elastic shock cord with adjustable tension was used to raise the seat. The pilot's seat was capable of fore and aft pivot movement about a fitting on the back of the seat to accommodate pilots of different heights.

Co-pilot's Seat: The co-pilot's seat was similar to the pilot's, but had the control handle on the right hand side. The co-pilot's seat supports were fixed to the floor, making it adjustable only in the vertical direction.

Navigator's Seat: The navigator's seat, made from riveted aluminum sheet, was located on the left side of the pilot's compartment aft of the pilot's position, and forward of the radio table. It was of the swivel type, capable of rotating 90° with a control handle under the seat that released it so that it could move from one direction to another.

Radio Operator's Seat: This seat, also made from riveted aluminum sheets, was located on the left side of the fuselage, opposite the liaison set. It was fitted with spring loaded plungers on the forward two wheels, and fit into holes into track installed on the fuselage floor. It was able to swivel fore and aft about 4.5 inches.

Seat Upholstery: Each bucket seat was provided with a spring cushion seat and a back It was covered on the outside with leather, and with duck on the inside surface. Cotton batting was used to fill the seat cushions, while kapok was used to fill the backs. The backs were held in place by a pocket that fit over the top of the seat, and four flaps that buttoned onto the seat.

Safety belt brackets were provided on the bucket seats for the installation of Type B-10 safety belts.

Radio Table: The radio table was installed on the starboard side of the fuselage between Station 198.5 and 252.5, and was made from aluminum alloy corrugated sheet covered by a flat aluminum sheet. It was supported by three channel legs on the inboard side, and was

Radio operator's seat. (Author/Pima)

Bunks. (Author/Pima)

fastened to the fuselage wall on the outboard side. The table had two additional leaves, and had a drawer installed in its forward end.

Sleeping Accommodations: Provisions were made for four chrome molybdenum frame bunks with cotton duck stretched tightly over the frame. They were located in the cabin and installed as follows:

Main bunk: The main was located to the left side of the fuselage aft of the bulkhead at Station 270.5. It could be lowered and used while the aircraft was in flight or on the ground. When not in use it could be folded up against the fuselage wall. It was attached to the fuselage frame by two fittings, and was supported by two legs inboard. Auxiliary bunks: The three remaining bunks, if installed, could only be used when the plane was on the ground. The right front bunk was supported similarly to the main bunk. The forward end of the upper rear bunk rested on the end of the bomb bay floor, and the aft end was supported by cables. The forward end of the lower rear bunk rested in a channel of the rear hinged floor, and its aft end rested on the forward end of the upper gunner's floor. These auxiliary bunks were carried as a special load, and were stored between Stations 371 and 432 by adjustable web straps.

Toilet: The toilet was located on the right hand side of the fuselage directly opposite the main entrance door. The receptacle was made of formed aluminum sheet, with a wooden birch seat and a corrugated metal top. The receptacle could become unlatched from the seat and cover for cleaning. After cleaning it was filled with one quart of disinfectant solution and one tablespoon of "Sanitor Special Disinfectant" per quart of water.

Toilet. (Author/Pima)

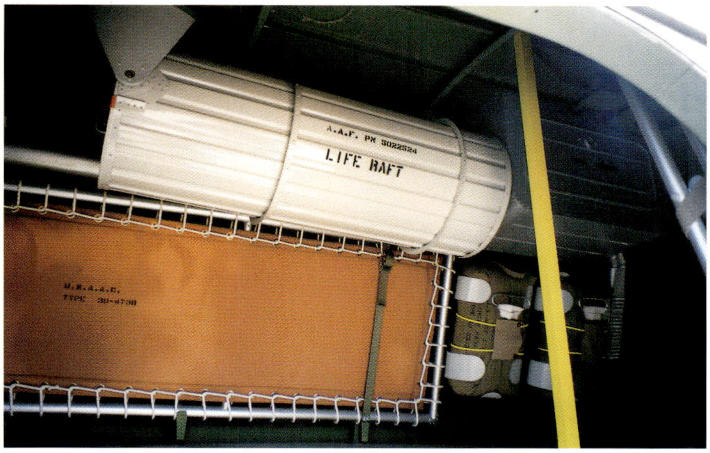

Life Raft Container. (Author/Pima)

Miscellaneous Fuselage Equipment:
The Type A-1 Life Raft Container was installed near the top of the fuselage, in a horizontal position, parallel to the centerline of the airplane, and directly opposite the main entrance door. A cable controlled latch suspended the aft end of the container, while the forward end was hung from brackets extending downward from the exterior access door.

The parachute brackets were located one on the left side of the fuselage adjacent to the lower escape hatch, and two on the right side of the fuselage aft of the life raft container. A container in which two parachutes could be stored could be placed under the forward end of the radio table. The pilot and co-pilot were provided with seat type parachutes, and no brackets were required in their compartment.

Three Type A-1 life preserver cushions were provided, with two located on the left side, and one on the right side of the fuselage beneath the floor, adjacent to the camera bay.

Warning Bells: Three warning bells were operated simultaneously by a switch on the pilot's electrical panel. The bells were located in the front gunner's compartment, bombardier's compartment, and in the cabin.

Thermos Bottle: A two quart Type A-35 Landers, Frary & Clark thermos hung in a support bracket was located on the forward face of bulkhead 270.5. A paper cup dispenser was located on the wall adjacent to the thermos.

A maintenance platform constructed of welded molybdenum chrome steel tubing held together by four pins on the joint fittings that allowed it to be broken down was stored upside down on the left fuselage wall aft of the main entrance door.

The miscellaneous container was located on the left side of the fuselage forward of the main entrance door, and contained the landing gear safety pins, tail mooring cable, and spare antenna wire.

The equipment box was located to the left of the rear upper gunner's left step, and contained the Zerk gun, trouble lamp, tool kit, two engine covers, and four bomb hoist rings.

The Flight Report Holder and Map Case were installed on the right side of the fuselage aft of the co-pilot's seat.

Glare Curtain, Window Shades, and Blind Flying Hood: The glare curtain was provided for night flying, and was attached at Station 164. The blind flying hood was attached to the pilot's compartment structure on the left hand side to simulate blind flying conditions. These features were installed on aircraft 36-343, 36-431 to -446, and 37-1 to -33.

The B-18 (left) had its drawbacks, particularly the nose configuration, which was confined and inefficient, and led to the B-18A (right). Douglas altered the nose design by having the bombardier and gunner change positions. The top position was moved forward, with the nose being extended into what was deemed a "shark nose" to accommodate the bombardier and bomb sight. The flexible forward-firing nose gun was relocated, and was mounted in a spherical ball turret located further aft and below. (USAF both photos)

Bombardier's Panel. (USAF)

Bomb Compartment. (USAF)

Bombing Equipment
Bombardier's Compartment:

Bombardier's compartment was located in the forward part of the aircraft below the pilot's compartment, and contained the Type M-1 bombsight, sighting window, and all bomb release controls. The bombardier's switch and instrument panel were mounted on the bulkhead at the forward end of the compartment. Before operating the bombsight it was necessary to raise the bomb sight window to the UP position by using a hand crank located in the upper right hand corner above the bombardier's window. The bombardier's electrical panel incorporated an indicator light for each bomb rack station to show the position of the bombs on the rack, and contained eight double throw release switches: one for each rack, one for each of the front and rear 2,000 pound bomb installations, one bomb release firing switch, and one bomb release master switch.

The controls for releasing bombs consisted of electrical selective release by the means of 1) Electrical contacts which will be automatically closed within the bomb sight, 2) A momentary contact switch manually operated, or 3) Manual salvo release of all bombs simultaneously.

There was a manual emergency release handle that was accessible to both the pilot and co-pilot located at the bottom of the engine control pedestal, and opened the bomb bay doors and allowed the release of all bombs in salvo.

Bombing Compartment
Useful Design Bomb Load:

1-2000lb. bomb	Type M-34, Mk-I, Mk-IMII, Mk-IMI or
2-1100lb bombs	Type M-33, Mk-III or
4-600lb. bombs	Type-32, Mk-IMI, Mk-III or
8-300lb. bombs	Type M-31, Mk-I, Mk-IMI or
20-100lb. bombs	Type M-30, Mk-I, Mk-IMI, Mk-IMII

B-18A being loaded with is complement of 20 100 pound bombs. (USAF)

Bomb Rack Releasing Mechanism. (USAF)

Bomb Racks: The bomb rack installation incorporated six individual racks. One contact of the firing switch released one bomb from each rack, or any combination of racks simultaneously.

Bomb Bay Doors

The bomb bay doors were hydraulically controlled, and formed the lower part of the fuselage when closed. There were weather strips made of gum rubber attached to the main keel to make a tight joint between the bomb bay doors and keel.

Opening: To open the bomb bay doors, a four way hydraulic valve was moved to the open position. A bungee assembly was attached to the bomb bay door operating cylinder arms, and attached to the bomb bay doors. The doors sprung open the moment the release was thrown. Another bungee system was incorporated in the door operating mechanism to open the doors in case of hydraulic system failure.

Closing: To close the doors, the four way valve was moved to the closed position and left there at all times, except when the doors were open. The bomb bay doors were held closed by hydraulic pressure, and a mechanical latch was incorporated in the operating mechanism to keep the door closed when the aircraft was on the ground and the hydraulic pressure had dropped.

There was an electrical and mechanical safety locking device incorporated in the bomb bay door's operating mechanism to prevent releasing the bombs before the doors were fully open.

Gunnery Equipment

There were three flexible .30 caliber Model M-2 Browning machine guns with muzzle stabilizers. Ammunition boxes were installed within easy reach of each of the gunners; and were loaded with 500 rounds for each of the two rear guns and 600 rounds for the front gun.

Bomb Doors. (USAF)

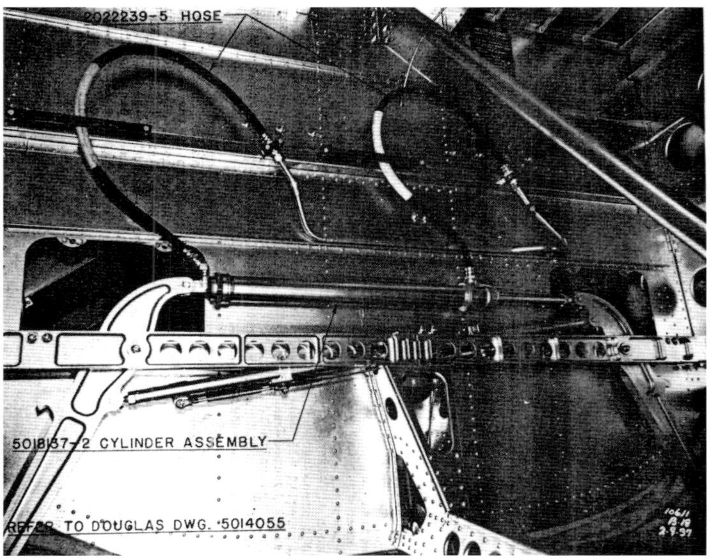

Bomb Bay Door Operating Mechanism. (USAF)

Gunnery Photo Gallery

Front Turret B-18. (USAF)

Front Turret B-18 stowed. (USAF)

Front Turret B-18A. (USAF)

Front Turret B-18B (arrow). (Pima)

Above: Rear Upper Turret manned. (USAF)
Right: Rear Upper Turret. (USAF)
Lower left: Rear Upper Turret stowed. (USAF)
Lower right: Rear Lower Turret (internal view). (USAF)

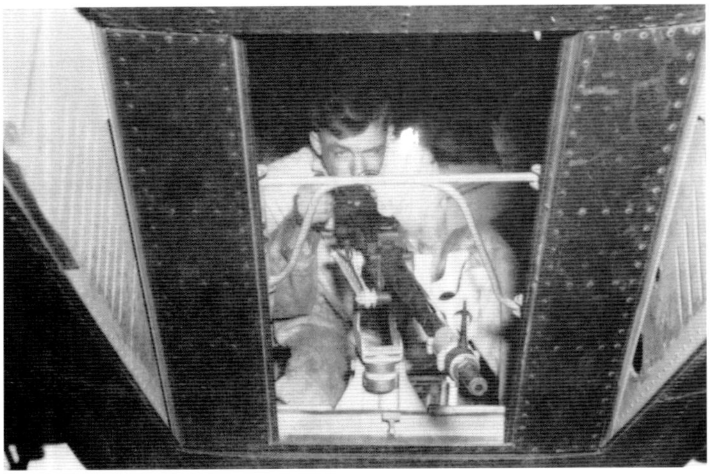

Rear Lower Turret (external view). (USAF)

Front Gun and Turret

The front gun turret was located in the nose of the aircraft, and was constructed of aluminum alloy frame. The gun was mounted on an aluminum alloy casting attached to the ring of the turret, and turned on ball bearings in a circular track. Flexible trailing windshields were attached above and below the gun as a wind break. A back rest with a retractable seat equipped with a Type A-3 seat belt attached to rings on the floor was mounted on the ring on the opposite side of the gun. To bring the gun into firing position, the latch adjacent to the spade grips was disengaged, and the right turret handle was moved until the safety catch pin on the handle could be removed. The gun could be rotated about its horizontal axis 195 degrees, and 7.5 degrees about its vertical axis without moving the turret. The gun was stowed with its barrel pointing vertically outside; the safety catch pin that prevented the turret from moving on the ring was engaged, and the latch was pulled up to engage the flange on the gun.

Rear Upper Gun and Turret

The fully retractable rear upper gun turret was located between Stations 473 and 493, and was constructed of transparent sheets fixed in an aluminum alloy framework. It was raised and lowered on four chrome molybdenum steel tubes attached to the fuselage by cables operated by a crank drive. When in its extended position the turret revolved through a full 360 degrees, but a latch could be set to prevent it from firing into the tail structure. The gun could rotate 360 degrees around its vertical axis and, in certain positions, 134 degrees around its horizontal axis. The gunner was supported on a canvas hammock seat attached to the turret, and by a back rest connected to the turret ring. Rings were provided on the floor for a Type A-3 safety belt.

When the turret was being retracted, a pin in the rotating handle was inserted into a stop on the track to hold the turret in place. The grip handles were then to be pushed down and aft until an aluminum sleeve clamped to the gun barrel jacket could be held by pawls in the top of the turret. A catch on the left hand side of the rotating handle was tripped, and the turret swung until it was held by the catch. The turret then could be retracted by turning the crank. When the turret was retracted, the gun barrel extended outside the aircraft.

Rear Lower Gun

The rear lower gun was mounted on the floor of the airplane at Station 432.5, and fired through a small arc to the rear under the tail. To bring the gun into the firing position, a catch was released on the window section of the door, the gun stop was raised, the hold down catches released, the gun placed in its firing position, the gun stop lowered and fastened, and the door was raised by means of a crank located on the left hand side of the fuselage at Station 432.5. The molybdenum steel gun stop regulated the firing angle of the gun to prevent hitting the bottom of the empennage and tail wheel.

To stow the gun, the gun stop was unfastened and raised, and the stowage latch engaged with the latch on the aluminum sleeve located on the gun barrel, the hold down latches fastened, gun stop lowered, and the door lowered, and the window pulled forward until held by a catch.

Ammunition Box Holders

There were six corrosion resistant steel boxes that were built to accommodate Type L-4 ammunition boxes. They were held by springs built into the top of the holders in two vertical banks on the left hand wall next to Station 432.5, and four in a single horizontal bank on the right hand side of the partition at Station 432.5.

Gun Cameras

A Type G-4 gun camera was installed on each of the aircraft's machine guns. Only the front gun mount needed a special adaptor assembly for the attachment of the camera.

The B-18 and 75mm Cannon Testing

In 1932 it was decided that .50 caliber was to be the standard aircraft gun, and no funds were to be expended on aircraft weapons over that size. Interest was renewed in larger caliber aircraft guns in December 1935 when the Chief of Ordnance ordered tests at Aberdeen Proving Grounds to determine the effect of high explosive instantaneous-fused projectiles of 37mm, 25mm, 20mm, and .50 inch caliber. The results were forwarded to the Air Corps Technical Committee, which formulated the characteristics of aircraft cannon that were forwarded to the Chief of Ordnance for approval. During 1936 the U.S. investigated the possibility of purchasing cannon manufactured by foreign countries, but negotiations were hindered by red tape, or unreasonable prices and conditions of sale. By March 1938 it had been determined that there was a demand for an aircraft cannon, that foreign purchase was unfeasible, and that various American designed cannon be developed. The Ordnance Committee recommended that a 75mm semi-automatic cannon that could fire a HE shell at 1,500 to 2,000 feet per second be developed. A meeting between the Air Corps and the Ordnance Committee determined on 1 November 1938 at Wright Field, Dayton, OH, that an existing Model 1897 75mm cannon and its control system was to be fitted into the first B-18 available for static tests at Wright. The DB-1 prototype was fitted with a fixed forward firing 75mm cannon with a shortened muzzle in the forward bomb bay. After static tests this aircraft was to be shipped to Aberdeen Proving Ground for further testing. However, the aircraft was also being used for radio tests at Wright, and would not be available to Aberdeen until April or May 1939. A production model B-18 was found to be available at Langley, VA, and transferred to Aberdeen. Here it was fitted with a 75mm gun and fire control system in a simple improvised experimental installation that would provide data on accuracy and effectiveness on hitting a target, and determine the final military mounting of a newly designed 75mm cannon and its fire control system, and the type of aircraft to carry it. A Model 1897 75mm cannon with limited elevation and traverse, shortened muzzle, and its telescopic sight and range finder was mounted in the forward bomb bay. The tests were first conducted on the ground and then in the air, and finally against aerial and ground targets, with the results being better than expected. The first concern was to determine whether the aircraft fuselage could withstand the stress of the firing in flight of a gun with such large recoil, but the fuselage and mounting suffered no adverse effects. The simple stereo telescopic sight proved to be an accurate range finder, and it was possible to fire accurately against towed aerial targets and ground targets. The average error at 3,000 yards was 50 feet, and at 1,500 yards it was 35 feet. The B-18 tests proved convincingly that mounting a 75mm cannon in an aircraft was practical from both an engineering and tactical standpoint, but that the development of a new 75mm cannon and fire control system was necessary, rather than improve the existing cannon and fire control system. The joint Air Corps and Ordnance Committed recommended that development of the cannon/fire control system be started immediately, and that the Air Corps initiate the design and manufacture of an aircraft to mount the cannon when facilities became available, but not later than the fiscal year 1942. The B-18 would pass further 75mm cannon testing on the XA-26B, which would carry the newly developed 75mm M3 that was initially designed for tanks, but was soon replaced by the lightweight T9 cannon.

The DB-1 prototype was later modified, installing a forward firing 75mm cannon with a shortened muzzle mounted in a fixed position in the forward bomb bay. It was used in feasibility tests for large caliber cannon in aircraft at the Aberdeen Proving Ground, and over Lake Erie. (USAF)

Landing Gear

Main Landing Gear

The hydraulic landing gear consisted of two independent units: Each gear was mounted under the nacelle, and so arranged that it could be folded up into the nacelle, leaving only the bottom of the tire projecting.

The main hydraulic landing gear consisted of two independent units with interchangeable Bendix wheels mounting 45x17-16, 10-ply tires. (Pima)

Main gear disassembled. (USAF)

The Bendix dual hydraulic 14x3 brakes were mounted on each wheel, with hydraulic pressure transmitted through the power brake control valve. The brakes operated independently by toe pressure applied to the rudder pedals. (USAF)

Oleo pneumatic shock absorber cylinders with a booster pump to provide pressure.

The Bendix wheels were interchangeable, right and left, mounting 45x17-16, 10-ply tires.

Brakes

The Bendix dual hydraulic 14x3 brakes were mounted on each wheel, with hydraulic pressure transmitted through the power brake control valve. The brakes operated independently by toe pressure applied to the rudder pedals. Hydraulic pressure of at least 500psi was transmitted to the brakes through the power brake control valve, each wheel operating independently of the other. To set the brakes for parking, the plunger type parking lock control located on the upper left corner of the control pedestal base was pulled; then the rudder pedals were depressed until the plunger moved out, after which the pressure was removed from the pedals and the plunger released. To release the brakes from the parked position, the rudder pedals were to be depressed to unlock the spring lock.

Wheel Stop

A wheel stop was attached to each landing gear upper truss. Once the landing gear was retracted into the nacelle, the wheel stop automatically engaged with the tire, and caused sufficient braking to stop the rotating tire motion caused by the slip stream pressure acting on the exposed portion of the tire.

Operation of the Main Landing Gear

Retraction

The landing gear latch was placed in the LATCH RAISED position by pulling the control handle back until it was held by the trip. A red warning light went ON as soon as the latch was raised.

The landing gear four-way control valve was moved to the UP position, and the wheels would retract when the hydraulic lines came under pressure.

After the wheels moved up into the retracted position, a hard stroke of the hand pump insured that the axles were up against the rubber bumpers in the nacelle.

The position of the wheels were to be checked occasionally, and if they were not against the bumpers they were to be pumped up with the hand pump, and the return valve control was placed in the midway position when the operation was finished.

The trip lever was moved aft, placing the safety latch in the SPRING LOCKED position.

Extension

The four way landing gear control valve was pushed to the DOWN position, which automatically placed the safety latch into the SPRING LOADED position; the wheels should have then extended.

When the landing gear was DOWN and LATCHED, a green light would go ON. The latch control handle was then to be moved down against the floor, placing the latch in POSITIVE LOCKED position, and it was held in place by a clip on the floor. The valve control was to be left in the DOWN position at all times when the landing gear was extended.

Note: If for any reason the landing gear could not be retracted or extended with engine pressure the hand pump had to be used, and required 35 seconds to retract and 30 seconds to extend.

Landing Gear Indicating System

This system consisted of signal lights and a horn. The Klaxon Type K-16 Horn was located on the left hand side of the fuselage beneath the instrument panel, and the red and green signal lights were on the right hand side of the fuselage, just forward and above the co-pilot's seat. When the gear was in the DOWN and LOCKED ONLY mode the green light would be ON. When the red light was ON, the gear was *not* DOWN and LOCKED. The horn would sound when the throttles were CLOSED, and one or both wheels were *not* fully DOWN and LOCKED.

Safety Latch

A safety latch was mounted on the forward face of the front spar in each nacelle, and was used to lock the landing gear in the extended position. The latch was controlled by a lever located forward of the landing gear four-way valve. The valve handle and latch lever were mechanically interconnected, making it necessary to manually raise the latch to the UNLOCKED position before operating the valve handle to retract the gear. A clip on the fuselage structure was used to hold the latch in the POSITIVE LOCK position, and was to only be used when the gear was fully extended. The latch lever was held in the UNLOCKED POSITION by the interconnecting mechanism. A trip lever for releasing the latch from the UNLOCKED POSIION was located immediately aft of the latch lever.

Two air-oil shock absorber struts were solidly clamped to the axle of each wheel, and connected together at there upper ends by rigid trusses. The hydraulic operating cylinder was attached near the center.

Safety Lock Pins

The safety lock pins, one for each landing gear, were used supplementary to the mechanical safety latch to lock the landing gear DOWN when on the ground. A red strip of cloth was attached to each pin, and served as a warning that the landing gear was locked,

The tail wheel was of the full swiveling type, and was designed so it remained in the trailing position during take off and landing. The shock absorber was of the oleo pneumatic type. The Hayes Industries 9x6 wheel mounted a 22x9x6, 8-ply tire. (Pima)

Tail gear disassembled. (USAF)

and could not be retracted without removing this pin. It was the pilot's responsibility to be sure that these pins were removed before taking off. If the aircraft was scheduled to be on the ground for a long period of time these pins were to be installed through the safety latch fitting. The pins were stowed in the miscellaneous container.

Tail Wheel

The tail wheel was of the full swiveling type, and was designed so it remained in the trailing position during take off and landing. In addition, a latch that was controlled by a lever on the control pedestal in the pilot's compartment could be used to lock the tail wheel in the trailing position. The shock absorber was of the oleo pneumatic type. The Hayes Industries 9x6 wheel mounted a 22x9x6, 8-ply tire.

Mooring

The aircraft was moored at three points: a ring at each wing tip, at rib 18, and at the tail skid. The wing rings that attached to ropes or cables were set into a slot, and partially retracted into the wing when not in use. The tie down cable for the tail skid was stored in the service box.

Empennage. (USAF)

Empennage

The horizontal and vertical stabilizers were all metal, and attached in alignment to the fuselage. The rudder and elevators were fabric covered, and were both statically and aerodynamically balanced. A trim tab was fitted into the trailing edge of each elevator, and could be adjusted during flight. The rudder had a similar tab, providing adjustment for engine torque with either or both engines running.

Wing

The wings were made up of the center section with two engine nacelles attached, and fastened to the center section were two outer panels that had detachable tips to give a wing span of 90 feet, and an area of 965 square feet. The fabric covered ailerons had an area of 48 square feet each. The right aileron was fitted with a trim tab to compensate for unsymmetrical loading. The wing flaps, with an area of 64.2 square feet each, were of the split trailing edge type, extending from just outboard of the fuselage to the aileron. A flap was formed by two sections: one in the center wing panel, and one

Wing Photo Gallery

A B-18 of the Squadron CO (two vertical fuselage stripes) of the 1st RS of the 34th BG comes in for a landing with flaps down. (USAF)

A good look at the upper wing surfaces of a B-18A piloted by the A Flight Leader (one vertical fuselage stripe) of the 18th RS. (USAF)

Outer wing panel. (USAF)

Wing center section. (USAF)

Landing light. (Author/Pima)

in the outer panel, with the two sections functioning as an integral unit. The flaps had a movement of 55-60 degrees, giving an increase in lift of 35%, and an increase in drag of 350%. The outer wing section between the nacelles and the fuselage was reinforced for walkway purposes. If it were necessary to walk on the outer wing panels or on top of the fuselage it was necessary to place padded boards on them.

Landing and Running Lights

Two 240 watt, 12 volt, Type A-8 landing lights were installed in the leading edge of the outer wing panels, pointing at a 10 degree angle outboard from the fuselage centerline. These lights were controlled by switches on the pilot's electrical control panel. The 20 volt, 50 candlepower Type B-1A red running light was mounted in the left hand landing light bay.

Engines

The B-18 was powered by two nine cylinder, one row R-1820-45 air-cooled, G-series radial engines. The G-series R-1820 replaced the F-series at the beginning of the war, and comprised the bulk of R-1820 production during the war. Studebaker was responsible for most of the G-series production, with the main assembly facility in South Bend, IN, supported by factories in Fort Wayne, IN, and Chicago.

The steel crankcase was made up of two sections split on the cylinder centerline. The cast aluminum cylinder heads were screwed and shrunk into the steel barrels. One inlet and one sodium-cooled exhaust valve per cylinder were actuated by push rods. The Wright developed supercharger was a gear-driven two-speed type, and replaced the inefficient General Electric type fitted in the F-series. The carburetor was a Bendix-Stromberg PD-12H3 two barrel injection type downdraft with automatic mixture control, and a four position manual mix control. The magnetos were either built by Bendix-Scintilla or American Bosch. Cooling was a constant problem in radial engine performance and the R-1820 was designed with more and deeper cooling fins, with the G-series adding one more cooling fin than the F-series. The engine was attached to the nacelle firewall at nine points by rubber fittings. The exhaust collector ring was composed of two separate sections; each made up of four segments.

The B-18A, -18B, and -18M were powered by the R-1820-53 series engine which differed from the -45 only in impeller size, and being specifically designed to use 100 octane gasoline, instead of the previously standard 87 octane. At the time the R-1820 was the largest successful single row radial engine, and during its development many improvements were incorporated that yielded higher horsepower.

View of left engine section. (USAF)

Engine accessories. (USAF)

Rear of engine cowling. (Author/Pima)

Rated Horse Power
930hp at 2,100rpm for take off
830hp at 2,100rpm at sea level
860hp at 2,100rpm at 3,200 feet
810hp at 2,100rpm at 10,300 feet

Engine Controls: The throttle and propeller pitch controls were grouped at the upper end of the control pedestal, and were situated so that each set could be operated individually, or together with one hand by either the pilot or co-pilot. Spring loaded manually adjusted brakes held these controls in any desired positions. The engine controls were operated by cables to three-arm bellcranks on the front face of the firewall. Forward of the firewall the controls were operated by push-pull rods.

The carburetor was the two barrel downdraft Holley type mounted on top of the supercharger section. The fuel mixture was fed through an annulus to the General Electric supercharger impellor or blower, and then fed to the cylinders. The General Electric two-speed controllable superchargers were inadequate for that time, as the GE design fell behind its European counterparts. The Eclipse Hand Electric Starter (Type C-21) was operated by two solenoid starting and meshing switches, one for each engine. The engines could be started by a hand crank that was stowed on the left wall of the cabin just forward of the entrance door. Ignition that fired two plugs per cylinder was provided by two Scintilla magnetos mounted on the rear cover. Lubrication was the standard dry sump, with one pressure pump and one scavenge pump. The generator system consisted of two engine driven Type E-5 generators, one mounted on each engine, which charged the batteries. Engine covers for each engine were stowed in the equipment box.

Propellers
The Wright R-1820s mounted a Hamilton-Standard Hydromatic constant speed 3 blade, fully feathering propeller.

Blade Length	6 feet, 6 inches
Height: Propeller hub (at rest)	8 feet, 6 inches
Clearance Propeller tips to ground (at rest)	24 inches
Clearance Propeller tips to fuselage	9 inches

Control Pedestal. (Author/Pima)

Engine Maintenance Gallery

Maintenance Platform. (USAF)

The ground crew of the 1RS service a B-18A in Panama. (USAF)

Engine maintenance of a B-18 of the 9BG at Mitchel Field in 1938. Note the O-46 observation aircraft in the background. (USAF)

Propeller Pitch Controls and Feathering

The propeller pitch controls were located on the upper part of the control pedestal adjacent to the throttle controls. They regulated the amount of oil furnished to the propeller to maintain a constant RPM predetermined by setting the control lever. The oil pressure exerted a force tending to rotate the blades toward lower pitch, and the coil springs and centrifugal action of the counterweights attached to the blades at the hub exerted a force tending to rotate the blades toward a higher pitch. A separate electric high pressure oil pump supplied propeller feathering, turning the propeller blade to 90 degrees pitch to prevent it from windmilling if the engine became inoperative.

Fuel System

The fuel system consisted of:
Left and Right Auxiliary Tanks: 86 gallons each
Left Main Tank: 215 gallons (including 57 gallon reserve)
Right Main Tank: 215 gallons
Left Bomb Bay Tank: 184 gallons
Right Bomb Bay Tank: 184 gallons
Total fuel capacity 1,170 gallons with fuel tanks to be emptied in the following order: 1, 2, 5, 6, 4 and 3.
Two AN4009 wobble pumps
AN4001-2 cross feed valves
Two C-2A strainers
Two R-600BE Type C-7 Pesco engine driven fuel pumps

Fuel Tanks

The left and right main (430 gallons) and left and right auxiliary fuel tanks (172 gallons) were filled at the filler necks located in the upper surface of the wing center section. The left and right bomb bay tanks (368 gallons) were filled at the filler necks that were accessible through the ports inside the bombardier's compartment by lines brought through the escape hatch. All fuel lines were made of aluminum alloy, except those that were less than 3/8 inches, which were copper. The tanks were constructed of aluminum alloy, with internal baffles riveted to the shell. All tanks were equipped with electrically operated liquidometers, and had water collecting sumps and drain cocks. The two main tanks were mounted in the wing between the center spar and rear spar on each side of the centerline of the aircraft. The left main tank had 20 minutes (57 gallons) fuel reserve. The two auxiliary fuel tanks were mounted in the wing between the front spar and center spar forward of the main tanks. The two bomb bay tanks were mounted in the bomb bay by a method

Hamilton Standard Constant Speed 3 Blade propellers measured 6.5 feet. (Author/Pima)

The two main tanks were mounted in the wing between the center spar and rear spar on each side of the centerline of the aircraft. (USAF)

Engine (nacelle) oil tanks. (USAF)

similar to that used to attach the bombs, and were jettisonable in flight.

Fuel Flow

The fuel flow was as follows. The fuel lines from the two main tanks converged at the outer tank selector valve, but there were two lines from the left main tank to the selector. The standpipe of one of the two lines was located high enough in the tank that the 20 minute 57 gallon reserve could not flow through it, and to tap this reserve, the other line had to be used via the selector. The fuel lines from the auxiliary and bomb bay tanks connected to the inboard selector valve. The two selector valves were interconnected. From the outboard selector valve, the flow was through the engine selector valve to the strainers. The strainer outlets were connected to the wobble pumps. The fuel supply lines then connected to the engine driven fuel pumps. The fuel pressure lines received pressure from the line connecting the fuel pump and carburetor in each nacelle, and transferred it to the pressure gauges mounted on the right hand side of the instrument panel. The fuel pressure warning lights operated when the fuel pressure was reduced to three pounds or less. The correct fuel pressure was between four and five pounds.

Oil System

The oil system consisted basically of an oil tank in each nacelle, an auxiliary tank in the fuselage, oil temperature controls, oil coolers, oil pressure gauges, and the necessary valves, lines, and fittings.

Oil Tanks: The main oil tanks were filled at their filler necks accessible through hand holes in the top of the engine nacelles. The auxiliary tank was filled at its filler neck accessible through the door on the right hand side of the fuselage. All units of the oil system forward of the firewall were interchangeable between right and left engines. The tanks were constructed of aluminum alloy sheet with internal baffles riveted and welded to the shell, and were held in place by four aluminum straps. The total capacity of the three tanks was 87 gallons.

Oil Coolers

Air was passed through the cooler by a scoop that extended below the engine section cowling. Butterfly valves within the scoop were

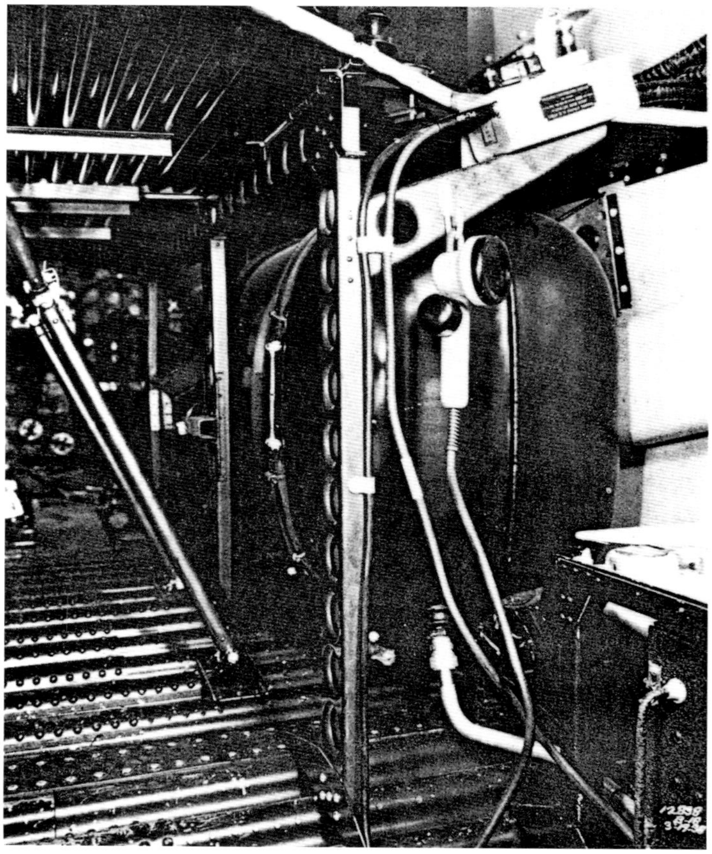
Fuselage oil tanks. (USAF)

controllable from the pilot's compartment, and were used supplementary to the automatic oil temperature control.

Oil Flow

From the nacelle oil tank the oil flowed through the "Y" drain and dilution valve to the engine. Leaving the crankcase, it passed through a Type D-5 temperature regulator on the oil cooler where automatically, depending upon the temperature, the oil was passed through the cooler, or was shunted directly back to the tank. When the quantity of oil in the nacelle tanks needed replenishing, oil could be pumped from the fuselage tank by a wobble pump; the desired tank or tanks was selected by moving a Type E valve that was located adjacent to the pump. The intake for the oil coolers was equipped with butterfly valves that were manually controlled by the pilot, and were used to supplement the oil temperature regulators.

Hydraulic System

General

Two Pesco high pressure engine-driven oil pumps supplied pressure to the hydraulic and automatic pilot systems, and the complete hydraulic system, including the automatic pilot, contained about nine gallons of oil. The oil was supplied to the system from the hydraulic fluid reservoir mounted just aft of the front gun turret on the left hand side of the aircraft. The filling capacity was 2.33 gallons, and the total capacity was 3.5 gallons, with a sight gauge located beside the reservoir. The oil for the hydraulic system passed to a pressure tank, and then continued through the system under pressure. The normal hydraulic pressure was between 600 and 800 pounds, and was used to operate the landing gear operating cylinders, the wing flap operating cylinder, bomb bay doors operating cylinder and the brakes.

Operation

The operation of the landing gear, wing flaps, and bomb doors was controlled by individual four-way valves of the piston type. The movement of the piston connected the proper ports of the valve to allow operation of the various units of the system. The valves for the landing gear and wing flaps were located in an assembly in the floor between the pilot's and co-pilot's seats. The bomb door valve was located just forward of Station 164 in the forward gunner's compartment, with remote controls adjacent to the bombardier.

On airplanes Nos. 37-458 to 37-594 inclusive, a manually operated selector valve located on the inboard side of the pilot's seat support allowed the arbitrary selection of the pump which was to supply the hydraulic system, at which time the other pump was automatically connected to the automatic pilot system. When the selector valve handle was in the forward (normal) position, the left pump supplied the hydraulic system, and the right pump supplied the automatic pilot. When the selector control was in the aft position the reverse was true.

Emergency Hand Pump

An emergency hand pump was located adjacent to the landing gear and wing flap four-way valves, and was so connected to the system that any hydraulic unit, including the brakes, could be operated by the hand pump. The hand pump could also be used to increase the pressure in the pressure tank when the by-pass valve (located adjacent to the pump) was changed from the normal SYSTEM to TANK position.

Pitot System

The Type C-1 Pitot Static Tube (electrically heated) extended forward of the leading edge of the right outer wing midway between Stations 232.7 and 250.6. The lines passed through the outer wing between the leading edge and front spar.

Flight Controls

The B-18 was equipped with complete dual controls of a type allowing the pilot and co-pilot to be seated side-by-side. Each man had his individual control column and a set of rudder pedals. The rudder, elevator, and aileron tab controls were located on the control pedestal forward and between the two seats, so as to be easily accessible to both the pilot and co-pilot without requiring abnormal reach. The controls were operated by cables and pulleys, with the cables of 3/16 extra flexible steel, and only anti-friction pulleys were used.

Control Column: Controlled the movement of the elevators and ailerons. It was constructed of aluminum alloy tubes and castings, and was attached to the fuselage by three roller bearings. Mounted at the ends of the control column were two hard black rubber coated aluminum alloy wheels. On the shaft of each wheel was a sprocket, around which a chain attached to the aileron cable turns. The cable passed over pulleys in the upper elbow, and down and over the lower ends of the upright tubes.

Rudder Control: The rudder cables ran along the aircraft centerline to the rear wing spar, where they turned along the right side of the fuselage just above the cabin floor into the tail compartment, where they diverged and connected to the rudder horn. The rudder pedal movement was limited by the torque tube arm striking fixed stops

Pilot's controls and instrument panels in 1937-458 to 37-594 aircraft. (USAF)

Co-pilot's control column. (Author/Pima)

on the fuselage structure. The rudder pedals were adjustable fore and aft to accommodate pilots of various heights, and the energy the pilot had to exert was decreased by an 11 pound counterweight in the rudder area.

Elevator Control: The elevator attached to a horn on the control column, and followed the same route as the rudder control cables to the tail. The forward and aft movement of the elevator control was limited by stops on the fuselage structure. An 8 pound counterweight decreased the energy required by the pilot.

Empennage Tab Controls

Rudder Tab: This tab was mounted in the trailing edge of the rudder, and was controlled by a crank on the control pedestal. It was actuated by a rod and drum assembly located in the main torque tube of the rudder via a push-pull rod attached to the tab. The rod was actuated by a cable running from a drum in the control pedestal down under the pilot's compartment floor to the to the rear wing spar, where they took the same route as the rudder cables.

Elevator Tabs: The elevator tabs were mounted in the trailing edges of the elevators. They were controlled by a crank on the control pedestal. An indicator behind the crank showed the position of the tab. These tabs were actuated by a rod and cable drum assembly located in the main torque tube of each elevator, and functioned in the same manner as the rudder tab.

Aileron Control: The ailerons were controlled by turning either wheel on the control column. The main aileron control cables connected to those in the control column, and ran along the centerline of the aircraft to a point under the radio compartment, where these four cables converged and attached to two main cables that attached to a four arm bellcrank. Two arms of this bellcrank attached to the aileron of one wing via push-pull rods that operated the aileron.

Aileron Tab Control: This tab was located on the trailing edge of the right aileron, and was operated by a crank on the control pedestal. The control cables for these tabs followed the same route as the right aileron controls out to a cable mounted on the rear face of the wing rear spar. This tab operated in the same manner as the rudder tab.

Control Locks

Provisions were made for locking the control surfaces when the aircraft was moored. Four felt-covered channel locks (two for the ailerons, one for the rudder, and one for the elevator) were used to connect the movable surfaces to the airplane structure. The appropriate surface was noted on top of each lock, as the locks were not interchangeable. The clips to which shock cord hooks were attached held the locks in place, and were located:

Aileron: lower surface of the wing on the rear spar
Rudder: left hand side of the vertical stabilizer
Elevator: left hand horizontal stabilizer fairings

Wing Flaps: The hydraulically controlled wing flaps had a movement of 55-60 degrees, giving an increase in lift of 35%, and an increase in drag of 350%. A position indicator was attached to the instrument panel in front of the pilot, and was connected by a flexible control to the hydraulic cylinder; with any movement of the cylinder being shown on the indicator. The flaps could be placed in any desired position between FULL UP and FULL DOWN by moving the wing flap hydraulic four-way valve to the NEUTRAL position when the desired setting was reached. It took about seven seconds to fully lower the flaps, and ten seconds to raise them from the FULL DOWN position with the engine driven pump supplying the hydraulic pressure. A hand pump was provided in case of emergencies, but required a much longer time to move the flaps. The wing flaps were never to be lowered when the indicated air speed was in excess of 112mph, nor was the IAS to be increased above 112mph with the wing flaps lowered.

Automatic Pilot: The Sperry Automatic Pilot Type A-2 was installed on aircraft nos. 37-458 to 37-594 inclusive, and provided control of the rudder, ailerons, and elevators maintaining directional,

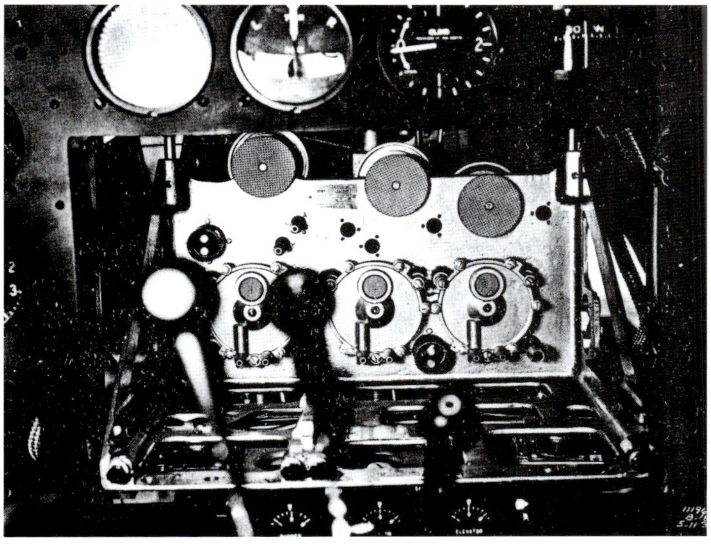

The Sperry Automatic Pilot Type A-2 was installed on aircraft nos. 37-458 to 37-594 inclusive, and provided control of the rudder, ailerons, and elevators, maintaining directional, lateral, and longitudinal stability. (USAF)

lateral, and longitudinal stability. It consisted of the following major components: gyro control, mounting hydraulic servo, speed control valves, oil pressure relief valve, oil pressure gauge, oil supply tank, oil pumps, oil filter, vacuum pumps, and vacuum relief valve. The two Sperry gyros provided datum lines from which the control surfaces could be hydraulically operated. These gyros were driven by vacuum of 4 to 5 inches supplied by a Pesco vacuum pump mounted on the angular drive of each engine. The controls were operated by oil pressure (120psi) supplied by either the right or left hand engine oil pump as selected by the selector valve. On aircraft equipped with automatic pilots using directional gyro control assemblies that were interconnected with special tactical equipment, the automatic pilot operating oil pressure would be adjusted to between 155 and 165psi.

Instruments

The pilot's instruments were grouped according to their use as follows:

1) Flight instruments to the left and center of the instrument panel
2) Engine instruments on the right side
3) Electrical instruments to the extreme left

The bombardier had a small instrument panel adjacent to the electric bomb control panel.

Communications Equipment

The aircraft was equipped with the following communications equipment:

SCR-187-A Liaison Set consisting of:
1) Type BD6 dynamotor mounted in the cabin floor adjacent to Station 351.5.
2) Type BC-224-A receiver mounted on top of the radio table near the wall.
3) Type BC-191-A transmitter mounted aft of the radio table on the right hand side of the fuselage.
4) Type J-5 flame proof key mounted on the radio table, and was easily accessible to the radio operator.
5) Type RL-30 antenna reel located on the right hand side of the fuselage just forward of the bulkhead at Station 270.5.

Type SCR-133 Command Set consisting of:
1) Type AD-12 Eclipse dynamotor mounted on the right hand side of the fuselage directly below the transmitter.
2) Type BC-AG-230 receiver mounted on a shelf on the right hand side of the fuselage immediately aft of the bulkhead at Station 270.5.
3) Type BC-AG-230 transmitter mounted forward of the receiver on the same shelf.
4) Type MC-125 control unit mounted in the pilot's compartment enclosure at the upper point of the "V."

Communications Equipment Gallery

Radio Operator's Station. (Author/Pima)

Antenna Reel and Command Radio. (Author/Pima)

Liaison Radio Transmitter and Receiver. (Author/Pima)

Interphone Station and Headphones. (Author/Pima)

5) 36B2528 antenna relay mounted on the ceiling of the cabin directly above the command set.
6) Type BC-AE-232 transmitter control mounted on the fuselage directly forward of the liaison set junction box.
7) Type MC-135 dual coil control mounted in the center top of the pilot's compartment near frame 164.
8) Type BC-AE-231 receiver control mounted directly above the dual coil control.

Radio Compass consisting of:
1) Loop mounted on top of the loop rotator.
2) Loop rotator mounted on a retractable base operating on two chromium plated tubes. Two doors on top of the fuselage opened to allow the loop to be extended outside. (The loop was not retractable on aircraft # 37-4 to 37-33.)
3) Receiver located on the floor at the base of the loop support tubes.
4) Two compass indicators mounted: one mounted on a bracket on the left hand side of the fuselage for use by the radio operator and navigator, and the other mounted on the pilot's flight instrument panel.

Type A-1 Marker Beacon Equipment consisted of:
1) Receptor mounted on the centerline of the lower surface of the aircraft, and was supported by a stub mast. The forward end of the antenna was fastened to a short mast at Station 120, and the aft end was attached to a similar mast at the main keel beams at Station 234.5.
2) 35D1422 marker beacon indicator located on the pilot's flight instrument panel.
3) 35B5645 marker beacon rectifier mounted on the left hand wall of the bombardier's compartment.

RC-15 Interphone Equipment consisted of:
1) Type BC-212 amplifier located under the shelf that supports the command radio.
2) Eight Type BC-13A jack boxes supplying each of the crew at their positions.

Antenna Array. (USAF)

De-icing shoe on the leading edge of the vertical stabilizer. (USAF)

3) Eight Type T-17 Microphones attached adjacent to each of the jacks.

Antennas
1) Liaison set had a trailing antenna that trailed downwards from a point to the right of the fuselage under the center section wing panel.
2) Command set antenna fastened to the rear radio mast and the lug on top of the vertical stabilizer. Both the receiver and transmitter used the same antenna.
3) Marker beacon antenna located on the underside of the fuselage, and was supported by two short masts.
4) Radio compass "T" antennae were a retractable loop type, and located between the two radio masts on top of the fuselage.

Compass Type D-4 mounted on a bracket assembly supported from the upper cover of the wing center section panel, and located forward of the navigator's seat in the navigator's compartment adjacent to the driftmeter.

Drift Indicator Gatty #5545 installed in the fuselage at Station 179.5 for use by the navigator.

Anti-icer and De-icer Equipment
Anti-icer
The propellers were equipped with alcohol glycerin anti-icers in a four gallon tank. The pump switches were located on the co-pilot's electrical panel and on the tank.

De-icer
This system consisted of Goodrich de-icing boots mounted on the leading edge of the wing, vertical and horizontal stabilizers, and two radio masts. The boots were inflated by air pressure from the exhaust side of a Type B-1 engine driven by a vacuum pump. The boots were inflated in five sections or stages by the rotation of the distributor valve, which rotated once each 40 seconds by an electric motor, and on every rotation all de-icing boots were inflated and then deflated in sequence. The system was supplied by a four gallon alcohol deicer tank located inside the fuselage cabin.

De-icer Alcohol Tank. (Author/Pima)

FIGURE 78 – BOILER INSTALLED IN ENGINE SECTION

Engine Boiler. (USAF)

Cabin Heating and Ventilating Equipment. (USAF)

Heating and Ventilating System:
The B-18 heating and ventilating system was unique, and consisted of:

Water Supply Tank: The tank, constructed of seam welded aluminum alloy sheets, was located on the right hand side of the forward part of the cabin. It was filled through a filler neck fitted with a cork stopper with a chain. There was a vent line on top that ran to the outside of the aircraft. There was a drain cock on the forward side of the tank to determine the correct water level.

Boiler: The Inconal steel boiler was installed in the right hand engine section, and acted as part of the exhaust collector ring. The boiler was arranged so that the exhaust heat from cylinders #3, #4, and #5 was used.

Radiator: This unit was installed above the water supply tank in the cabin. It was constructed of copper, except for the brass cover. There were 16 54-inch coiled tubes inside the radiator, with thin copper sheets running between the coiled tubes that helped to conduct the heat from the steam in the tubes to the incoming air.

Heating and Ventilating Duct: This duct attached to the aft end of the radiator, and ran from Station 100 to 465.5 on the right hand wall of the fuselage. The forward portion of the duct (to bulkhead 270.5) was rectangular, while the aft portion was circular. Manually adjusted, simple sliding type aluminum alloy outlet ducts branched off from the main duct for the radio operator, upper gunner, and lower gunner. The duct for the pilot and co-pilot was equipped with a butterfly valve operated by a knob below the instrument panel on the right hand side. (See Compartment Color Gallery for photos)

Operating Principal: When the steam shut off valve was opened, water from the supply tank ran into the boiler, where the exhaust heat converted it into steam. The steam left the top of the boiler, and was conducted through an insulated duct to the radiator, where it passed through coils over which air taken in from a retractable scoop was circulated. As the steam lost its heat to the cold air, it condensed and ran down into a tube on the lower side of the radiator. This tube connected to a regulating valve that allowed the water to return to the supply tank to recirculate, but shut off the escape of steam. The air, after it had been circulated through the radiator, passed through warm air ducts into main compartments of the aircraft. The amount of heat was controlled by a valve in the water supply line located forward and below the auxiliary oil tank. Allowing more water to go to the boiler tended to raise the temperature; while lessening the amount lowered the temperature. If the compartment became too hot, this valve was to be closed to stop the steam from going to the radiator, and allowed cold air to circulate through the system.

Oxygen Equipment
A complete set of oxygen equipment was supplied for each crew member and consisted of:

Oxygen Cylinders
Six Type E-1 oxygen cylinders were located:
One, held in place by braided cloth strap on the right hand wall of the front gunner's compartment forward of Station 85.75, was for the use of the front gunner and the bombardier.

Four mounted side-by-side on the right hand wall of the cabin between Stations 371.75 and 412.25 were for the use of the pilot, co-pilot, radio operator, and rear lower gunner. They were held in place by two brackets by two clamps on each cylinder, and could be adjusted by tightening and loosening wing nuts.

One on the left hand side of the fuselage just forward of Station 473 for the use of the upper rear gunner.

Oxygen Regulators
The Type A-6 oxygen regulators were automatic, and indicated the cylinder pressure. The six regulators supplied each cylinder, and were located at:

Front gunner and bombardier were supplied by a two-way shut-off valve on the left hand side of the fuselage.

Pilot's regulator was located on the left hand side of the fuselage adjacent to his seat.

Co-pilot's regulator was mounted in a similar position as the pilot's on the right hand side of the fuselage.

Navigator was supplied by a two-way shut-off valve supplied from an extension from the co-pilot's cylinder.

Radio operator's unit was located on the left hand side of the fuselage near his seat.

Rear lower gunner's unit was mounted on the right hand side of the fuselage near his seat.

Rear upper gunner's regulator was located on top of his cylinder.

Oxygen Cylinders. (Author/Pima)

Fire Extinguisher. (USAF)

Fire Extinguishers
Engine Type A-12 CO_2 Pressure System: The CO_2 cylinder was located on the left hand side of the fuselage adjacent to station 100 in front of the gunner's compartment, while the control panel was located on the floor of the pilot's compartment to the right of the hydraulic hand pump and valve assembly, and was covered by a door. The lines ran from the cylinder from the control panel, and then aft in the right hand fuselage tunnel to the wing tunnel. From this point one line went to each nacelle and engine section, where they circled the engine, with one tube discharging into the carburetor air intake.

Hand Operated: There were two hand operated Type A-2 hand extinguishers, with one located just aft of the lower escape hatch in the bombardier's compartment, and the other mounted on the aft face of the right hand side of the bulkhead at Station 270.5.

Pyrotechnics
The pyrotechnic installation included:
Two Type A-3 parachute flares installed in the vertical position in the tail compartment.

Four Type M-6 parachute flares carried on brackets on the left hand side of the aircraft between the upper turret and Station 513.5.

Signal pistol Type M-2 located in a holster on the floor to the left of the pilot's seat.

Signal Flares. (Author/Pima)

Signal flare brackets for carrying signal flares were mounted on the left hand side of the fuselage between the pilot's and navigator's seats. The upper forward brackets carried ten M-10 or M-11 flares, while the forward lower brackets held ten M-10 flares. The aft brackets held six M-9 flares.

Photographic Equipment

Provisions were made for the installation of a Type T-3A or Type K-3B Camera, a Type B-2 Intervalometer, a Type A-2 View Finder, a Type A-1A Spare Magazine, and a Shutter Induction Coil Box. There was a folding seat installed to the rear of the view finder for use of either the camera operator or the rear lower gunner through the floor in the rear cabin at the machine gun installation.

Tow Target Equipment

A Type C-4 tow target windlass was installed in the camera bay of Aircraft 37-10 to 37-33. The tow line tube extended from the camera bay to a point under the rear lower gun door.

Making the Fuselage and Wing Watertight

The exterior seams and joints formed by the fuselage covering were assembled without any sealing compound. After assembly, the interior surfaces of the fuselage between Stations 120 and 270.5 and the outer wing sections were made watertight by a brush application of Lionoil tinted with Prussian Blue. The Lionoil application was to be generous, so as to flow into and seal open crevices and between the rivets. All joints and seams in the pilot's compartment windshield frame were sealed with Fostoria Titeseal, and the seal of the windshield frame to the fuselage deck was sealed with a strip of Canton flannel impregnated with a Soya bean oil solution/varnish mixture.

There were three watertight compartments in each outer wing panel that provided for emergency water landings. There were venting and purging lines extending from these compartments to the cabin. Immediately after a water landing, the crew was instructed to close the outlets of the purging lines with corks provided adjacent to the outlets. All six wing compartments were to be checked for leaks by operating a hand-held bilge pump on each connection.

There were several instances of "amphibian" B-18s. Capt. C.D. Meadowcroft and his crew, flying a 90th Bomb Squadron Bolo based at Zandery Field, Surinam, were forced to ditch in the Caribbean after one engine failed. The crew clambered out of the aircraft and floated in rafts until a Dutch patrol boat picked them up. Five days later the B-18 was still floating, and was considered a hazard to navigation; it was finally towed into port, where it was washed down

Camera Installation (in floor). (USAF)

and spent its career donating parts. During a night training mission from Hollywood, FL, a B-18B piloted by Al James ran out of fuel at 500 feet, three miles off the Florida coast. James ditched the bomber, tearing off the bomb bay doors, with several of the unarmed depth charges sinking to the bottom without exploding. The six man crew was rescued by a Coast Guard patrol boat that also towed the still floating B-18 to Hollywood Beach. On approaching the beach, a bomber crewman attempted to drop the gear by hand, but only one landing gear would drop, and it was left to bob in shallow water. A week later it was towed into Biscayne Bay, and finally dragged onto the Pan American-Grace seaplane ramp, where the aircraft was cannibalized for parts. Another "floater" was a 40BG B-18 based at Mitchel Field, LI, NY, that ditched into the Atlantic, and finally had to be sunk by naval gunfire, as it posed a threat to shipping lanes. Still another example of the B-18 "seaplane" occurred on 9 July 1943 when 37-594 ditched after dusk due to engine trouble off Kanon Point, Aruba. The crew was rescued after floating for several hours on the bomber, which continued to float, and finally had to be sunk by the Navy. In early 1943, a 1st Observation Squadron B-18 piloted by Squadron CO Capt. Nestor Cole was flying a long Atlantic coastal patrol out of France Field, Canal Zone (CZ), when a combination of nightfall, a tropical storm, and loss of instruments occurred. Cole was forced to fly on the deck trying to reach the coastline, but finally became lost and low on fuel, and was forced to ditch. In the darkness of night, unknown to Cole, he had ditched only a 100 feet or so offshore. When Cole ordered the crew to get into life rafts, a crewman at the aft of the aircraft yelled that he had jumped onto land from the rear fuselage door! The B-18 had been pushed by the wind and waves, rear first, onto the beach, where the crew could easily exit. The next morning a search aircraft easily spotted the B-18 and its crew on the beach, and they were rescued, while the B-18 was salvaged, and was reported to have flown again.

Exterior Finish and Markings

In March 1935, the AAC authorized unpainted aluminum surfaces on tactical aircraft as an economic—not weight savings—consideration. Previously, the typical AAC aircraft was clad in a blue fuselage with yellow wings and tail. Stripping old paint required approximately 175 man hours on pursuit aircraft, and up to 400 hours on bombers, plus the time and expense per aircraft for repainting. The weight savings was only about 25 pounds per pursuit, and 80 pounds per bomber aircraft. There was a delay in implementing this order, as there was a surplus of blue and yellow paint that had to be used before the natural metal scheme could be adopted, and manufacturers were not allowed to deliver unpainted aircraft until 1937.

In available photos, all prewar B-18s appear to have natural metal finishes. The exterior aluminum alloy surfaces of the wings, fuselage, and empennage were cleaned after assembly was completed, and left without a finish. However, to eliminate glare, the deck of the fuselage forward of the pilot's compartment was given one coat of primer, followed by two coats of Pine Green lacquer. Effective on aircraft 36-343 onward, the inboard half of each nacelle was given one coat of primer, followed by two top coats of flat Bronze Green lacquer shade #9. The landing gear and tail wheel parts, and the exterior surfaces of the metal strips on both front and rear turrets, were given one coat of zinc chromate primer, followed by one coat of aluminum spar varnish. Any exposed steel hydraulic lines were given one coat of primer, and finished to match the adjacent surface.

Individual parts in the ailerons, elevator, and rudder frame assemblies that were not covered with aluminum (e.g. fabric covered) were given one coat of zinc chromate primer before assembly, while aluminum covered and aluminum alloy parts were given a dip coat of Lionoil before assembly. The completed control surface frame assemblies were given, as a final finish, a double coat of aluminized zinc chromate primer applied over the initial coating. The outside fabric covered surfaces, except the rudder, were given three hand brushed coats of yellow semi-pigmented nitrate dope.

The 1938 Maneuvers provided for the first large scale test of temporary water soluble paint on natural aluminum aircraft. Since this was a spring exercise, green paints prevailed over grays. It was at this time the Air Materiel Command first standardized their paint colors with numbers: 25=White, 26=Sand, 27=Light Blue, 28=Sea Green, 29=Dark Blue, 30=Dark Green, 31=Dark Olive Drab, 32=Neutral Gray, 33=Black, and 34=Rust Brown.

Aircraft Camouflage Paint

In May 1939, Gen. Arnold requested that the Air Materiel Command research the permanent application of camouflage paint to combat aircraft. The AMC protested that the added cost and time, increased weight and drag of the paint, and the difficulty of painting aluminum left temporary water soluble paint as the painting solution. However, in August, Arnold ordered the AMC to conduct a study of camouflage paint and paint schemes, and a number of studies and reports were issued over the next year, culminating in a preliminary findings report in September 1940. Permanent camouflage number colors were standardized: 41=Dark Olive Drab, 42=Medium Green, 43=Neutral Gray, 44=Black, 45=Insignia Red, 46=Insignia White, 47=Insignia Blue, and 48=Identification Yellow. The study found that paint; particularly the widespread Dark Olive Drab, did not adhere properly in all areas, and faded signifi-

In available photos, all prewar B-18s appear to have natural metal finishes. The exterior aluminum alloy surfaces of the wings, fuselage, and empennage were cleaned after assembly was completed, and left without a finish. (USAF)

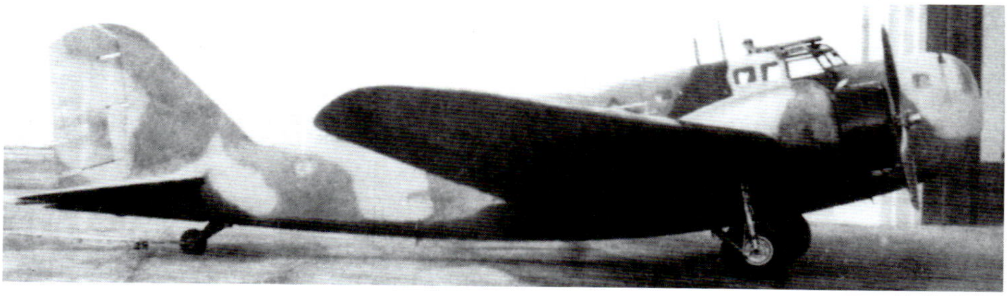

The 1938 Maneuvers provided for the first large scale test of temporary water soluble paint on natural aluminum aircraft. Since this was a spring exercise green paints prevailed over grays. This 7BG out of Hamilton Field, CA, was painted in a water-based dark olive drab and neutral grey, with black undersides and red cowlings. (USAF)

cantly. Neutral Gray, the other predominant color, fared better due to its light shade after being painted over aluminum. The AMC also found that the changing background conditions in flight negated any benefit of permanent camouflage paint.

Anti-submarine Camouflage Paint

On a clear, sunny day a lookout on a U-Boat was able to see an Olive Drab/Neutral Gray (OD/NG) B-18 approaching at nine to ten miles, giving the U-Boat ample time to dive. A more effective camouflage scheme was needed, and in April 1942, a commercial grade of Oyster White was tested at Mitchel Field, NY. Oyster White was a creamy yellow white that was sprayed to cover the exterior of a PBY. A few B-18s and other ASW aircraft were sprayed Oyster White or Oyster Gray (white with a black tint). When the 1st Sea Search Group was formed, further color tests were conducted on a B-18, B-17, and B-34, and it was concluded that a "just off white" color be accepted.

However, the Air Materiel Command was reluctant to change its Olive Drab/Neutral Gray policy until it conducted further tests. This scheme had been designed to be used at altitudes above 10,000 feet, while ASW searches were being conducted at 300 to 3,000 feet. In early September 1942, Proving Ground Command engineers conducted tests at Langley, and found white to be effective on head-on approaches, but too bright from above or the side. The PGC engineers recommended that Oyster White be discontinued, and suggested that, pending further tests, NG upper surfaces and White under surfaces be adopted. Meanwhile, the standard OD/NG remained in service, while the PGC conducted tests during late September and October evaluating the standard OD/NG, the newly recommended NG/White scheme, all White, and Haze paint (a recently developed paint technique that was supposed to render an aircraft invisible when sprayed in different thicknesses and shadowing). The camouflage schemes were tested from under every lighting condition and angle of attack, and NG/White was the proven winner, and the newly formed AAF ASW Command (formed from the 1st Bomber Command on 15 October) supported the decision. The first B-24s were painted in the scheme, but a few of the older B-18s and A-29s used Navy Blue Gray instead of the Neutral Gray with a wavy demarcation line.

In February 1943, the Anti-submarine Command again reversed its policy, and directed that OD remain as the color of the upper surfaces, dropping NG. The reason for this was that the AAF bombers coming off the production lines needed to be sent to ASW modification centers to have NG painted over the existing upper OD. This added weight to the bomber, but also would have to be repainted again when the ASW bombers reverted back to strategic bombers. The NG under surfaces were to be painted White. The painting pattern was to follow the RAF Coastal Command's, with the lower surfaces to be painted white, with the White/OD demarcation line being a wavy line with a height of 12 inches up and down, with the White extending over the leading and trailing edges of the wings.

Finally, in June 1943 the Air Materiel Command Engineering Division tested a variety of camouflage color schemes at Wright Field, OH, to determine the most effective one for anti-submarine aircraft. The 1st Sea-Search Attack Group provided five B-18s (37-464, -465, -561, -574, and -621) to be painted the prescribed test color schemes. After the aircraft were painted all the test flights were made on days with unlimited ceiling, normal haze conditions, and widely scattered or no clouds. Observers on the ground judged the relative effectiveness of the camouflage scheme of each aircraft flying from N-S and E-W directions at altitudes of 2,000, 4,000, 6,000, and 10,000 feet; first from directly overhead, and then at one mile from the observation area. The most effective camouflage scheme was determined to be:

Top surfaces: Shade No.41 Dark Olive Drab
Side surfaces: Shade No.43 Neutral Gray
Bottom Surfaces, leading edge, and front view: No.47 Insignia White

The Pima Air & Space Museum's B-18B is restored to 4th Anti-submarine Squadron markings, with a Dark Olive Drab fuselage and Neutral Grey undersurfaces. However, many ASW units painted the undersurfaces of their aircraft the recommended Insignia white. (Author/Pima)

In January 1920, new specifications established the colors, size, and location of the National Insignia wing star. The five-point white star containing a red circle was circumscribed inside a larger blue circle. Since the B-18 had tapered wings, the specification was changed, with Insignia's aft end placed at the forward edge of the aileron at an area that the width was 3/4th of the wing chord, and the center of the Insignia was placed at 1/8th the span of the wing. (USAF)

Only a few B-24s were painted in this newly recommended scheme, as on 9 July 1943, the "U-Boat emergency" was over, and the Navy was to assume ASW duties, with the AAFAC disbanded on 24 August 1943.

Markings
"U.S. Army"

In a Technical Order of 15 October 1926 it became mandatory to display the words "U.S. ARMY" in 24 inch letters under the wings and four inch letters on the fuselage. The letters were to be either black or white, which ever contrasted best against the aircraft's background color. In October 1942, all "U.S. Army" underwing markings were to be removed from all aircraft.

National Insignia

In January 1920, a specification established the colors, size, and location of the wing National Insignia, which was a five-point white star containing a red circle that was circumscribed inside a larger blue circle. The position of the star was placed at the ends of both sides of the wings, with one point of the star facing forward. The original diameter of the Insignia was to be 60 inches, and placed at the very tip of the wings, but this specification was altered on wings with ailerons. The B-18 had tapered wings, and the specification was changed, with the Insignia's aft end placed at the forward edge of the aileron at an area that the width was 3/4th of the wing chord, and the center of the Insignia was placed at 1/8th the span of the wing. In July 1940 the AAC Board recommended that one National Insignia be removed on top of one wing, and another be removed on the underside of the opposite wing, but added a larger Insignia on each side of the rear fuselage. All Insignia were to be painted with flat paint, and as war threatened, any colors that detracted from the camouflage or were too apparent were deleted, particularly the red circle (i.e. Japanese "meatball") in the center of the Insignia.

The 1926 specifications directed that rudders were to be painted with 13 alternate horizontal stripes: seven red and six white, with one blue vertical stripe forward of the 13 horizontal stripes. The rudder stripes were also ordered to be removed in July 1940. (USAF)

In July 1940 the AAC Board recommended that one National Insignia be removed on top of one wing, and another be removed on the underside of the opposite wing, but added a larger Insignia on each side of the rear fuselage. All Insignia were to be painted with flat paint, and as war threatened any colors that detracted from the camouflage, or were too apparent, were deleted, particularly the red circle (i.e. Japanese rising sun "meatball") in the center of the Insignia. (USAF)

Rudder Markings

As per specifications issued on 1 November 1926, rudders were painted with 13 alternate horizontal stripes of equal width, seven red and six white, with one blue vertical stripe forward of the 13 horizontal stripes. The proportions were 1/3 blue vertical stripe to 2/3 horizontal stripes. The rudder stripes were also ordered to be removed in July 1940.

The November 1937 Aircraft Marking Standardization

On 15 November 1937, the first standardized system for squadron colors, command markings, and designators (call letters) was adopted.

Squadron Colors: White, yellow, and red (and blue for a fourth squadron) were the only permitted squadron colors. These colors were to be painted "a suitable depth of the front portion of the engine nacelles."

Command Markings: The previous diamonds, chevrons, and spurious bands were replaced by five inch stripes in the squadron colors (with black replacing white on aluminum fuselages, and white replacing blue on blue surfaces). The Squadron Commander's aircraft had two vertical stripes located about midway on the rear fuselage: the "A" Flight Leader had one vertical stripe; the "B" Flight Leader had one stripe slanted downward from the top of the fuselage aft to the belly; and the "C" Flight Leader had one stripe slanted upward from the aft belly forward to the top of the fuselage. (See photos in wing section of this chapter)

Designators

The 1937 standardization placed the designators that identified the unit the aircraft was associated with on the tail, and above and below the left wing. The first letter indicated the type of unit, and the second denoted the number of that unit, with a number to identify the individual aircraft in that unit. For example, BG over 38 would designate the 38th bombardment (B) aircraft in the 7th Bombardment Group (G is the 7th letter in the alphabet), and would be the radio call sign for this bomber. These letters were generally black on light colored tails and yellow on darker surfaces, and were to be of a height to be readily seen from 150 yards.

In May 1940 these designators were revised so that all active AAC units would be assigned an individual code, with individual organizations being identified by a unit number rather than a letter. Originally, the AAC had anticipated that there would be 24 tactical Groups available by the end of June 1941, which would conveniently correspond to the 26 letters in the alphabet. However, the

After 1937, the new specifications replaced the double diamond banding signifying a Squadron Commander with two vertical stripes. (USAF)

The 1937 standardization placed the designators that identified the unit with which the aircraft was associated on the tail, and above and below the left wing. The first letter indicated the type of unit, and the second denoted the number of that unit, with a number to identify the individual aircraft in that unit. For example, BG over 38 would designate the 38th aircraft in the 7th Bombardment Group (G is the 7th letter in the alphabet), and would be the radio call sign for this bomber. (USAF)

41 Group Program of May 1940 provided for 15 more groups than alphabet letters. Thus, the aircraft number was placed on top of the unit number, followed by the unit type (e.g. B=Bomb Group, P=Pursuit, R=Reconnaissance, T=Transport, W=Wing, etc.) For example, 7 over 18 R would designate the 7th bomber of the 18th Reconnaissance Squadron.

In August 1941, the final form of the radio call designator was based on the aircraft's serial number, but did not become wide spread until 1942. The letters were to be at least four digits, consisting of the last digit of the fiscal year ordered, plus three numbers of the serial number. Also effective in August 1941, the yellow designators were to be repainted in black to diminish their visibility, and many units deleted their unit designator numbers, leaving only the serial number. Also squadron colors, emblems, and command stripes were to be eliminated.

Interior Color Schemes

All interior metal surfaces exposed to view, other than in the pilot's and radio operator's compartments, were finished with aluminized spar varnish. The finish in the pilot's and radio operator's compartments' exposed metal surfaces were finished as specified in two coats of Pine Green (over a coat of primer), or one coat of Light or Dark Grey Duco. The instrument panels, instruments, and control wheels and handles were finished in dull black. The fabric covering over the soundproofing in the pilot's and radio operator's compartments were finished with acetate dope. All corrugated metal flooring, other than that in the soundproof compartment, was given a final coat of aluminized spar varnish. The soundproofed compartment's corrugated metal flooring was painted with Pine Green Duco. Structural parts that were already coated in Lionoil were given a sealing coat of primer before the green lacquer was applied.

Markings of Lines

Lines, such as fuel, oil, etc., were marked in accordance with Spec. 98-24/105, and banded at frequent intervals for the purpose of identification. Where a single color was used to identify a line it was applied in a 1/2 inch wide band approximately 3 inches away from each joint or connection. Where two colors were specified, they were applied side-by-side in 1/2 inch wide bands located the same as specified for single colors.

Banding:

1) Fuel lines and connections	Red
2) Oil lines and connections	Yellow
3) Fire extinguisher lines	Brown
4) Airspeed lines	Black
5) Oxygen lines	Light Green
6) Vacuum lines	White and Light Green
7) Supercharger lines	Orange
8) De-icer lines	Light Blue and Light Green
9) Automatic pilot lines	No banding
10) Floatation equipment lines	Light blue
11) Manifold pressure	White and Light Blue
12) Steam (heating) lines	Light Blue and Black
13) Purging (bilge) lines	Light Blue and Yellow
14) Exhaust analyzer	Light Blue and Brown

The hydraulic lines were marked with triple banding, consisting of a Yellow band bordered on each side with Light Blue banding.

5

The B-18 in the Continental U.S. 1937 to War's End

Introduction

After its acceptance, the B-18 entered the mundane world of pre-war testing and endless days of boring base life; punctuated by exercises and reviews, and by isolated incidents that made life interesting. The B-18 was "the" American bomber when it was introduced, and set several records, but when the B-17 came on board the B-18 became increasingly outdone by its favored four engine rival, which garnered more and more of the records and headlines. Nonetheless, by 7 December 1941, 112 of the 220 production B-18s and B-18As were the most numerous American bomber to be stationed overseas.

Deliveries for Testing

On 23 February 1937, the first production B-18 (36-262) arrived at Wright Field, followed on the 28th by the DB-1 prototype. The next three aircraft were sent, one each, to Chanute Field, IL; Aberdeen Proving Grounds, MD, and Lowry Field, CO, for testing and evaluation. The next three aircraft were sent to the 7th Bombardment Group at Hamilton Field, CA, and soon after that trio was followed by another 30 aircraft. The next 94 B-18s were distributed among the 5th Bombardment Group at Hickam Field, HI; the 19th Bombardment Group and 38th Reconnaissance Squadron at March Field, CA; the 18th Reconnaissance Squadron, at Mitchel Field, NY; and the 21st Reconnaissance Squadron at Langley Field, VA. Several B-18s from the initial factory order eventually were delivered to the 2nd Bombardment Group.

Aircraft #37-34 was completed as the DB-2 prototype, incorporating a new powered nose turret and completely modified clear windows in the bombardier's compartment, which was located in the lower nose section. The DB-2 was delivered to Wright Field on 8 November 1937 as a flying test photo laboratory to test various new cameras (see photo), but its factory modification was found wanting, and the aircraft was restored back to a stock B-18 arrangement, and transferred to the 18th Reconnaissance Squadron at Mitchel Field, NY.

The AAC Puts the B-17 into the Headlines

While the B-18 was put into large scale production, the Air Force was eager to flaunt its new B-17, and demonstrate what it could do in a blatant attempt to show that B-17 quality was superior to B-18 quantity. The 2nd Bombardment Group had a dozen Flying Fortresses by March 1937, and its commander, Lt.Col. Robert Olds, and Major Barney Giles flew the first one to Bolling Field on 9 March to put it on display for four days for media and public inspection. On 16 May, Olds led four B-17s over 20 eastern cities in 15 states during a much publicized 11 hour flight. The flaunting continued

The first production B-18 (36-262) waits at Clover Field before its flight to Wright Field, where it arrived on 23 February 1937 for service testing. (USAF)

The DB-2 was delivered to Wright Field on 8 November 1937 as a flying test photo laboratory to test various new cameras, but its factory modification was found wanting. (USAF)

when six B-17s were given the featured role in an air review that the GHQ Air Force put on for the American Legion Convention in New York. Later that year a B-17 was flown to California to be placed on exhibit for the well attended Golden Gate Exposition at Treasure Island, San Francisco.

1937 Joint Exercise to Attack the *Utah*

Joint Exercise No.4, which has been commonly referred to as the search and bombing of the battleship *Utah*, took place on 13-14 August 1937. The exercise was conceived in fall 1936, when President Roosevelt wished to know how effective AAC land-based aircraft would be against an invading enemy fleet. Maj. Gen. Frank Andrews submitted a list of factors to Malin Craig and the War Department that should be considered in drawing up the parameters for the exercise. The main issue was that the exercise be held on the American East Coast, where the facilities, communications, and weather were better than the West Coast. However, on 10 July 1937, Roosevelt, Craig, and Adm. William Leahy met and issued a secret memo stating that the exercise would be held on the West Coast beginning noon 12 August, and ending at noon the next day, using the battleship *Utah* as a target, with no bombs being dropped after dark on the 12th. The old battleship *Utah* (BB-31) was launched in 1909, and was used as a mobile target ship between 1932 and 1941. She subsequently was hit by a torpedo and rolled over and sunk during the Pearl Harbor attack. The search area would cover an area of 100,000 square miles, extending out 300 miles into the Pacific west from San Francisco and Hamilton Field on the north, and San Pedro Bay, near the Los Angeles aircraft factories, in the south. The choice of the West Coast came as a surprise to the AAC, as during that time of year there were daily heavy fog conditions extending 300 to 500 miles out to sea that would hamper air search, and could hide the *Utah*. However, the Navy had dictated the biased terms of the exercise, and the AAC had no other option but to accept.

Gen. Delos Emmons was assigned as the Commander-in-Chief of the AAC forces, the defending "Blue Forces," which was comprised of 30 to 34 B-10s, four B-18s of the 7th and 19th Bombardment Groups, and seven B-17s of Robert Olds' 2nd Bombardment Group. There would also be about 30 Navy patrol aircraft available for reconnaissance, but they were limited in beginning their search until the dictated noon exercise starting time on the 12th. Also, the Blue defenders would have no ships available except for rescue purposes. The attacking "Black Force" was to be made up of two battleships, one aircraft carrier, and nine destroyers, but was ultimately limited to the *Utah* and several destroyers during the exercise. The bombs were 50 pound Navy Mark VIII water bombs in a light metal case that the ACC had never used previously. When the AAC requested the water practice bombs from the Navy it was told there was a shortage, and did not receive them until just a few days before the exercise. So in preparation for the exercise the AAC used powder filled practice bombs on stationary targets shaped like battleships, and on some small radio-controlled boats.

Emmons set up bases on the coast, and also inland, so that if the coast bases were fogged in he could fall back on the aircraft at the inland bases. As expected, at noon on the 12th, fog covered the entire West Coast, and extended westward 200 miles out into the Pacific. With no clue to the *Utah's* position, Emmons correctly guessed that the *Utah* would enter at the northwest corner of the search area and make a feint toward San Francisco, then steam toward Los Angeles, and ordered the Navy search planes to that quad-

The battleship *Utah* was the enemy during the exercises on 13-14 August 1937. The exercise was conceived in fall 1936, when President Roosevelt wished to know how effective AAC land-based aircraft would be against an invading enemy fleet. In the exercise the B-18 took a back seat to the B-17. (USN)

rant. At 1537 the Navy patrol planes discovered the *Utah* 300 miles southwest of San Francisco, and by the time the message was forwarded through 1st Naval District HQ to Emmons at Hamilton Field it was 1600 hours. Emmons dispatched the B-17s and B-18s, but held back the B-10s, as they had neither the speed nor range to intercept the *Utah*. The lead B-17 was piloted by Capt. Caleb Haynes, and had Gen. Frank Andrews and Col. Robert Olds on board, along with a young Lieutenant named Curtis LeMay, who was serving as Blue Force navigator. Several minutes after the bombers took off the Navy recon plane radioed that Black Force was 40 miles further east than first reported, and another report confirmed the sighting. LeMay used the Navy's reported positions to calculate the intercept point for the bombers flying over the fog. When Haynes' lead B-17 dropped down through the fog at 700 feet there was only empty ocean and fading daylight, and the Blue Force had to turn back; there were now faced with trying to find San Francisco through not only the fog, but also darkness. Fortunately LeMay found the city, but could not land at Oakland or the Air Corps' Mather Field at Sacramento, as they had no landing lights, and the Blue Force landed at Sacramento Municipal Airport at 2300. Once on the ground it was found that the Navy reconnaissance reports were inaccurate, which the Navy defended as by "only one degree," but which actually meant an error of 60 miles. Andrews and his staff were incensed, as they believed that the Navy error was intentional. In his biography *Mission with LeMay*, LeMay maintained that the Navy willfully gave the wrong or misleading coordinates:

"It might be charitable to assume that an incorrect position could be relayed to us on the first occasion quite unwittingly; or maybe, say, half-wittingly. But not twice in a row."

After fueling their aircraft, the AAC crews bought their dinners at nearby food stands and grabbed some sleep on a hangar floor, hoping to vindicate themselves the next day.

The next morning the West Coast was again fogged in, and the Navy recon aircraft that were to shadow the *Utah* during the night had lost the battleship. The Navy recon flights scheduled for the morning were also grounded by the fog, but Andrews ordered them up anyway. By 0900 Andrews decided to send up his B-17 and B-18 aircraft through the fog and hope for the best. By 1100 a Navy plane spotted the *Utah*, and LeMay made the calculations for a last chance intercept. But this time Andrews did not trust the Navy's coordinates and recalculated, dead reckoned, and then hoped, and ordered the B-17s to drop through the overcast for a look at 285 miles from the coast. When the B-17s, flying in line abreast, broke through the fog Andrews saw the *Utah* with its crew relaxing on deck. Soon there was pandemonium below, as the sailors raced for the hatches to escape the oncoming water bombs. The B-17s scored three direct hits and many near misses, but all the B-18 crews could do was cheer over the radio, as they lagged far behind the faster Fortresses.

The Navy claimed that the exercise, though apparently successful, did not prove anything, as the AAC bombers exploited the situation by using a sneak attack, and the *Utah* did not have a chance to maneuver! Andrews challenged the Navy to fend off another announced attack from altitude with evasive maneuvers in open seas the next day. At 1155, in clear weather, seven B-17s from the 2nd Bombardment Group and three B-18s from the 19BG intercepted the zigzagging *Utah* from varying altitudes ranging from 8,000 to 18,000 feet, and five minutes later there were 37 water bag hits on the battleship, 23% of the total bombs dropped.

After the Joint Air Exercise, the Navy insisted for security reasons the results be kept secret, and they were buried, as if that act could keep Navy ships safe from air attack. The Navy even kept the results from reaching higher Navy echelons, as the Senior Naval Commandant at San Francisco in his 31 August 1937 confidential memo to the Chief of Naval Operations made no mention of the second day's attack, and of the first said: "...The ceiling over the

B-18s from the 19BG were to take part in the first day of the exercise, but were unable to keep up with the faster 2BG B-17s, and did not intercept the *Utah*. However, on the second day three B-18s BG and seven B-17s scored 23% direct hits on the battleship. (USAF)

Utah was 400 feet. No bombing was possible." Nearly four years and five months later the *Utah* lay shattered on the bottom of Pearl Harbor, the victim of a Japanese sneak air attack.

In his report to the Army War College in October, Andrews described Joint Air Exercise No.4 as not only a great victory for the GHQ Air Force and airpower, but also demonstrated the superiority of the B-17 over any other GHQ bombers (e.g. B-10s and B-18s). Not only was the Exercise conducted against the Navy; apparently, the B-17 faction was out to sink the B-18.

Gen. Andrews and his staff had been eager to prove that the B-18 did not meet the AAC's requirements, and that a formation of B-17s compared favorably, not only in cost, but also in performance to a formation of B-18s. 1Lt. Leonard Harmon of the Engineering Department at Wright Field prepared extensive tables and charts comparing the speed and range for both bombers with different loads of fuel and bombs during the series of attacks. On D-Day minus One at 0600, with the *Utah* over 900 miles at sea, each B-17 could carry 2,200 pounds of bombs, and the B-18 1,200 pounds to attack the battleship. But if the B-18s were sent out on this mission, they would be unable to fly the second mission of the day due to their slower speed. If they passed on the first mission and flew the second mission, in which the *Utah* was closer, it was calculated that they could carry 3,400 pounds of bombs, as compared to the B-17s' 4,000 pounds. Attacks on the next day found the *Utah* much closer to the coast, and consequently less fuel and more bombs could be carried by both bombers. For the first attack at 0600 the next day (D-Day), with the target 350 miles west, the B-17s carried 8,000 pounds of bombs and the B-18s 4,400 pounds. For the second attack at 1300 the B-17s carried 8,800 pounds, and the B-18s 5,300. During the third attack at 1800, only 85 miles out to sea, the B-17s again carried 8,800 pounds, and the B-18s 6,200 pounds.

From these figures, the B-17 faction at Langley demonstrated that it would require 50 squadrons of B-18s to match the results of 31 squadrons of B-17s. With their cost figures of $176,000 per production B-17 and $103,000 per B-18, the B-17 faction deduced that it would cost $79 million for a fleet of superior B-17s, as compared to $77 million for an equivalent fleet of dated B-18s. As described earlier, these figures did not persuade Gen. Malin Craig of the War Department to procure any more B-17s, "except for experimental purposes until international conditions indicated a need for them." In the meantime, Craig reasoned that the B-18 "fulfilled reasonable requirements at a justifiable cost."

The B-17 and the Liner *Rex*: 12, May 1938

On 12 May 1938, Gen. Andrews commenced maneuvers that would give the B-17 an undeniable public relations boost over the B-18. With the cooperation of the Italian luxury liner *Rex*, Andrews sent out three B-17s to intercept the liner as it was 725 miles off New York. Under intense newspaper and radio coverage orchestrated by Lt.Col. Ira Eaker, head of the AAC's Information Division, the three B-17s left Mitchel Field under the command of Maj. Vincent Meloy. On board were a radio announcer and sound engineers from NBC, and a newsman and photographer representing the major news services. Ace navigator 1Lt. Curtis LeMay computed the expected position of the *Rex*, and estimated the time of interception. After flying through heavy overcast and rain showers the B-17s arrived exactly on time, and circled the liner while the passengers waved. The next day newspapers carried a dramatic front page photo of two B-17s flying over the *Rex* at smoke stack height.

"Bomber Rushes Son to Save His Mother"

While the B-17 was grabbing headlines with record setting flights, the B-18 had to settle for lesser acclaim. Always looking to justify their warplanes, the AAC often publicized humanitarian missions flown by its aircraft. On 20 November 1937, AAC Public Relations released the following story to accompany the above headline:

"An emergency flight November 20th to Red Bluff, CA, where the mother of Pvt. Cleveland Knox lay critically ill, probably saved her life." Pvt. Knox, stationed at Hamilton Field, learned that his mother needed an immediate blood transfusion, and "a B-18 twin engine bomber was wheeled out" at 11PM, and flown to Red Bluff in the darkness. The transfusion was administered, and the "physicians stated that the soldier's mother showed immediate improvement." The poignant story ended: "Leaving Pvt. Knox behindÖthe plane returned to Hamilton Field, arriving three hours after taking off on the 300 mile round trip."

On 12 May 1938, Gen. Frank Andrews commenced maneuvers that would give the B-17 an undeniable public boost over the B-18. With the cooperation of the Italian luxury liner *Rex*, Andrews sent out three B-17s to intercept the liner when it was 725 miles off New York. The next day newspapers carried this front page photo of two B-17s flying over the *Rex* at smoke stack height. (USAF)

After its introduction the B-18 took part in speed and range test flights. Shown is the navigator of an 88RS using an octant. (USAF)

Testing and Training

B-18 Speed and Range Test Flights

When the first two B-18As were delivered to the 73rd Attack Squadron at March Field, CA, they were flown on extended navigation qualification flights. The first cross country navigation mission was flown 4,500 miles to Washington, DC, by Capt. Ernest Lawson, and this flight was followed by a 5,000 mile flight to Middletown, PA, flown by Lt. A.W. Tyler.

Scheduled attack navigation qualification flights were then flown by the 95th Attack Squadron from March Field, CA, to such areas featuring difficult terrain as mountainous Bishop, CA, and St. George, UT; rugged coastal Monterey, CA, and desert Phoenix, AZ. The requirements of attack navigation were more exacting than cross country navigation, demanding a 300 mile flight with only a 30 second margin in the ETA, while flying no higher than 500 feet during the entire flight, utilizing hills and valleys for cover. Before take off a flight plan was made in duplicate (one for the pilot and one for the check pilot), consisting of main check points, compass courses, distances, and ETAs. During the flight the data was entered for ground speeds, actual times of arrival at check points, and elapsed times. After the flight this data was correlated, and checked against the original flight plan to determine the results of the tests.

In February 1937, Col. Follett Bradley, Chief of Staff GHQ Air Force, and co-pilot James Hodges flew a B-18 from Randolph Field, TX, to Langley Field, VA. Tail winds at 9,000 feet helped the aircraft make the flight in the record time of 5:40, at an average speed of 275mph.

During B-18 testing, it was determined that the following flight maneuvers were to be prohibited: loops, spins, rolls, dives, vertical banks, stalls, Immelmanns, and inverted flight. The Operation and Flight Instructions issued 1 July 1939 prohibited flying faster than 212mph, and not to operate with flaps at airspeeds in excess of 112mph.

B-17 Speed and Range Test Flights

The Air Corps Newsletter continually hyped the B-17, and publicized its "remarkable" speed and range: five hours from Kelly Field, TX, to Langley Field, VA; five hours from Miami to Langley; Wright Field, OH, to Langley in 1:45; etc. On 6 January 1938 Col. Robert Olds attempted a well publicized cross country flight, flying from Langley to March Field, outside Riverside, CA, a distance of 2,317 miles. Olds flew in the face of strong headwinds, and landed in California 13 hours and 27 minutes later. However, on his return flight, favored by tailwinds, he arrived back in Langley in 11 hours and one minute.

Probably the most publicized B-17 flight to demonstrate the long-range capability of the B-17 occurred in February 1938, when the State Department requested a goodwill flight to Buenos Aires for the inauguration of Argentina's new president. Col. Olds, in command of six B-17s, left Langley and flew to Miami for servicing. The B-17s then flew directly to Lima, Peru—2,695 miles in 15 1/2 hours—and after a seven hour layover flew 2,200 miles in less than 12 hours to an airfield outside Buenos Aires. That night Olds delivered a letter from President Roosevelt to the Argentine president, and the next day he led a formation in an impressive flight over the inauguration ceremonies. Olds was awarded the Distinguished Flying Cross for the flight, and the 2nd Bombardment Group collected the Mackay Trophy. During the next two years the AAC flew numerous well publicized B-17 flights to South and Central America, while the B-18 fell further and further out of the limelight.

The B-18 Tries to Find a Niche

Anti-Aircraft Cooperative Missions

The B-18 was used for cooperative missions with anti-aircraft units, mainly daylight tracking and firing on towed targets. However, night missions were conducted using aircraft equipped with muffled engines and night camouflage paint, or combinations of these, to determine their effectiveness against night anti-aircraft defenses.

B-18 as a Transport

In January 1940 a test mission was conducted to test the feasibility of using the B-18A as a troop transport. A battalion of 342 enlisted men and 12 officers was loaded into 19 B-18As of the 7th Bombard-

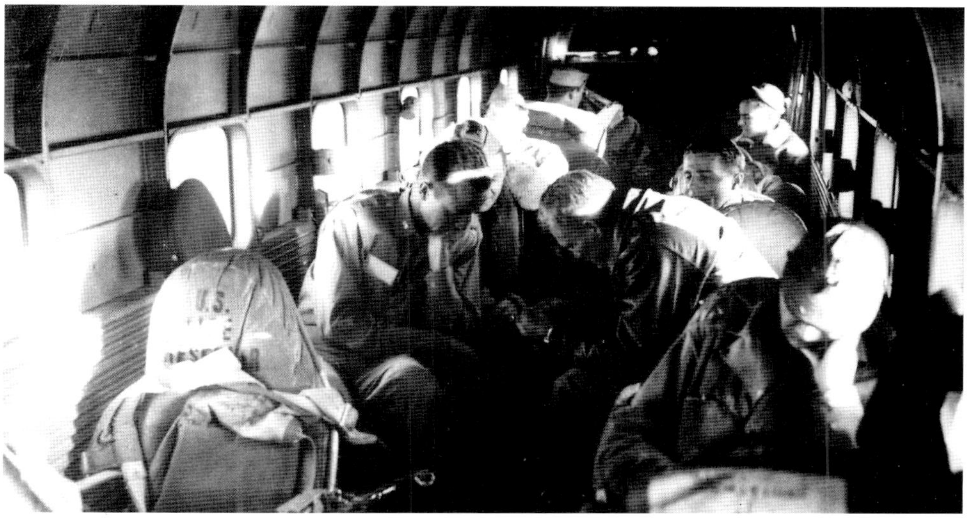

In January 1940 a test mission was conducted to investigate the feasibility of using the B-18A as a troop transport. A battalion of 342 enlisted men and 12 officers were loaded into 19 B-18As of the 7th Bombardment Group, and 19 more B-18As from the 19th Bombardment Group. Each aircraft carried four crewmen and ten infantrymen, along with 5,000 pounds of equipment. (USAF)

ment Group and 19 more B-18As from the 19th Bombardment Group. Each aircraft carried four crewmen and ten infantry men, along with 5,000 pounds of equipment. At 0900 the transports left the 7BG's home base at Hamilton Field, CA, near San Francisco, for the 19th's home base at March Field, near Los Angeles. The next day the battalion was transported back to Hamilton Field.

B-18 Tests the Norden Bombsight

The Navy had conducted impressive bombing tests against the USS *Pittsburgh* using the new gyro stabilized, mechanically synchronized bombsight invented by Carl Norden. Gen. Benjamin Foulois wanted to procure the bombsight for the AAC, but the Navy had classified the sight as top secret, and would not allow the AAC to talk with Norden. Nonetheless, Foulois was able to acquire a few of the sights to be used in courses at the Air Materiel Command at Wright Field by a few selected personnel to learn of its maintenance and operation. The bombsight was so secret that extreme measures were taken to store and safeguard it, and this significantly delayed their distribution and testing. It was not until May 1936 that the older model Norden bombsights that had been stockpiled in safekeeping at Wright Field were cleared for equipping bombers in Hawaii, and then to other Continental air stations.

Meanwhile, by 1935 Norden greatly improved the performance of his sight by developing an automatic pilot to use with it. Before the AAC could service test the improved Norden bombsight, it had to wait until the Navy had tested it and received approval for procurement. Once the Navy finished its trials, Lt.Col. Robert Olds tested it for seven weeks, and recommended that procurement and installation of the automatic flight control equipment (AFCE) be expedited on all bombing and reconnaissance aircraft, giving the B-17 priority, of course. It was not until mid-1937 that the GHQ Air Force permitted the AAC to acquire enough of the new bombsights to equip all its B-17s and a squadron of B-18s for service testing. The Air Materiel Command installed the new Norden bombsight on a B-18 and tested it, and turned the aircraft over to the 2nd Bombardment Group on 27 May 1938. It was the AAC's intention to replace the earlier automatic pilots installed on its B-18s with the

In the pre-war the Norden Bombsight was so secret that extreme measures were taken to store and safeguard it, and this significantly delayed their distribution and testing. This photo shows that the censor crudely obscured the bombsight's position in the front of this B-18A. A close inspection of the photo shows the man on the left to have his pistol drawn, while the bombsight was being transported in a canvas bag carried by the man in the middle. (USAF)

newer compatible AFCE type to be used with the new Norden sight, and for Douglas to install the new Norden sight on new B-18s on the assembly line. However, the equipment was not available for general installation until early 1939.

Bombardier Training

In late 1937, the 19th Bombardment Wing of the GHQ Air Force, March Field, CA, conducted progressive bombardment training for its crews in its new B-18s. The operations were conducted by Maj. Grandison Gardner, Group Operations Officer, and directed by Group Commander Lt.Col. Harvey Burwell. The training included all phases of the ground handling of high explosives: hoisting and loading, and then the air fusing, carrying, aiming, and dropping of bombs.

In December 1938 Arnold Skogstad, 2nd Wing Commander, decided to test drop a large number of demolition bombs of various sizes on Plum Tree Island, near Langley Field, in Chesapeake Bay. A squadron of nine B-17s and two squadrons of B-18s flew two bombing missions under Col. Carl Connell from Mitchel Field, and the larger capacity B-17s again confirmed that size did matter, and so did superior performance.

In the 1930s, each bomb group was responsible for training its own bombardiers, who were enlisted men. After war began in Europe in September 1939, Gen. Arnold prepared a 16 week bombardier program at Lowrey Field, outside Denver, CO, to meet anticipated personnel requirements. Gen. Delos Emmons, CO of the GHQAF, wanted the school to be under his control, but in July 1940 Gen. George C. Marshall put Gen. Hap Arnold in charge of the program. Fred Anderson became CO of the Lowrey bombardier school, and the first class of 18 candidates made up of pilot training washouts graduated in October, followed by 52 in November, and 50 more in December. Once they graduated, these bombardiers were given an additional ten week service test class to qualify them as instructors. In 1941 Arnold expanded the program to Barksdale Field, LA, and Ellington Field, TX. The first class of combat bombardiers was not graduated until May 1941. By the end of 1941 only 206 bombardiers had graduated, which was far short of the 1,367 planned. In December 1941 and February 1942 new schools were opened at Albuquerque, NM, and Midland, TX, to meet the new bombardier requirement, which was increased to 5,590. However, in March 1942 the requirement was raised to 14,000 bombardiers, and then to 22,400 in July 1942. To meet this incredible demand, Arnold established the Flying Training Command (FTC) on 23 January 1942, which opened eight new schools in New Mexico, Texas, and Arizona. Because of the demand for combat aircraft, many aging B-18s were pressed into service as bombardier training aircraft. Kirtland AFB, Albuquerque, NM, was a major bombardier training center, opening in February 1942 with an initial training aircraft complement of 28 B-18s and 50 AT-11s.

The bombardier's course was 12 weeks, with the first three weeks in preliminary ground training; the fourth through ninth weeks inclusive consisting of ground and air training, and the tenth through twelfth weeks inclusive consisting of air training to include tactical bombing and reconnaissance missions. For air training, the classes were divided into teams of four students who would train with the Norden bombsight in B-18s or AT-11s crewed by a pilot, co-pilot, and instructor. The Norden bombsights were so highly classified that they were stored in a guarded vault. When the students were scheduled to fly a practice mission they had to get clearance to enter the vault area to be issued a bombsight that was to be carried in a canvas case. When they left the vault, two students per bombsight were accompanied to the aircraft by an armed guard. The first training missions were dry runs carrying no bombs, in which the students became familiar with the Norden bombsight and sighting the target. With each dry run the student became more proficient, resulting in the bomber's vulnerable straight and level run over the target becoming shorter and shorter. The next step was to divide the teams into pairs for practice runs with practice bombs. One cadet would be placed in the nose with the instructor, while the other cadet was in the rear with the automatic camera to take photos of his classmate's hits and misses. The 100 pound sand filled practice bombs had 2.5 pounds of black powder packed in the nose that would explode in a big puff of smoke. After each run the student would note the results, which would be entered on the official 12-C form after landing.

In 1942 Flying Training Command (FTC) opened eight new schools in New Mexico, Texas, and Arizona. Because of the demand for combat aircraft, B-18s were pressed into service as bombardier training aircraft. Kirtland AFB, Albuquerque, NM, was a major bombardier training center, opening in February 1942. Pictured is a training officer accompanying a trainee carrying the Norden bombsight in a canvas bag. (USAF)

In the pre-war years the Army Air Corps PR corps issued a number of dramatic photos to publicize its aircraft. Shown is a flight of B-18As flying over the Statue of Liberty in August 1939. (USAF)

The May 1938 Maneuvers were the largest to date. Pictured (from left) is Maj.Gen. Frank Andrews, CO of the GHQ Air Force, Brig.Gen. Arnold Krogstad, CO of the 2nd Bomb Wing, and Brig.Gen. Delos Emmons, CO of the 1st Bomb Wing. (USAF)

Air Maneuvers and Reviews

The B-18 had been granted favored status and production contracts over the objections of top AAC brass. To prove their contention that the B-17 was the better aircraft, AAC officers Lt.Gen. Frank Andrews and Lt.Col. Robert Olds, in particular, showcased the B-17 as often as possible. The B-18 and B-17 vied for recognition in air reviews and exercises, but the competition proved to be one sided, as the four engine B-17 was truly a "Flying Fortress," and was breathtaking in sight and sound as it roared across the sky. In exercises and maneuvers the B-17 far outclassed the B-18, so that by the beginning of 1940 it was never considered to be a rival except in quantity.

Florida Field Exercises of March 1938

The tactical units of the 2nd Wing deployed to Florida in mid-March 1938 for a series of field exercises aided by the Florida National Guard, which provided the air units with supplies and living quarters. The exercises were under the command of Col. Henry Claggett, new Commander of the 2nd Wing. The HQ and tactical units from the 2nd Bombardment Group, with three B-18s, arrived at Orlando, while the 21st Reconnaissance Squadron (at Tampa) and the entire 8th Pursuit Group (at Sarasota) arrived from Langley. The 17th Pursuit Squadron arrived at Tampa from Selfridge, while the entire 9th Bombardment Group (at Lakeland) and the 18th Reconnaissance Squadron (at Tampa) arrived from Mitchel Field. The 18th and 21st Recon Squadrons were equipped with the B-18. The Wing's equipment and personnel were transported in their respective units' B-17s and B-18s. The first Sunday was declared Visitor's Day at each of the fields by Claggett, with the aircraft on display, and the crew standing by to answer questions. It proved to be a popular event, with 20,000 people turning out in Tampa alone. The Florida training exercises consisted of pursuit aircraft interceptions of approaching bombers, gunnery and bombing practice, and photo recon missions. To simulate ocean going targets a mixture of oil and aluminum dust was spread over the water's surface that formed into a condensed slick to be attacked.

The Air Maneuvers of May 1938

The War Department ordered the "concentration and maneuvers of the General Head Quarters Air Force during the month of May." Participating in these maneuvers were 468 officers and 2,380 enlisted men flying 131 aircraft. Of the 131 aircraft, there were 34 B-18s, 56 A-17 attack planes, 8 B-17s, 30 pursuit aircraft, and 3 miscellaneous aircraft. The 19BG(HQ) and 30BS from March Field,

A General Headquarters (GHQ) B-18 at Orlando during the Florida Maneuvers of March 1938. (USAF)

CA, and the 7BG and 9BS from Hamilton Field, CA, each flew five B-18s to the northeast.

The GHQAF under Maj.Gen. Frank Andrews, the 1st Wing under Brig.Gen. Delos Emmons, headquartered at March Field, CA, and the 2nd Wing under Brig.Gen. Arnold Krogstad, headquartered at Langley Field, VA, were to move from their various bases throughout the U.S. from 1 through 11 May to bases in the northeastern U.S. From 12 through 17 May "intensive training was to be held, with all units of the Air Force participating." The remainder of May was to be "taken up with Command Post Exercises, staff training and critique, and the movement of the units from their temporary airdromes back to their home bases." The purpose of the maneuvers was a training exercise to develop a warning and defense system, with the attacking planes reporting their positions and the defenders intercepting them.

Fort Bragg Exercises of October 1938

Beginning 10 October 1938, Brig.Gen. William Bryden, CO of Fort Bragg, NC, directed a well publicized, large-scale exercise to demonstrate air warning and air defense against an enemy offshore aircraft carrier air attack. Brig.Gen. Fulton Gardner, Commander of the 4th Coast Artillery District, directed the "Blue Force" in defense of Airdrome A at Fort Bragg, NC, and had an interceptor force under Lt.Col. William Kepner, consisting of a squadron of P-36s and two squadrons of PB-2As. "Black Force," based at Langley Field, VA, was commanded by Brig.Gen. Arnold Skogstad, and comprised of a bombardment squadron each of B-17s, B-18s, and B-10s; a reconnaissance squadron of B-18s, and an attack squadron of A-17s. AAC press releases gave the B-17 the "starring" role in the exercises in words and photos.

Beginning 10 October 1938, Brig.Gen. William Bryden, CO of Fort Bragg, NC, directed a well publicized, large-scale exercise to demonstrate air warning and air defense against an enemy offshore aircraft carrier air attack. This staged photo is interesting, as it shows ground crew of the 19BG/1W at March Field loading 100 pound practice bombs into a B-18 at March Field while wearing gas masks during a simulated gas attack. (USAF)

The exercises were scheduled to begin on 10 October, and run for six days; eight hours per day, divided into four hours by day and four hours by night. Newsmen led by Hanson Baldwin followed the action on an 8x16 foot map, with red lights showing the reported enemy locations, and green lights showing friendly aircraft and airfields. Interception of the bombers was reported on the map via rapidly blinking lights. The action was reported over loudspeakers.

A typical attack consisted of the reconnaissance B-18s flying ahead of the attacking force to reconnoiter, report on the weather en route and over the target, and then take photos during the attack. Two hours later the "attack" aircraft flew out east over the Atlantic, and then turned back to the west to cross the coast between Wilmington and the North Carolina-Virginia border to represent an enemy carrier-based aircraft attack. A squadron of A-17s led the attack by flying on deck to lay a smokescreen to blind AA gunners. It was found that the smoke screen was ineffectual, and the low-flying A-17s were vulnerable to machine gun fire and attack from pursuit interceptors. The high flying bombers arrived from a different route, flying over the coast, with the B-17s arriving first at 25,000 feet and dropping their bombs; followed by the B-18 and B-10 squadrons flying at 12,000 feet to drop their bombs. The warning network was found to be effective, and the bombers were usually intercepted by the pursuits, and the anti-aircraft fire was judged to be decreasingly effective as the bombing altitude increased.

Another part of the exercise was attacking a target during an extensive blackout covering over one quarter of the state of North Carolina, including 65 cities and towns and 21 counties. Six B-17s led by Lt.Col. Robert Olds, and accompanied by a contingent of newsmen, flew east over the ocean at 12,000 feet and returned inland, expecting a total blackout, but uncovered automobile headlights along the approach streets and highways to the target led the B-17s to the target, which was darkened, and hypothetically declared bombed into ruins.

The July 1940 Presidential Inspection

After France fell, and with the European war looking bleak for Britain, President Roosevelt was contemplating asking Congress for a large increase in airpower appropriations. To impress FDR, the AAC planned a three day inspection and an aerial review at the end of July 1940, with all Langley Field personnel and all available B-17s and B-18s readied on the Langley ramp for a 1445 arrival by FDR on the first day. The President arrived, and quickly left the review without comment, but the three day event was widely documented by the press and movie news media as an impressive demonstration of American airpower.

First Army Maneuvers August 1940

On 18 August 1940 six officers and 48 enlisted men left Langley Field, VA, in three B-18s for Mitchel Field, NY, to set up Headquarters for the 1st Composite Bomb Group, which was to participate in maneuvers with the First Army. The next day nine B-18s arrived from Barksdale Field, and were followed a day later by eight B-17s from March Field. Once all the aircraft arrived 15 missions were scheduled, but poor weather cancelled the last five mis-

Reviews and inspections were the order of the day, as Europe was at war. Pictured is a line up of aircraft at Langley Field in September 1940 for the inspection of Latin American officials. In the foreground are P-40s and a P-36 of the 8PS, while a large number of B-18s of the 2BG are seen across the runway. (USAF)

sions of the first day, and all the next day's missions. On the third day the aircraft had to be prepared for their turn around flights back to home base, and the maneuvers ended in a demonstration of how airpower could be cancelled by poor weather.

Dedication of Washington National Airport

In late summer 1940 the new Washington National Airport was completed, and was to be dedicated on September 25th by FDR. On 15 September 1940 the 2nd Bombardment Group Langley Field flew a large number of B-17s and B-18s, along with pursuit aircraft, to set up a mass flyover on the 25th. The 2BG's aircraft were to be joined over the next several days by aircraft from across America. However, poor weather cancelled the event, which was rescheduled for the 28th, and in clear weather FDR was impressed by the airpower display, and commended the 2nd Bombardment Group for its role in the show.

FDR 1941 Inauguration

On 18 January 1941 the 2nd Bomb Group's B-18s out of Langley, supplemented by nine 22nd BG B-18s out of Bolling, formed the B-18 contingent in the mass Inaugural aerial review. The aircraft took off with other Army aircraft, including the impressive B-17, at 1100, and flew several passes over the Capitol and returned to base.

Trouble on the Way to Another Exercise

On 20 March 1941, Lt.Col. Harold George led 12 2BG B-18s out of Langley at 0530 heading for exercises near Miami. After about an hour and a half George's aircraft suddenly went into a climbing turn with all the controls, except the ailerons, jammed. The crew quickly put on their parachutes and awaited orders to bail out. George gained back some control, and when the aircraft neared Fort Bragg, NC, the six crewmen bailed out; leaving George and his co-pilot, Maj. Donald Lyon. The decreased weight of the departed crew fur-

Five B-18As of the 21RS based at the Miami Municipal Airport fly over the Intercoastal Waterway during Army Day 1940. (USAF)

In 1941 the 21RS took part in sea searches off the Florida coast. These aircraft carried the National Insignia on each side of the nose to signify Neutrality Patrol. (USAF)

ther improved control, and George moved to the rear to assess the damage, and found that the left elevator was missing. He and Maj. Lyon experimented with the controls, and concluded that they could land the aircraft, and safely landed at Pope Field, near Fort Bragg. The only casualty was a crewman who sprained his ankle on his parachute landing. The crew was transported back to Langley via transport plane. Afterward they were inducted into the Caterpillar Club, reserved for aircrew men who had survived a bail out. Soon there appeared a tongue-in-cheek set of "Instructions to All Passengers in B-18 Aircraft," which follows:

"In case of engine failure or other minor trouble, such as the loss of a wing or wings, loss of propeller, fire, etc.: the following procedure should be carried out:
a) Remove all loose radio equipment and tools.
b) Ask pilot for Form 1 and fill out same
c) Send radiogram to Corps Area Headquarters requesting permission to make an emergency jump.
d) Check altitude and position, being sure to include this information in the above mentioned radiogram.
e) Make a list of the best telephone numbers in area.
f) Notify the pilot that you are ready to jump.
g) JUMP
After leaving the ship proceed as follows:
a) Count ten (It may be necessary for some passengers to carry a slide rule to accomplish this. If necessary it will be included in the bundle of spare radio parts and tools carried).
b) Pull rip cord. This is quite essential.
c) The usual procedure here is for the parachute to open.
d) If step b or c or both, are omitted, immediately upon landing the passengers will proceed to the Post Operations Office, secure and fill in Form I1131 (Request for Sympathy), and mail same to the Chief of Chaplains, U.S. Army, Washington, DC. This will be accompanied by an Unsatisfactory Report on the parachute used.
e) The Form 1, radio, spare parts, tools, etc., will always be carried by the passenger on his jump.
f) In same cases the Booklet, 'How to Swim in Three Easy Lessons;' will be found to be very helpful."

B-18 in Show Business

Irving Berlin and the Air Force Song, "Wild Blue Yonder."

In 1937 Brig.Gen. Hap Arnold decided that the Air Corps needed a song to create a musical identity to equal the Navy's famed "Anchors Away." *Liberty Magazine* sponsored a contest, and received over 600 entries, with the winner being Robert Crawford, who introduced the Army Air Force Song, "Off We Go (into the Wild Blue Yonder)" at the September 1939 Cleveland Air Races. Among the other hopeful composers were an unknown Meredith Wilson, who much later put the "*Music Man*" on Broadway, and the celebrated Broadway composer Irving Berlin, who was flown around in a B-18 by the Air Corps as a creative incentive. Even though the famed composer did not win the contest, he used this experience to help Moss Hart bring his play "*Winged Victory*" to Broadway in November 1943. Hart went on to produce the play as a movie in June 1944, and Hart spent at least two months living among aviation cadets, masquerading as a private, in order to get material for his play, and asked Berlin, his friend and collaborator, for advice.

Test Pilot

The 1938 MGM film *Test Pilot* starred Clark Gable, Myrna Loy, and Spencer Tracy. The movie was the story of test pilot Gable, who decides to train ACC cadets after losing friend Tracy in a crash of an experimental bomber. Paul Mantz, renowned aviation expert and aerial cinematographer, was in charge of the aviation scenes in the film, which had large scale and authentically produced aerial sequences that make it a classic aviation film. The AAC, wanting to impress Congress in the lean prewar years, was eager to participate in the film, and in the final scenes hundreds of aircraft were

The 1941 Paramount movie *I Wanted Wings*, based on Bierne Lay's bestselling book, starred Ray Milland, William Holden, and Wayne Morris, and was produced with the support of the War Department to encourage recruitment into the military as war approached. The bomber sequences featured the B-17, but there was a sequence that prominently showed a B-18 at Randolph Field. This studio photo shows a B-18 surrounded by cameras and light reflectors. (Paramount/author)

filmed at March Field, CA, including a number of B-18s of the 19BG. The AAC was particularly interested in introducing the capabilities of the YB-17s to the world, and for the filming Lt.Col. Robert Olds, CO of the 2BG, led 12 YB-17s across country from Langley Field, VA, and were featured in the movie. The movie was one of the biggest hits of the year, and was nominated for three Academy Awards, including Best Picture.

Bombardier

The 1943 RKO movie *Bombardier*, starring Pat O'Brien and Randolph Scott, was the story of the "rigorous training received by the men behind the Norden Bombsight." The heavily propagandized movie was filmed in six weeks by Director Richard Wallace at Kirtland Field's Bombardier Training School at Albuquerque, NM. Although B-18s were the primary bombardier training aircraft at Kirtland, B-17s were transferred to Kirtland and starred in the movie. However, there are several background shots of parked B-18s in several outdoor scenes.

The 1943 RKO movie *Bombardier*, starring Pat O'Brien and Randolph Scott, was the story of the "rigorous training received by the men behind the Norden Bombsight." The movie was filmed at Kirtland Field's Bombardier Training School, Albuquerque, NM. B-17s were featured in the film, even though B-18s were the major AAF bombardier training aircraft. A B-18 (arrow) can be seen in the background. (RKO/author)

I Wanted Wings

The 1941 Paramount movie *I Wanted Wings*, starring Ray Milland, William Holden, and Wayne Morris, was produced with the support of the War Department to encourage recruitment into the military as war approached. The movie was based on Bierne Lay's bestselling book of the same title, and followed three young ACC cadets through flight training at Randolph and Kelly Fields, where producer Arthur Hornblow decided to make the film on location. The film featured a large number of AAC aircraft, and had numerous exciting aerial sequences filmed by Paul Mantz. It won the 1941 Academy Award for Special Effects, and was Paramount's top grossing film of the year. Again, while the bomber sequences featured the B-17, there was a sequence which prominently showed a B-18 at Randolph Field.

Canadian Rescue Operation

On 18 November 1940 a RCAF B-18 Digby #749 piloted by F/O J.G. Richardson was flying a routine night time war patrol flight from Newfoundland when the weather closed in the Gander Newfoundland Airport, forcing Richardson to head toward Montreal to find an open airfield along the way. About 200 miles from Montreal, and unable to find a field, the bomber iced up, and was almost out of fuel. About midnight the pilot ordered the crew to bail out near Megantic, ME, but there were only five parachutes available for the six crewmen. Richardson and the radioman, as the two lightest members of the crew, volunteered to jump with one chute. Once the Digby did not arrive the RCAF requested help in the search from the 2nd Bombardment Group, which dispatched six of its B-18s, and three more from the 18th and 41st Reconnaissance Squadrons. The nine aircraft arrived at St. Hubert Airport, Montreal, early in the morning of the 20th, but poor weather prevented an immediate search.

The RCAF had limited quantities of 100 octane gasoline for use by the AAC B-18s, and it had to be hand pumped into the fuel tanks used for take off, while the other tanks were filled with 80 octane fuel used for cruise. The search of a 30x50 mile area was commenced the next day, and soon three parachutes were reported, but poor weather stopped the search. The next day the search area was reduced to 8x12 miles, with three aircraft each flying a three hour search pattern that spotted the crashed aircraft. But again the weather closed in, and the search aircraft were recalled, and were grounded for the next two days as heavy ice formed on the parked

B-18s. The ground crews tried to remove the ice on the aircraft by beating it with ropes and rubber hoses, but the remaining ice needed to be thawed. RCAF Lockheed Hudsons were removed from a hangar, and two B-18s were towed inside to be de-iced at four hour intervals.

On the 26th, eight days after the crew had bailed out in freezing weather, landing in rugged country; four B-18s were ready to resume the search. The three chutes that were spotted five days earlier were again found, but by that time two of the Canadians had been rescued by land search, and another wandered into a lumber camp. It appeared that the Richardson/radioman parachute tandem descended safely, but drowned when they landed in lake. The other missing man was declared lost.

The B-18s returned to Montreal, and were placed two at a time in a hangar to be de-iced in preparation to return to Langley. Two B-18s were unable to return with the others, as during the search one experienced a broken starter, and the other a sheared fuel pump drive shaft. The parts had to be ordered by telegram, and sent air express to Burlington, VT, then trans-shipped by a Canadian Airline to Montreal. The American Customs Agent at Burlington held up the shipment, as it did not have an export license! While official teletype communications bandied back and forth across the border the weather closed in, and the parts had to be transported to Montreal via railway, during which time the parts were lost for a day.

B-18 in Radar Jamming Testing

After Dr. Ivan Getting had developed the top secret experimental SCR-584 microwave anti-aircraft gun-laying radar system at the MIT Radiation Laboratory, Vannevar Bush at the Harvard Radio Research Laboratory needed to determine the system's vulnerability to jamming, so that if it could be jammed the necessary modifications could be made. Bush appointed Dr. Winfield Salisbury to conduct the jamming tests, much to the chagrin of Getting, who felt that his SCR-584 was jam-proof. Salisbury devised an airborne jammer, and had it installed on one of the several B-18 s made available to Harvard and MIT for test purposes. Salisbury and Getting met for the test on Deer Island, in Boston Harbor, where a SCR-584 equipped truck was set up. Salisbury was in contact with the B-18 pilot and the jamming crew on the B-18, which flew out to sea, and then turned toward the island. Salisbury asked the truck SCR operator to tell him when he picked up the B-18 on his set so he could tell his crew to use the jammer. Meanwhile, Getting had climbed on top of the SCR truck for a better view, proclaiming that his set was unjammable, and that this test was a waste of his time. Salisbury asked Getting to get away from the antenna and come down from the truck, but the adamant inventor refused. Salisbury instructed the B-18 jammer to begin, and the antenna on top of the truck immediately swung wildly in all directions, knocking Getting off the truck. Salisbury rushed over, thinking that he had killed Getting, but found him to be OK, though extremely angry. Getting insisted that it was an equipment malfunction that caused the antenna gyrations, not jamming that caused the failure, and demanded another test. Salisbury agreed to another test, which never happened, as Getting was always too busy. Both men made their reports and, after analyzing the results, Getting wisely improved the SCR set, and it was put into operational use.

B-18s in Photo Reconnaissance

The Fairchild F-1 and Beechcraft F-2 (UC-45) were the major prewar photo recon aircraft, but a number of bombers, and observation aircraft such as the B-18, Lockheed A-29 Hudson, and B-34 Ventura, or O-47 were employed in special projects where their physical and performance specifications, and even their availability and base location, were the criteria for their use. The DB-2 was delivered to Wright Field on 8 November 1937 as a flying photo laboratory to test various new cameras, but its factory modification was found wanting, and the aircraft was restored to a stock B-18 arrangement, and transferred to the 18th Reconnaissance Squadron at Mitchel Field, NY.

In the prewar, Maj. (later Brig.Gen.) George Goddard was the head of Army Air Materiel Command photographic research at Wright Field. Goddard was active in general camera and lens development, particularly the K- and T- series of cameras and high resolution metrogon lenses. After refining the film, lenses, and techniques used in daytime recon photography, Goddard made major contributions in the development of night aerial photo-reconnaissance. He took the first night aerial photo on 24 November 1925, using the K-1 camera using flash bombs he developed. Continuing his nocturnal photographic tests in the late 1930s, Goddard used a reliable B-18A (-102) flown by 1Lt. Marcus Cooper to circle his hometown of Rochester, NY, dropping his newly developed magnesium photographic "flash bombs." The 45 pound cylindrical bombs, identified by two black bands, were developed with the cooperation of the Eastman Co. of Rochester, NY. The flash bombs exploded with a millions of candlepower flash lasting but one-sixth second, and culminating in a thunderclap on the ground. By using a photo cell and electric synchronization device, the flash was coordinated to the shutter of an automatic 8x10 negative camera that was mounted to shoot through the bomber's floor. At 5,000 feet the bomb illuminated an area of about five square miles below, and was so blinding that it concealed the position of the bomber from

Col. George Goddard (left) used a B-18A flown by 1Lt. Marcus Cooper (right) to circle his hometown of Rochester, NY, to drop his newly developed magnesium photographic "flash bombs." (USAF)

AA batteries. The flash bomb allowed accurate observation of enemy movements at night, and as Goddard stated, "(they) marked the end of day that military leaders prayed for darkness or for a few extra hours of daylight."

By June 1942 there was a pressing need for strategic aerial photo-reconnaissance, and the Second Air Force was training pilots and technicians at Lowrey Field in B-18s, Beechcraft F-2s (UC-45), and North American F-4s (B-25).

B-18 Considered for the Doolittle Raid

Three weeks after Pearl Harbor, RAF Chief-of-Staff Peter Portal met with U.S. General of the Army Air Forces Hap Arnold and Air Marshal Arthur Harris of the RAF Bomber Command. In the course of their discussions Portal asked when Japan could be bombed, and the use of aircraft carriers carrying bombers was broached. During this time a Navy Captain, Francis Low, had seen Army bombers at Norfolk, VA, using an airfield marked for Navy aircraft carrier take offs and landings, and suggested the use of carrier-borne bombers to Adm. King. With FDR eager to strike back at Japan, the bomber/carrier strike force concept soon developed into the First Aviation Project, the first bona fide Army-Navy cooperative effort. Gen. Jimmy Doolittle was assigned to determine which Army bomber would be able to take off within 500 feet with a 2,000 pound bomb load to be carried over 2,000 miles. He had recently been sent by Arnold to B-26 transition school to calm apprehensive trainee pilots, and assure them that the B-26 did not deserve its "Widow Maker" reputation. So Doolittle already knew that the B-26 required too much take off space, and that left the B-25, B-18, or B-23, all of which required modifications and extra fuel tanks to have the required range to reach Japan. The limiting factor between the three aircraft was the 75 foot wide take off area on the carrier deck, which eliminated the B-18 and B-23, which had a 90 foot wing span, compared to the B-25's nearly 68 feet. Perhaps 22 feet of wingspan prevented the B-18 from a niche in history.

B-18 with the Airborne

On 26 June 1940, the 48-man Parachute Test Platoon was formed from 200 volunteers from the 29th Infantry Regiment at Fort Benning. The Test Platoon was commanded by 1Lt. William Ryder, and underwent intensive physical training and 40 hours of instruction regarding the parachute and its packing at Lawson Field. After this initial training, the unit and its three B-18 bombers moved to Fort Dix, NJ, where there were two 150 foot parachute practice jump towers similar to those erected at the 1939 World's Fair. After spending ten days training on the towers, the unit flew back to Benning for the final phases of its training, which would be from their B-18s. There were five jumps scheduled: the first two jumps would be individual jumps, followed by three mass jumps in which a full plane load of troops would jump from the plane in quick succession. After Ryder had explained the jump schedule, he told the unit that a B-18 was about to fly over the field at 1,500 feet, and a dummy was to be thrown out so the men could see a demonstration of a jump from an aircraft. The dummy's parachute did not open, and as it crashed into the ground so did the confidence of the trainees. Less than an hour later the B-18 again flew over the assembled trainees, and a real live paratrooper made a perfect jump. On 16 August 1940 the unit made its first individual jumps, followed by the second jumps the next day. It was during these jumps that two rules of parachuting were put into effect that are still in operation today. The first rule was for a parachutist standing in the aircraft doorway to look out over the horizon and not down to the ground. The second rule was to have the jumper keep his chin tucked to his chest so as not to hit his face on the parachute's connector links when jumping out the door. A tradition, not a rule, was also established during this time, as the yell "Geronimo" was introduced when jumping by Pvt. Aubrey Eberhardt. The next three jumps were to be mass jumps, with 12 trainees packed into the B-18 cabin with their bulky T-4 parachutes on their backs. After the aircraft slowed over the jump zone, the jump master gave the hand signal to jump, and the men squeezed out the narrow door in quick succession while the other trainees watched the jump from below. The fifth and last jump was to be a mass jump for the benefit of the Brass from Washington, including Secretary of War Henry Simpson and Chief of Staff Gen. George Marshall. The entire platoon was to jump from three B-18s flying in a column. Once they had landed, the platoon was to collect their weapons' containers and attack an objective located near the field. Except for one jumper who landed on top of a hangar the exercise was successful, and the unit would establish a parachute training unit, and many of its members would soon volunteer for the 501st Parachute Regiment.

During the German airborne campaign during the invasion of Crete at the end of May 1940, troop reinforcements were brought in by Ju-52 transports once the German paratroops secured an airfield. This new practice influenced American airborne thinking, as the 550th Infantry Airborne Battalion was formed, neither as a parachute nor glider unit, but as a combat unit transported by aircraft to be deployed after the aircraft landed. The 550th and 501st Parachute Battalions were ordered to Panama in July 1940 to conduct joint maneuvers there in response to the possible threat of large German national populations in Central and South America. On 28 July,

On 26 June 1940, the 48-man Parachute Test Platoon, commanded by 1Lt. William Ryder, was formed from 200 volunteers from the 29th Infantry Regiment at Fort Benning. (USAF)

After spending ten days making training jumps from 150 foot parachute practice towers at Fort Dix, NJ, the unit flew back to Benning for the final phases of its training, which would be from their B-18s. There were five jumps scheduled: the first two jumps would be individual jumps, followed by three mass jumps, in which a full plane load of troops would jump from the plane in quick succession. (USAF)

Company C of the 501st left New Orleans by troopship to join the 550th at Fort Kobbe, near Howard Airfield, Panama, located on the Pacific side of the Panama Canal. Meanwhile, a three day reconnaissance of five Central American countries surrounding Panama was flown by a B-18 that landed at or near the capitals of the countries to be inspected. Two 501 Parachute Battalion lieutenants dressed in civilian clothes were to reconnoiter the best parachute and air landing locations, and then draft a plan for possible airborne operation in case of any German insurrection. But as German military fortunes in Europe declined so did their influence in Latin America, and the plans never needed to be realized.

In the fall of 1940, the 550th Parachute Battalion and C Company of the 501st took part in the Army's first combined parachute/air landing tactical training exercise in Panama. C Company was carried by 74 B-18s and four C-39s based at Howard Field, and parachuted on the airfield at Rio Hato, establishing a perimeter there. These aircraft turned around immediately after the drop, and returned to Howard Field. After landing, and with their engines running, leading elements of the 550th embarked, which landed in two waves on the airfield, and then secured and enlarged the drop perimeter. As the 550th landed at Rio Hato, C Company paratroopers assisted the 550th in unloading their equipment from the B-18s, and the aircraft took off again, and returned an hour later with the remainder of the battalion. By mid-afternoon the perimeter was expanded and secured well beyond the airfield, and the maneuver umpires declared the exercise a total success. As a result Gen. Marshall directed that a second air landing battalion, the 88th, be formed.

Soon after Pearl Harbor, the restrictions on Army troop strength were removed, and in March two central Army training headquarters were established: the Army Ground Force and the Army Air Force. The Army Ground Force then established the Airborne Command at Fort Benning, which would organize and train the parachute and air-landing forces, and control allocation and training of such AAF transport or glider aircraft; coordinating training and operating procedure for airborne operations. A few days after the Army Ground Force was established, the Army Air Force established the Air Transport Command at Stout Field, near Indianapolis, where it was to train the pilots who would fly the aircraft and gliders used by the airborne troops. When the Air Transport Command was established it had only 56 aircraft, including a small number of B-18s, but mainly consisted of its replacement and sister, the C-47 Skytrain. The Skytrain had a 6,000 pound payload and one jump door, and in June 1942 a second troop carrier was introduced, the C-46 Commando. The Commando had a 10,500 pound payload and two jump doors, but because of its large capacity would be used as a cargo carrier, and was used only once in combat, during the jumps across the Rhine in March 1945.

6

The B-18 in Hawaii

Despite the proven superiority of the B-17, at the outbreak of the war, B-18 and B-18As were the most numerous American bomber to be stationed overseas, with 112 of the 220 production bombers stationed offshore. The first day of the war would expose the B-18 to the Japanese attacks, not only at Hawaii, but also in the Philippines.

On 7 December 1941, the air contingent of the Hawaiian Air Force (7th Air Force) of the Hawaiian Department had a personnel strength of 754 officers and 6,706 enlisted men, the majority of which were based on the island of Oahu. The Hawaiian Department was commanded by Lt.Gen. Walter Short, who directed all non-aviation Army personnel and activities on Hawaii. Maj.Gen. Frederick Martin reported to Short as both the commander of the Hawaiian Air Force, and as the Hawaiian Department Air Officer. In this capacity Martin directed all Army aviation personnel and activities. The Hawaiian Air Force had been activated on 1 November 1940 at Fort Shafter, and its personnel were assigned to the 18th Bombardment Wing stationed at Hickam Field, the 14th Pursuit (as fighter units were designated at the time) Wing at Wheeler and Haleiwa Fields, and the 86th Observation Squadron, along with several P-40s of the 44th Pursuit Squadron at Bellows Field. Personnel were also assigned to the 17th Air Base Command at Hickam Field, to the 18th Air Base Command at Wheeler, or to the Hawaiian Air Depot at Hickam.

Maj.Gen. Martin and the Hawaiian Air Force were responsible for defending the Hawaiian Islands and the U.S. Pacific Fleet from air attack. Martin had arrived in Hawaii the day after the activation of the Hawaiian Air Force, and was an experienced pilot, tactician, and veteran commanding officer, having led the 3rd Bombardment Wing at Barksdale Field, LA. On his arrival in Hawaii Martin found that the Army and Navy had developed a rivalry that seriously impaired inter-service cooperation. In order to strengthen the Air Force's position in Hawaii, Gen. Arnold instructed Martin to try to have his commander, Lt.Gen. Short, and the Navy's commander, Adm. Husband Kimmel, resolve this problem. Martin was not the man for the task, as he had a serious chronic ulcer that had required

An impressive PR photo of all the 33 B-18s of the Hawaiian Air Force in April 1940 flying over Honolulu's Royal Hawaiian Hotel, with Diamond Head in the background. (USAF)

Hawaiian High Command (Front: left to right): Lt.Gen. Walter Short, Hawaiian Department CO, visiting British Lord Louis Montbatten (center), and the Navy CO Adm. Husband Kimmel. (Rear: left to right): Maj.Gen. Frederick Martin, CO of the Hawaiian Air Force (right), and RAdm. Patrick Bellinger, CO of Navy air units. (USAF)

surgery, and forced him to curtail many official duties and engagements that made him appear aloof to the men under his command. Lt.Gen. Short was an old line infantry man who was obsessed with sabotage, and the invasion of the Hawaiian Islands. To combat a Japanese invasion, Short insisted on a six to eight week infantry training program for aviation personnel in July 1941 that was to be to the detriment of their more important aviation training. While trying to placate both Kimmel and Short, the ailing Martin gave into Short's infantry predisposition, and his obsession with enemy sabotage. On the day of the Japanese air attack, Short's fixation would have serious consequences, as American aircraft were parked closely together to protect them from sabotage by Japanese living on Oahu. Also anti-aircraft ammunition was stored in a guarded central location, but out of reach for immediate use to protect the vulnerable aircraft from air attack.

Martin-Bellinger Report 31 March 1941

In 1941 Generals Short and Martin recognized that the AAC was unable to perform the long-range reconnaissance necessary to protect Hawaii from a sneak air and sea attack, and requested modern aircraft in larger numbers. In a study of the situation Martin and his Navy opposite number, RAdm. Patrick Bellinger, drafted the historic report that took their names. The report described how a sneak attack could occur, and the measures necessary to prevent it. The report concluded that the crux of prevention was in 360 degree long-range air reconnaissance extending 1,000 miles out to sea. Both men realized that with the available equipment this reconnaissance objective could not be carried out on a daily basis, and could only be implemented with advanced knowledge that war was impending.

At the time the Hawaiian Air Force had 33 B-18s and 12 B-17s available, while the Navy had 69 long-range PBY flying boats. The PBYs operating with PATWING 2 were considered obsolescent, and were flown with inadequate personnel and facilities. Since the B-18s were already considered outdated, and had relatively short endurance, it was decided that the Hawaiian Air Force be responsible for the short-range reconnaissance, which meant circling within 20 miles of the islands, while the long range Catalinas would be better suited to fly extended patrols far offshore. However, Adm. Kimmel rejected this proposal, as he wished to use his PBYs to cover the fleet once war had begun, and if they were used on daily patrols they would soon wear out, and be useless when war began. In late 1941, Hawaii was a mid-Pacific fortress defended by the powerful Pacific Fleet and its aircraft, the Hawaiian Air Force, with its hundreds of bombers and fighters, and the Army with 25,000 well trained infantry soldiers. America had spent millions of dollars of its defense budget on its Pacific bases, extending from Dutch Harbor in the Aleutian Islands to the north, and then on those bases swinging south and west, extending through Midway, Wake, Johnston, and Palmyra Islands. At the time the Navy had too few PBYs and crews available to adequately patrol the entire 360 degree search area, and Kimmel, wishing to conserve those aircraft, decided that the only possible threat to Hawaii were the Japanese held islands far to the south. He allowed only a limited number of PBY patrols in that direction, and in hindsight it is now well known that the Japanese carrier attack was launched from the north.

In 1941 the Hawaiian Department and Navy held some exercises and full alerts on Oahu, but generally they were regarded as a supposition, and were not taken seriously, not only by enlisted personnel, but sadly by their officers. It was the prevailing opinion from the Army's General Short and the Navy's Admiral Kimmel on down through their senior commanders that there would be an ample warning of any Japanese attack, which would allow the initiation of long-range reconnaissance, establishment of communications between the Army and Navy, manning of the aircraft warning cen-

The 5th Bombardment Group, as the 5th Composite Group was the first bomb group in Hawaii. In March 1938 the unit was redesignated as the 5th Bombardment Group and, during that year, received its B-18s, which were distinguished by their red engine cowlings. (USAF)

ter, the arming and deployment of available aircraft, and dispersal of others. The U.S. fleet would then sortie toward the oncoming Japanese and join in a battle covered by a sky full of American aircraft.

Hawaii Receives the B-18

As the 5th Composite Group, the 5th Bombardment Group served in Hawaii at Luke Field, on Ford Island, during the placid prewar years, flying Keystone Bombers of the B-3, B-4, B-5, and LB-6 versions. By 1937 their Keystones were in such poor condition that they were loaded on barges and dumped offshore, and were replaced by the newer Martin B-12s. In March 1938 the unit was redesignated as the 5th Bombardment Group, and during that year received its B-18s, and later received the supplementary B-17s. The 5th was composed of: 23rd Bombardment Squadron, 31st Bombardment Squadron, 72nd Bombardment Squadron, and the 4th Reconnaissance squadron.

The 11th Bombardment Group was activated on 1 February 1940, and was equipped with a mix of B-18s and B-17s. The 11th was composed of: 26th Bombardment Squadron, 42nd Bombardment Squadron, 58th Bombardment Squadron, and the 50th Reconnaissance Squadron.

Hickam Field

When Hickam Field was completed in October 1939, it was not only the largest and most important base in Hawaii, but also the largest and most modern in the entire Air Corps arsenal. The AAC's previous Hawaiian base at Luke Field was located on Ford Island, and was shared with the Navy, which was given the entire island after the Air Corps left. The 2,200 acre Hickam site was purchased in 1935 to be the site of a bombardment wing, and an air depot capable of complete servicing of all Army aircraft in Hawaii. The base was named after Lt.Col. Horace Hickam, a distinguished officer who died on 5 November 1934 in a flying accident in Texas.

With the activation of the Hawaiian Air Force on 1 November 1940 at Fort Shafter, bombardment and pursuit groups were organized into separate wings, with the 18th Bombardment Wing stationed at Hickam and the 14th Pursuit Wing at Wheeler. On 13 May 1941 the AAC assigned 21 B-17Ds of the 19th Bombardment Group to fly from March Field, CA, to Hickam Field, and the Hawaiian Air Force had more heavy bombers than any other overseas base, signifying the War Department's concern about the possibility of a Pacific war. After their arrival the B-17s began a rigorous training program, while the B-18s continued their debilitating reconnaissance patrols. This training program continued into Fall, when the War Department decided to send air reinforcements to the Philippines by dividing the strength of the Hawaiian Air Force, and then allocating fewer aircraft to it. On 5 September 1941, nine Hawaiian AF B-17s and 75 crew members, commanded by Maj. Emmett O'Donnell, left for Clark Field on Luzon, Philippines, which was to be the first flight of land-based bombers to cross the central Pacific. During 1941 the remaining Hawaiian B-17s were kept on standby, while the expendable B-18s continued to depreciate on daily patrols.

Hickam Field ground crews began repairing the two B-18s on the right immediately after the attack, while a totaled B-17 sits without its rear fuselage on the left. (USAF)

7 December 1941

The first Japanese attackers—nine torpedo aircraft—appeared over Hickam Field at 0755, heading for Pearl Harbor. Shortly afterward 17 dive bombers attacked the Hawaiian Air Depot and the hangars for about ten minutes. Several minutes later nine Zero fighters strafed the hangar line from the southeast, and were followed by nine more Zeros from the southwest. At 0915 a second attack, lasting about eight minutes, was carried out by 27 aircraft that bombed the parked bombers, the base service buildings, and the large barracks. Eighteen Zeros strafed individual dispersed parked aircraft, and then hit the barracks, service buildings, and Post Exchange.

The attacks on Hickam destroyed the majority of the B-18s (11 of 33 were in commission after the attack) and B-17s (four of 12 remained in commission). The aircraft repair facilities, and the aircraft undergoing repair there, were destroyed. Hangar #11 was totally destroyed, and Hangars #7 and 15 received extensive damage. Over 50% of the aircraft maintenance stock was lost, and there was extensive damage to the ordnance building, affecting munitions supply The Hawaiian Air Depot, which included administrative and engineering offices, was immobilized. The main 3,000 man Hale Makai barracks, called the "Big Barracks," was damaged, as

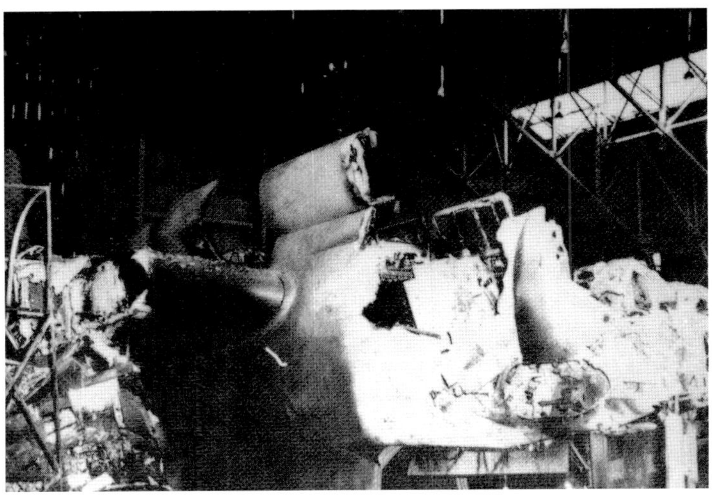

A demolished B-18 in its hangar. (USAF)

This B-18 of the 5BG HQ (see fuselage HQ emblem) was undergoing maintenance when it was damaged in its hangar. (USAF)

was the base's water system, technical buildings, and the Hawaiian Air Defense Buildings.

The first Japanese attacks were strafing runs on the hangar flight line that set a number of B-17s and B-18s ablaze. The aircraft were closely packed due to Gen. Short's sabotage mania, and once one caught fire the others were also doomed. During the attack an order was given to disperse the aircraft, and brave men rushed along the flight line under a hail of Japanese bullets to move the endangered aircraft. A B-18 was taxiing when it was hit by strafing Japanese aircraft that shut down an engine. The pilot revved the good engine, pulling the aircraft forward to one side, and then he slammed on the brake on that side. This action forced the other wing up, and the skilful pilot was able to waddle the B-18 across the flight line to relative safety. Some of the aircraft were threatened by fire, and quick thinking crewmen disconnected several engines and pulled them away so they could be used again. The 50th Reconnaissance Squadron had two B-17s and four B-18s parked on the ramp, but only one B-18 was salvaged. Hangar 11, housing several B-18s, was completely destroyed, killing almost all of the 11th Bomb Group's armament and maintenance personnel. Brig.Gen. Jacob Rudolph, CO of the 18th Bomb Wing, had scheduled a flight for 0800 that morning for a number of young pilots who had not completed B-18 training. There were 24 men in the hangar moving the bombers outside for their 0800 take off when the Japanese attacked. The attack killed 22, and seriously injured the other two. A Master Sergeant changing a B-18 tire on the ramp vanished under a direct bomb hit.

Among the many stories of individual heroism, a number of sources relate the story of a "mild-mannered" clerk, carrying a .30 caliber machine gun, who climbed into a B-18 parked on the apron in front of the hangars. He tried to mount the machine gun in the nose, but found that the gun was unusable; undaunted, he braced the gun against his shoulder to fire. As he fired a Japanese fighter came in low and strafed the B-18 with incendiary shells and set it on fire. The private did not try to escape, but continued to fire until he was enveloped by flames.

Fate of the Oahu B-18s

On the morning of the attack there were 33 B-18s on Oahu, including 32 at Hickam, and one based at Bellows, and of the 33, only 21 were listed as in "commission" that day. The bomber at Bellows was in fact assigned to Hickam, but was based at Bellows with the field's observation aircraft, and was used to tow targets for the pursuit aircraft that were temporarily stationed at that field. On the 7th this aircraft was on the island of Molokai, and flew back to Wheeler during the attack.

After the attack of the 33 B-18s, 13 were destroyed or in need of extensive repair, and nine were repairable within days. Of the 21 listed in commission before the attack, 11 were listed as still in commission.

Fate of the Oahu B-18s on 7 December
5th Bombardment Group

Group #	Serial #	Status 8 Dec 41	Notes
80	37-1	In commission	Hole in left wing and rudder
81	37-2	In commission	
31	36-339	In commission	
62	36-342	In commission	
63	36-329	In commission	

Bomb craters outside Hangar 7, which stored a damaged B-18 seen in the background. (USAF)

This B-18 of the 18BW was totally destroyed, and 22 pilots waiting to complete their training that morning were killed in the hangar. (USAF)

64	36-433	In commission	Left wing, vertical fin and rudder damaged
92	Unk.	In commission	
2	36-333	Repairable	Right wing and elevator damaged
83	36-310	Repairable	Right engine damaged, many bullet holes
33	36-438	Repairable	Left wing, ailerons and elevator damaged
35	36-337	Repairable	Right propeller damaged
36	36-334	Repairable	Vertical fin and right tire bullet holes
3	37-4	Destroyed	
82	37-3	Destroyed	By fire
32	37-11	Destroyed	
37	37-19	Destroyed	Tail, tail wheel, right engine & tire damage
11th Bombardment Group			
50	36-327	In commission	#2 & 3 engines replaced and main fuel tank
52	36-336	In commission	
53	37-20	In commission	Patch bullet holes
94	37-5	Repairable	
54	36-328	Repairable	Replace all wheels and repair empennage
77	36-436	Repairable	New vertical stabilizer, rudder & right aileron
78	36-437	Repairable	Extensive repair required
79	37-15	Repairable	Repair fin, right wheel, elevator & engine
76	36-288	Repairable	Complete repair at Air Depot
93	37-6	Repairable	Extensive damage
4	37-7	Repairable	Extensive metal repair and right gear
51	36-335	Destroyed	Destroyed by fire
92	36-270	Destroyed	Total loss
95	37-12	Destroyed	Cannibalized for parts

Hawaiian Air Force

1	?	Destroyed	Total loss

18th Bombardment Wing

1	?	Destroyed	Total loss

Post Attack

The first patrols after the attack were conducted by two B-17s, two B-18s, and four A-20s. Capt. LaVerne Saunders, the provisional CO of the B-17s, led a three plane search (but one B-17 aborted), and was joined in the patrol by Capt. Russell Waldron, provisional CO of the B-18s. The parked aircraft had been locked, and the keys were stored in the armament section, which had received a direct bomb hit. The air crews had to shoot or pry off the locks, and then hand load the bombs, as much of the bomb loading equipment had been destroyed. Waldron's two B-18s of the 31st Bombardment Squadron took off at 1330 toward the northwest, inadequately armed with six 100 pound bombs and two .30 caliber machine guns. On their return the search aircraft were greeted by nervous and intense "friendly" anti-aircraft fire, and Waldron's B-18s had to stand off while he broke radio silence to ask for the shooting to stop so they could land. But it was nearly an hour before the B-18s, running low on fuel, were finally cleared to land. After the attack a total of 48 search sorties were sent out using aircraft of various types (bombers, pursuits, and observation) to find the Japanese aircraft carriers.

The next day, and for the next several anxious days, the B-18s, B-17s, and A-20s flew numerous more organized, but unsuccessful patrols. In the predawn of the 8th at 0400, 2Lt. Trent took the repaired B-18 #94 out on a patrol conducted out to 350 miles. To reach this range the aircraft was overloaded by over a half ton with fuel in the bomb bays, reducing the bomb load. Soon, the 72nd Bombardment Squadron had its remaining B-17s transferred to the 23rd Squadron, and was assigned several of the repaired and undamaged B-18s. Hawaii's B-18s were more suited to "search and attack" sorties, and the B-17s were held in reserve as a "striking force" in case the Japanese returned. While more B-17s arrived from the mainland, the B-18s of the 72BS flew about 70% of the "search and attack" missions from Oahu. These missions, taking off at dawn, usually covered a sector from 10 to 20 degrees, and extended to about 300 miles from the island. To extend the endurance and range to 500 miles, half of the bomb load was replaced by two bomb bay fuel tanks. The missions usually lasted 8 1/2 to 10 1/2 hours, depending on the weather and winds. On the return leg to base unpredicted weather fronts either had to be skirted, or if fuel was running low had to be penetrated, many times in the dark.

By spring 1942 the Japanese threat subsided, and the B-18s were relegated to flying aircraft parts from one island to the next and hack status, flying brass, such as in April 1942, when Maj.Gen. Clarence Tinker was given a widely publicized conducted tour of Oahu in a B-18. An unusual mission for the B-18 occurred in late April, when the huge volcano, Mauna Loa, erupted on the Big Island, and threatened the town of Hilo and its harbor installations with lava flows. The initial reconnaissance flights to observe the extent of the eruption and the volume and flow of the lava were made by a B-18 of the 72nd Bombardment Squadron. Until the eruption subsided on 12 May several bombing runs were made on the lava flow by 72BS B-18s, and also by other Army and Navy aircraft wanting to get in on the fun. B-18s 37-1 and 37-2 were converted to transports, and 37-2 was used in the Gilbert-Marshall Island Campaign as an "armed transport," with its turrets modified and fitted with .50 cal. machine guns. One-by-one the aging B-18s based in Hawaii gave way to the B-17s and B-24s, and faded from the picture first, succumbing to the wear and tear of patrolling, and then to the eventual lack of spare parts.

7

The B-18 in the Philippines

With the collapse of France and Holland in mid-1940, the situation in Europe also jeopardized the situation in East Asia, and particularly the American status in the Philippines. In the late 1930s America had prepared the classified *War Plan Orange 3*, that garrisoned the Philippines with a small force that was to defend only the approaches to Manila until the Pacific Fleet could arrive six months later. *Orange 3* meant that the U.S. would cede its Pacific territories to Japan without much of a struggle until help arrived. When Maj.Gen. George Grunnert assumed command of the Philippine Department in May 1940, his Philippine air defense consisted of a small number of obsolete and cast off B-10 bombers and P-26 pursuit aircraft. He immediately began to lobby to augment his inadequate air defenses, which would be responsible to ward off possible aerial and naval attacks. The War Department decided not to abandon the Philippines, and in late October 1940 Grunnert received two Pursuit squadrons: the 17th from Selfridge Field, and the 20th from Hamilton Field. These squadrons were equipped with 40 (later 57) Seversky P-35 fighters that had been diverted from a Swedish export order. Grunnert continued his barrage of requests to the War Department to bolster his air defenses. He requested two more pursuit squadrons equipped with more modern aircraft, a bombardment squadron, more airfields, and an early air warning system. In March 1941 he received 31 crated Curtiss P-40B Tomahawk fighters, but by late May 1941, the B-10s still were the only frontline bombers available in the Philippines. Finally, in June 1941 the *SS Washington* arrived in Manila with a shipment of Hawaiian 5th and 11th Bomb Group B-18 hand-me-downs with many flying hours on them. These bombers had been dismantled and shipped in crates, and then trucked to Nichols Field for assembly to be supplied to the 28th Bombardment Squadron. Grunnert upgraded his air command by creating the Philippine Department Air Force, and appointed Brig.Gen. Henry Clagett as the Chief of Staff. Clagett was an old line Army officer who believed that the Japanese would have to be stopped offshore, and the B-18s would be the only offensive weapon available until the B-17s arrived.

Clark Field, located 60 miles north of Manila, summer 1941. At the time, the Philippine Department Air Force was composed of a single unit designated as the 4th Composite Group, which was based there. (USAF)

A damaged 28BS B-18 at Clark Field after the attack. Two large holes can be seen at the top of the rear fuselage above the National Insignia. (USAF)

On 24 July 1941 the Japanese invaded French Indo-China (Vietnam), and captured strategic air and naval bases there that threatened the Netherlands East Indies and Malaya. President Roosevelt was furious over this overt aggression, and two days later ordered all Japanese assets in America to be frozen, and reorganized all American and Filipino forces in the Philippines into the USAFFE under Douglas MacArthur. Several days later Army Chief of Staff George Marshall announced that it would now be American policy to defend the Philippines, and that the territory would have the highest priority for military equipment and the construction of airfields. Clagett informed MacArthur, based on estimated Japanese air strength, that he would need 27 pursuit squadrons, 18 light bombardment squadrons, and 30 heavy bombardment squadrons. However, there were only ten airfields available, and 56 would be required to accommodate the number of squadrons Clagett requested. In their assessment of the Japanese threat to the Philippines, the War Department realized that it would come from their bases in Formosa that were out of range of the B-18s. To meet the Formosan threat, Clagett received nine longer range B-17s in mid-September, and MacArthur was given priority over all B-17s coming off the Boeing production line, with 26 scheduled to arrive in October and 33 in December.

At the time, the Philippine Department Air Force (PDAF) was based at Clark Field, which was located 60 miles north of Manila. The PDAF was composed of a single unit designated as the 4th Composite Group, which was made up of the 28th Bombardment Squadron, a medium bomber unit, and the 2nd Observation Squadron. The 4th CG also had three pursuit squadrons (the 3rd, 17th, and 20th), and the Headquarters Squadron at Nichols Field, which was located about six miles south of Manila. With only two major airfields, the Philippines were extremely vulnerable to air attack and, although plans were laid for the construction of other fields, an operational airfield to accommodate a Group typically took about a year to complete, and construction was further impeded by the lack of engineers in the theater.

The 19BG arrived in the Philippines in August and began training immediately, and then stepped up training in September and October 1941, as its squadrons were being equipped with newly arriving B-17s. Its 14th and 93rd Squadrons at Del Monte, Mindanao, and the 28th and 30th squadrons at Clark Field were each equipped with eight B-17s, with the Headquarters' Squadron at Clark also receiving three. With the arrival of more B-17s, the worn B-18s were demoted to a transport and reconnaissance function.

The arrival and scheduled arrival of the large air reinforcements required MacArthur to again upgrade his air organization, and on 4 November, Maj.Gen. Lewis Brereton was named to head the newly created Far East Air Force (FEAF), comprised of the 5th Bomber Command and 5th Interceptor Command. On his arrival in the Philippines, Brereton's Pan Am Clipper was given a spectacular greeting by 42 P-40s and P-35s flying in a large V, followed by the B-17s and B-18s based at Clark Field. At the same time the defense of the Philippines had also been upgraded with the implementation of the new *Rainbow 5 War Plan*, that was now to defend all the Philippine Islands. By the end of November 1941, the Philippine Department Air Force had a complement of 5,609 men, including 669 officers and 4,940 enlisted men, supplemented by 1,500 men of the Philippine Army Air Corps and its 60 obsolete aircraft. Augmenting the *Rainbow 5 Plan* was a second pursuit group (the 35th) that was scheduled to be equipped with 50 P-40Es by the end of 1941 as part of MacArthur's scheduled delivery of 240 P-40Es. On 28 November, MacArthur was informed that the Japanese representatives in Washington had broken off negotiations, and war was believed to be imminent. MacArthur did not believe that the Japanese would be able to launch an air and sea attack on the islands for four months, but to be safe Brereton dispatched 16 of his 35 B-17s south to Del Monte Airfield on Mindanao, out of range of a Japanese air attack from Formosa. The pursuit aircraft, B-17s, and B-18s had arrived in the Philippines with their shiny natural metal finishes, and Del Monte had only one spray gun available, so the painters worked around the clock to paint the fighters, Flying Fortresses, and then the Bolos in camouflage olive drab. The lack of bomb blast revetments and natural cover caused the denuding of local coconut plantations for fronds to camouflage dispersed aircraft.

In the meantime, it was decided that the pilots of the 3rd, 17th, and 20th Pursuit Squadrons needed to sharpen their night interception aptitude, and Maj. Orrin Grover, CO of the 24th Pursuit Group, enlisted the B-18s of the 28BS at Clark and the 200th Coast Artillery to commence the night training at 0400 on 2 December. The 0400 time was chosen to minimize the risk to the valuable P-40s, so their pilots could land more safely in the light of dawn. The training procedure was to have six P-40s take off and form up into three trailing two-ship elements to search for the single B-18 that was flying an oval north-south pattern over Clark, illuminated by the 200th CA searchlights. As soon as the lead element found the bomber, the leader and his wingman would close for a simulated stern attack, and then veer off and radio for the next element to close. The lead element then fell in behind the third element, and in turn would attack again. This procedure continued until daylight, when the P-40s would land, and the exercise was critiqued over breakfast. On the third morning a P-40 element was approaching the illuminated B-18 when a "P-35" screamed down between them and the bomber. At the debriefing, the "P-35" incident was discussed and investigated. It was found that there were no other friendly aircraft in the area, and it was supposed that the bogie could have been a Japanese intruder. The next day, just after midnight SCR-270B radar sited at Iba intercepted an unidentified aircraft 50 miles offshore, and a P-40 was dispatched to check out the contact, which was lost after 25 miles. Later that night the scheduled B-18 night interception exercise was conducted, and while waiting for the B-18, the P-40 leader noticed aircraft navigation lights below. Since no friendly aircraft were in the air, an element was sent to investigate but did not make contact. Grover consulted Col. Harold George at the 5th Interception Command for instructions to deal with any future contacts, and George told Grover to act defensively if the bogies were intercepted offshore, but to attack if they were in Luzon airspace. On 6 December, six fully armed P-40s flown by the group's most experienced pilots took off on the regular B-18 night interception, in hopes of encountering the bogie, which did not appear. On 7 December, the last B-18 night exercise was carried out by fully armed P-40s, but again the bogie did not appear.

On 7 December the total operational U.S. air strength in the Philippines was 35 B-17s (two full squadrons of eight each at Del Monte, and 19 at Clark Field, including eight each with the 28th and 30th Bomb Squadrons, and three with the HQ Squadron), 18 B-18s (two at Nichols, three at Del Monte, three at Neilson, and ten at Clark), and 54 pursuit aircraft. At 0400 on the 8th Maj.Gen. Lewis Brereton called Col. Eugene Eubank at Clark Field to inform him that the Japanese had attacked Pearl Harbor, and to standby to organize a bombing mission. At dawn Brereton called again, and summoned Eubanks to come to his Headquarters at Neilson Field, outside Manila. Eubanks left Clark under the command of Maj. David Gibbs, and he and his staff took off in a B-18 for the meeting. At the meeting Brereton informed Eubanks and his staff that MacArthur had decided that American aircraft weren't to attack until fired upon, and the B-17s were to be held back from combat. Eubanks headed back to Clark still without orders to take offensive action. Meanwhile, Japanese warships had been spotted off the north coast, and Japanese flying boats were observed over the north islands, and Gibbs had ordered the B-17s and pursuit escorts into the air. By the time Eubanks returned at 1030 most of the 19 B-17s (4 B-17Cs and 15 B-17Ds) and their 24 P-40B escorts had returned to base. By noon rumors of approaching Japanese formations circulated, and pursuit aircraft were sent up to patrol, leaving Clark defended by one squadron of fighters that were in the process of refueling. At the same time MacArthur released the B-17s, and Brereton was finally able to issue orders for the B-17s of the 19th Bombardment Group to bomb targets on Formosa. A staff conference was held concerning the mission, and ordered a photo-reconnaissance mission for two B-17s that were waiting for cameras to be delivered by a B-18 from Nichols Field. While operational orders were being finalized, the B-17s were being fueled and loaded with bombs, and the remaining P-40s were being readied to take off when the Japanese appeared over the field at 1240. The surprise was complete, as the native Filipino forward air observers and the new SCR 297 sets had failed to issue a warning of the 54 Japanese bombers approaching in two Vs at 18,000 feet. The two waves of bombers dropped their bombs in train and left the field, its buildings, and aircraft smashed and ablaze. Soon Japanese strafers appeared low over the field, shooting up the parked bombers that had escaped damage from the bombing. All but two or three of the bombers on the field were destroyed, but those others were damaged. Four P-40s got into the air, but were quickly outnumbered and destroyed. After 45 minutes the Japanese attack was over, and Clark was no longer a viable tactical base. There were 100 dead and 250 wounded, and not one flyable aircraft remained. Two or three of the 19 B-17s were made flyable over the next several days from cannibalized parts, but these aircraft were never in condition to serve in combat. Nine of ten of the B-18s on the field were destroyed, and the hangars and shops were destroyed. During the attack on Clark Field a B-18 was able to take off under fire, and flew to the secondary Rosales Airfield, 100 miles northwest of Nichols Field, where it managed to land, but had to be written off, as it was shot full of holes. The Philippines B-17 striking force had been halved, and would be based out of Mindanao; they would have to use a repaired Clark Field as a staging area, which would add to the distance to Japanese targets, and add to the wear and tear on the aircraft and their engines. At Nichols Field two of the B-18s stored in Hangar 4

The B-18A of the 28th Bomb Squadron (5BG) Commander (two vertical stripes on the fuselage) at Clark Field. (USAF)

The only official caption on this photo was "B-18As from Nichols Field, P.I." (USAF)

were destroyed. At the end of the day there were five B-18s in commission in the Philippines, but there was the possibility of repair of several more damaged B-18s. That night the Japanese again attacked Nichols Field, and added three more killed and 15 wounded, Hangar 4 was completely destroyed, and a B-18 and O-52 were damaged.

Eager for revenge, plans were made to launch an attack against the large Japanese convoys approaching the Philippines at Vigan and Aparri on the morning of the 10th. Every aircraft suitable for combat was gathered: at Del Monte: the 14th and 93rd Bomb Squadrons each had one B-17C and seven B-17Ds available, which were to be escorted by 21 P-40Es from the 17th Pursuit Squadron at Nichols Field, and 22 P-36s from the 34th Pursuit Squadron from Del Carmen. Three B-18s at Neilson Field were assigned to the 27th Bomb Group (L), which had been awaiting the arrival of its A-24s (the AAC version of the Navy SBD Dauntless). There were also two B-18s based at Del Monte with the 5th Air Base Group, but one had been dragged off into the trees to serve as the communications center, with its radio tuned to the Clark Field frequency. The other Del Monte B-18 was used for reconnaissance, having checked the Davao area for Japanese activity. In the end, only five B-17s from the 93BS were available to make the attack, using Clark Field as a staging point to refuel and load 20 100 pound demolition bombs. The B-17s took off at 0530, and ineffectively bombed the Japanese from 12,000 feet to carry out the first American bombing mission of World War II. That day the 14th Bombardment Squadron at Del Monte sent three B-17s up to Clark Field on a mission that would become famous for Capt. Colin Kelly's bombing of a Japanese warship before he lost his life trying to keep his crippled aircraft flying while the crew bailed out.

With the virtual loss of the B-17 force, the B-18s took on added importance as transports, flying between the bases in Luzon (Clark, Nichols, Nielson, Iba, and Del Carmen) and Del Monte, on Mindanao. The slow, 160 to 170mph aircraft could either leave Luzon late in the afternoon and head for Del Monte 650-700 miles to the south, and arrive after dark, then refuel and return to Luzon before dawn; or, they could take off after dark from Luzon and return in the morning light. Either way they were exposed to Japanese fighters and tropical storms. To keep the remaining B-18s in the air, the beleaguered personnel cannibalized spare parts and engines removed from written off aircraft.

On 12 December Maj. David Gibbs was assigned to take over the command of Del Monte from Maj. Emmett O'Donnell, who was reassigned to Clark Field. Gibbs decided to fly to Del Monte in one of the B-18s, but before he left a weather report warning of intense thunder storms on the route was received. The Met man suggested that Gibbs leave six hours later at midnight, but he left at 1800, and his B-18 never landed at Del Monte.

The 27th Bombardment Group (L) had arrived in the Philippines in November, but there had been problems in collecting their A-24s, which were shipped separately. On 17 December the Group's A-24s had arrived in Brisbane, Australia, and the 20 pilots moved through rubble-strewn roads from Nielson Field to Nichols Field, where they were allotted a C-39 transport and two B-18s to make the trip to pick up their aircraft in Australia. While taxiing for take off the aircraft had to be guided by flashlight around the bomb craters and rubble. A ground crewman at the end of the blacked out runway flashed a light, the engines were run up, and the overloaded aircraft took off. The slower C-39, with less range, had left first, and had to land at Del Monte and Tarakan, Borneo, while the B-18s

A B-18 at Del Monte, with a C-39 in the foreground. Del Monte, on the southern island of Mindanao, was 700 miles to the south of Manila, and was a safe haven from Japanese attack for several weeks. (USAF)

flew to Darwin, on Australia's north coast, via a stop at Tarakan. On their return to besieged Luzon, all three decrepit aircraft were loaded with desperately needed .50 caliber ammunition and retraced their route, but were forced by a storm to land at Koepang, Borneo, and remained there over Christmas. While at Koepang, the B-18s were ordered to return to Australia and join the other 27th BS pilots in Brisbane. Meanwhile, the 27th pilots took a Quantas flying boat to reach their waiting A-24s in Brisbane. However, the A-24s were not new, had defective or missing instruments, and were useless, and the 27th continued to be the Squadron without aircraft.

In the morning of 19 December Brig.Gen. Clagett, who had been just relieved of his command of the 5th Interceptor Command, and his staff left Nichols in a pair of B-18s for Del Monte, where four of the remaining B-17s were hidden under trees awaiting to take Clagett to Australia that night. As they arrived at 1600 the aircraft were fired on by AA of the 5th Air Base Group, who were nervous and trigger happy, as they had just suffered their first Japanese air attack earlier that day. Fortunately the B-18s landed safely, but just after they shut down their engines four undetected Japanese Zeros attacked. They strafed the field and set one of the B-18s on fire, and hit two of the B-17s, which suffered only superficial damage.

As the Japanese were closing in on Clark Field, the men there continued to cannibalize and salvage wrecks, and send valuable parts and engines down to Del Monte on the almost nightly B-18 run. But by 24 December the situation became untenable, and the personnel at Clark were ordered to withdraw to Bataan with as much equipment and munitions as possible. The remaining personnel, and P-40s and P-35s, were ordered to evacuate to Del Monte, which had managed to escape serious Japanese attack. Once at Del Monte, further orders were received from Australia for higher ranking officers to evacuate. Several pursuit pilots from Clark were left off the evacuation list, as was Brig.Gen. William Sharp's G-3 operations officer, who was assigned to maintain communications between Del Monte and Australia. There was a damaged B-18 on the field that was being repaired by Maj. Emmett "Rosy" O'Donnell, 14BG CO, and later a B-29 Group general. O'Donnell and the men were able to complete repairs on the aircraft and its engines, but the bomber lacked sufficient fuel capacity for the trip. Four fuel drums were jury-rigged into the bomb bay, and a hole was cut into the fuselage floor so a man sitting on top of a drum could pump gas from these drums with a wobble pump into the B-18 wing tanks as pilot O'Donnell directed the fuel transfer. After the bomber was lightened by removing its armament it took off at 0300 on 15 January and safely arrived in Australia.

Once the Philippines were evacuated, the remaining four B-18s were moved to Java to continue their service as courier and transport aircraft. On 2 February 1942 Maj. A.A. Straubel was commander of the 7th Bombardment Group, whose B-17s were fighting a losing battle for Java. Straubel was called to a conference in Jagjakarta, and left in one of the old B-18s that had flown the 27BS pilots to Australia to pick up their A-24s. After the conference, the three man crew was ordered to fly to Soerabaja to deliver four FEAF personnel there. One of these passengers was communications expert Col. William Murphy, who Gen. Brereton assigned to revamp the inadequate air warning system. Paradoxically, the approaching B-18s were not warned that Soerabaja was under its first enemy air attack, and Japanese fighters shot down the bombers about 30 miles west of the city. Straubel and his co-pilot, 2Lt. Russell Smith, escaped from the crash, but after numerous attempts were unable to pull the trapped passengers free from the flaming wreckage, and both men succumbed the next day in a Dutch hospital from serious burns suffered in the rescue attempt.

Maj. A.A. Straubel was the commander of the 7th Bombardment Group. On 2 February 1942 his B-18 was shot down over Java. (USAF)

In April 1942, when the B-25s of the 3rd Attack Group that had just recently reinforced Del Monte were ordered to evacuate to Australia, the ground crews found that they did not have enough long range fuel tanks. The B-18s came to the rescue, as tanks were salvaged from out of action B-18s and installed in a B-25 piloted by the remarkable Capt. Paul "Pappy" Gunn, who in the prewar had been the manager of Philippines Airlines, and was to become famous during the war for his innovations and leadership. After the fall of the Philippines, a small number of USAAF bombers and transports retreated to Australia in January and February 1942, where they formed the first American Australian-based unit, the 21st Troop Carrier Squadron. Three B-18s were recorded as having been based in Australia (see B-18s in Australian Service section).

8

The B-18 in Alaska

During the summer of 1934, Gen. H.H. "Hap" Arnold led ten cutting edge Martin B-10 bombers on an ambitious round trip flight from Bolling Field, Washington, DC, to Fairbanks, AK. After the flight Arnold realized the strategic significance of Alaska in a Pacific War, and recommended the construction of air fields there. But it was not until Spring 1939 that Congress approved $4 million to build a cold weather test facility near Fairbanks, and preliminary clearing and construction on the site began in the summer of that year. The field was named for Maj. Arthur Ladd, who had been killed in a fighter in South Carolina in 1935 (this was in keeping with the AAC policy of naming air bases after notable deceased airmen). The rudimentary facility was completed, and the first aircraft landed there on 4 September 1939, but Alaska needed an operational airbase.

By early 1940 America had spent over $225 million on Hawaiian bases, but only S1.5 million to defend Alaska. With the rise of the Japanese threat in the Pacific, U.S. military planners finally decided that it was imperative to defend Alaska. Since the area was so large and harsh, they decided correctly that Japan could not logistically sustain a large land operation there. It was their opinion any enemy assault would be primarily by air power, and abetted by small naval units supporting small scale land attacks, and this would happen only if the U.S. Navy had lost control of the North Pacific. The planners had two defensive options: to build a number of military bases throughout Alaska, or build a large primary base near Anchorage that would provide a strong, mobile force that would protect, support, and supply the existing Navy bases at Dutch Harbor, Sitka, and Kodiak, and newly built forward Army airbases. The latter option was chosen, and on 8 June 1940 work was begun on Alaska's largest military facility, located four miles northeast of Anchorage, just north of the Alaskan Railway. The air base portion was named Elmendorf Field, after Capt. Hugh Elmendorf, a noted AAC officer of the interwar years who had crashed while flying an experimental YP-25 pursuit fighter on 13 January 1933. When completed, the base was to have a 7,500 x 200 foot East-West concrete runway and a 5,000 x 150 foot North-South runway, with revet-

On 8 June 1940 construction was begun on Elmendorf Field, northeast of Anchorage. It was to become Alaska's largest military base,, and when completed, the base was to have a 7,500 x 200 foot East-West concrete runway, and a 5,000 x 150 foot North-South runway, with revetments for 40 bombers and 80 fighters, and three permanent and one temporary hangar. (USAF)

ments for 40 bombers and 80 fighters, and three permanent and one temporary hangar. Construction was begun on extensive barracks and messing facilities, a repair sub-depot for major repairs, underground fuel and oil storage, and complete communications service. Construction continued in earnest until November, when the severe winter weather brought construction to a halt, but by that time most of the runway construction was well under way. On 12 December 1940 the War Department gave the name Fort Richardson to the military complex that surrounded Elmendorf.

In the meantime, on 22 July 1940 Col. Simon Buckner, a vociferous advocate of building Alaska's defenses, took command of the Alaska Defense Force. Despite being an infantry general, Buckner realized that defense in Alaska would be dependant on airpower, and the ability to strike first. On 12 August Maj. Everett Davis and two enlisted men arrived at Merrill Field in a old B-10B, and set up a command post there in a one room cabin. Davis would serve as the senior AAC officer in Alaska, but would have difficulties in accomplishing anything, as his duties were only defined as "Gen. Buckner's representative," and he had no authority. Also, Davis had little administrative help until November, when an officer and nine more enlisted men arrived.

During that summer Davis made many survey flights, searching for future airbases and air routes, and by October Buckner was able to submit a list of potential airfield locations and routes to Lt.Gen. John DeWitt, who was headquartered at the Presidio, San Francisco, and was responsible for the defense of the Western U.S. and the Territory of Alaska. The number of suitable airfield sites was slim: Anchorage, Fairbanks, Nome, Big Delta, and Juneau. DeWitt determined that the best air route to Alaska from the contiguous U.S. was via the interior, rather than along the coast, which was chronically socked in by bad weather. Later Lend Lease aircraft to Russia were delivered via this route, and the Alaskan Highway followed it. In September Davis moved his headquarters to

Col. Simon Buckner (right), a vociferous advocate of building Alaska's defenses, took command of the Alaska Defense Force in July 1940. Buckner's commander was Lt.Gen. John DeWitt (left), headquartered at the Presidio, San Francisco. DeWitt, who was responsible for the defense of the Western U.S. and the Territory of Alaska, hindered the defense of Alaska by insisting that he administer its defense from California. (USAF)

The first B-18 crew landed at Elmendorf on 19 February 1941. (USAF)

Elmendorf, and on 1 December 1940 he was named the CO of Elmendorf, and he and his small staff finally had the authority to execute AAC air defense plans.

In April 1940 two B-18s landed at Fairbanks Airport, bringing a contingent of officers to survey and expedite the continued construction of Ladd Field, Fairbanks. The group was led by Maj. Dale Gaffney and Maj. Edward George, the Quartermaster for Alaskan Projects. As part of the Gaffney airfield survey more land was cleared, soil was tested, and soon construction gathered speed, and by August the runway had been paved; on 4 September Ladd Field became an operational airbase. Maj. Gaffney made the first landing in an O-38 observation plane, and he would remain on as the commandant of the Fairbanks facility.

Also during September Buckner, accompanied by Capt. Ralph Parker, Alaskan Naval commander, sailed on the gunboat *Charleston* along the Aleutian Island chain to the naval installation at Dutch Harbor. The trip convinced Buckner that the only way to defend the Aleutians was to build air bases there, and fend off the enemy invasion with superior mobile airpower. But as the Navy had responsibility for the Aleutians, Parker insisted that the defense of the islands and security of the North Pacific should depend on warships supported by naval aircraft.

As winter approached the runway at Elmendorf was completed, and on 8 November 1940 a Douglas OA-5 Pelican was the first aircraft to land there, and soon after, on the 26th, two B-17Bs that were undergoing winterization testing at Ladd Field landed on the runway. The landing was challenging, as there were no runway markers, except cut spruce trees that were used to mark the edges. Since there was no control tower, an old radio located in a B-10 bomber parked along the runway was used to communicate landing instructions. The temporary hangar and three permanent hangars were not ready, and the ground staff had to work outdoors to maintain aircraft. The 73rd Bombardment Squadron (M) at McChord Field, WA, was scheduled to be the first bomber unit to be deployed to Alaska, and on 19 February 1941, after a two day flight, a B-18A piloted by Lt. Joe Schneider and co-piloted by Lt. Frank O'Brien arrived at Elmendorf as an advance organizational unit. Two days later the 18th Pursuit Squadron arrived at Seward by ship with their

In late May 1941 six 36BS B-18As led by Maj. William Eareckson made a stopover at McChord Air Base, Tacoma, WA, on their way to Alaska. During the stopover, the B-18 crews posed for this publicity photo. (USAF)

P-36s in the hold. The Curtiss Hawks were unloaded and transported by rail to Elmendorf. The 28th Composite Group, with three squadrons (the 36BS, 37BS, and the 34PS), arrived in Alaska on 23 February after being activated at March Field in February. The latter two squadrons were detached from the 28th CG, and the remaining 36BS was joined by the 18PS and 73BS already in Alaska to make up the new 28th CG under Maj. Donald Titus. The 34PS arrived at Seward by ship on 22 February, and the next day the Headquarters Squadron of the 23rd Air Base Group, and advanced elements of the 36BS, arrived at Elmendorf. This reorganization went smoothly due to the excellent organization of Maj. Davis. On 3 March the Alaska Defense Force was redesignated as the Alaska Defense Command, and the talented and deserving Davis was recommended for promotion to Lt. Colonel, but was denied. On 14 March the ground echelon of the 73BS disembarked at Seward, followed two weeks later by 73BS CO Capt. Jack Donohew, who arrived with eight B-18s after an uneventful three day flight from McChord. The ground echelon of the 36BS arrived at Seward on 17 March, followed on 26 May by six B-18As led by Maj. William Eareckson. The B-18As had been winterized at Sacramento Air Depot and flown for a publicity stop at McChord, and then touched down at Prince George, British Columbia, White Horse, Ladd Field, and Elmendorf. These were to be the last aircraft that Maj. Davis would receive in 1941, as Alaska continued to be the proverbial and geographically appropriate low man on the totem pole as far as AAC priorities were concerned. Davis had to manage with his obsolete 15 B-18s and 20 P-36s, and the 2,087 men on hand in June. Under the outstanding direction of engineer Capt. Benjamin Talley substantial progress was being made on Elmendorf, as most temporary buildings were completed, and foundations were laid for the permanent quarters and administration buildings. A large 60 x 4,980 foot section of the runway had been paved, and one of the three large permanent hangars was completed.

By July only 10 of the B-18s and 15 of the pursuits remained in service, mainly due to accidents, and the lack of maintenance facilities and parts. Davis named Eareckson to manage the training of the "cheechako" (greenhorn) Alaskan pilots. In order to take advantage of the long arctic summer days and lack of aircraft, Eareckson divided the training schedule into two shifts to prepare his pilots for Alaskan-style flying. The first group awoke early and flew until 1100, and then relinquished their aircraft to the second group, who slept late and flew until dark. Eareckson initially called on the celebrated Alaskan bush pilots, who were of great help in tutoring the inexperienced AAC pilots, who became adept at Alaskan operational flying through hours of practice missions using instrument flying and dead reckoning navigation. Nonetheless, both Buckner and Davis were concerned about the ability of their small obsolete air force to ward off any resolute enemy attack.

During 1941 one of the responsibilities of the B-18s was to fly support for the 1st Photographic Squadron out of Grey Field, Fort Lewis, WA. During the summer of 1941 the Squadron was assigned to photograph potential airfield sites, and the route of the Alaskan Highway. While returning the photo squadron personnel and equipment back to Washington after its summer assignment, a B-18 piloted by Lts. Edward Clark and Gene Yarborough had its engines quit near Prince Rupert. The heavily loaded aircraft lost altitude, but Clark was able to land it on a sandbar at low tide. The passengers and equipment were removed before the tide returned, and Clark and his crew tried to waterproof the aircraft, which was well known for being able to float. After the tide came in the Royal Canadian Air Force sunk a barge under the floating aircraft, and when the tide receded again the barge was sitting on the bottom, cradling the B-18 on its deck. The water that sunk the barge was pumped out, and on the next high tide the barge with the B-18 on board was floated and towed to Seattle, where it was repaired at the Boeing factory.

During the summer of 1941 the 1st Photo Recon Squadron was assigned to photograph potential airfield sites on the Alaskan coast. This 1PRS B-18 had its engines quit near Prince Rupert, and the heavily loaded aircraft lost altitude, but the pilot was able to land it on a sandbar at low tide. The B-18, known for its buoyancy, was floated on a barge at high tide and eventually repaired. (USAF)

As fall 1941 arrived tensions between Japan and America increased, but the Alaska Defense Command remained low on the AAC and Navy defense priority lists, even though there were now nearly 35,000 men in Alaska. Brig.Gen. Carl Spaatz, Chief of the Air Staff, had realized that Alaska would be a theater of air, rather than ground operations, and recommended that an Alaska Department be established under an Air Corps general. Gen. DeWitt, not wishing to reduce his authority, petitioned the War Department, and he was to retain control of the Alaska Defense Command for the next two years. Davis, still a Major, and as Chief of Aviation of the Alaskan Defense Command, had insufficient command authority over the air units. To give him more control the Air Field Forces, Alaska Defense Command was created on 29 May. This department gave him control over training, maintenance, planning, and the implementation of defense plans. However, the 23rd Air Base Group and associated air support units continued under the control of the local commander at Fort Richardson, and thus Davis had no immediate control over the ground units that supported his air operations. Fortunately two members of Arnold's staff, Col. Frank Kennedy and Lt.Col. Harold Clark, made an inspection tour in August, and found that Davis' situation was unacceptable, and again recommended that the Alaska Defense Command be made a separate theater of operations under an AAC general, and that Davis be given more authority. Again the first recommendation was refused, but the deserving Davis was finally promoted to Lt. Colonel and CO of the Alaska Defense Command, and given authority over the ground units. In this change, Maj. Eareckson was named commander of the 23rd Air Base Group, replacing Maj. John Davison, who was named Davis' assistant. Maj. Norman Sillin replaced Eareckson as CO of the 28CG, and Maj. Donald Titus took Davis' old post as Chief of Aviation, Alaska Defense Command.

During 1941 construction at Elmendorf continued, and by year's end there were two 5,000 foot paved runways and two 400 x 4,000 foot paved parking areas. Mechanics no longer had to work outdoors in subzero cold and then eat and sleep in tents, as the three large arched hangars were completed, and all personnel were housed in barracks. Even though he had more authority and more runways, Davis had only six of his original 15 B-18s in service: five had been damaged in accidents, and the others were out of commission for a number of reasons, mainly attributed to hard use. The P-36s of the 18PS were experiencing similar problems due to lack of spare parts, and several accidents involving structural failure. In mid-August two more of the B-18s and half a dozen P-36s were grounded for lack of spark plugs that became readily fouled in the damp Alaskan weather. Navy Air was in even worse shape, operating a Kingfisher at Sitka and a Grumman Duck at Kodiak, their only aircraft in service.

By the end of 1941, Elmendorf's two main runways and three large hangars were completed, and all personnel were housed in barracks. However, conditions were harsh, and only six of the original 15 B-18s were still in commission. (USAF)

On 22 June 1941 Hitler began Operation *Barbarossa*, the invasion of Russia, and rapidly advanced toward Moscow. On 3 July Gen. Marshall, fearing a Russian collapse, issued an alert that the Japanese under its Tripartite agreement with Germany might choose to capture Russian Pacific bases. The alert demonstrated the inadequate state of Alaskan preparedness, as it took four days for the

The B-18 aircrews flew many tedious hours without tangible result, but they did gain important training flying hours over water; an experience that would be of immeasurable value later while flying in other aircraft in the notorious weather in the Aleutians. (USAF)

Alaska Defense Command to inform all the bases in its command of the warning. During the alert the question of which service had the authority over Alaskan off shore patrols again surfaced. Since the Navy air units were more than stretched thin, Navy CO Adm. Ralph Parker approved the transfer by Buckner and Davis of the B-18s of the 73rd and 36th Bomb Squadrons to Nome to patrol the Bering Sea. The aircrews flew many tedious hours without tangible result, but they did gain important training flying hours over water, an experience that would be of immeasurable value later while flying other aircraft in the notorious weather in the Aleutians. After the B-18s accumulated 5,000 hours over the inhospitable Bering Sea the worn bombers were withdrawn back to Elmendorf. Despite Parker's approval of the B-18 Bering patrols, his superiors stressed that it was the Navy's duty to patrol over water, and the Army's over land. Army man DeWitt, protecting his turf, or water in this case, ordered Buckner to continue his Alaskan off shore patrols. In response the Navy dispatched more PBY-5 patrol aircraft to Alaska, deploying Washington state-based VP-41 and VP-42 to Alaska. Soon it became apparent that the focus of the Japanese threat was to the islands in the Aleutian chain, where the Navy had established its presence at Kodiak and Dutch Harbor Naval Stations. However, these stations were located on the eastern end of the Aleutian chain, and the western islands of Adak, Kiska, and Attu were vulnerable. DeWitt and Buckner concurred with Parker and the Navy that the Navy was responsible for the defense of the Aleutians, and the AAC would be responsible for the rest of Alaska. However, the Army wished to build a base on Otter Point, on Unimak Island, which was located northeast of Dutch Harbor. Both the Navy and War Department opposed this base, but DeWitt insisted that it would protect Dutch Harbor, and sent Davis out to survey the site. Finally, in late November the Otter Point base was approved.

The attack on Pearl Harbor had little immediate effect on the Alaskan Defense Command, other than a frenzied alert the day of the attack at Elmendorf and Ladd, and then for several days thereafter, as the outer bases were notified. On the 8th Buckner sent up six B-18s and 12 P-36s to patrol, and also to protect them from being caught on the ground in case the Japanese attacked, but this precaution eventually lapsed, and patrols were reduced. With war a reality, Buckner and Davis realized that their meager air force was helpless against any attack, and petitioned the AAC Chief of Staff, stating that they would rather have one bombardment squadron than an infantry division, as "the enemy is always at a disadvantage when he is on the water and we are on the land." After the Hawaiian attack, all military dependants and construction workers were evacuated from Alaska, and there was a concerted restriction of information from the media. Fortunately, nothing threatening would occur in Alaska until June 1942, when all of Alaska went on high alert as the Japanese attacked Dutch Harbor and captured the Aleutian islands of Attu and Kiska.

After Pearl Harbor Gen. Arnold and his staff recommended that Buckner be replaced by an Air Corps general, who would be more familiar with air operations, but again DeWitt, supported by his long time crony Gen. George Marshall, maintained that "instead of unity of command, it was essential that the most cordial cooperation be maintained." Again the personality factor prevented a unified Alaskan command, which would plague the conduct of the upcoming Aleutian campaign. However, Lt.Col. Davis' command improved when his personnel were released from its many administrative duties and base defense, which was correctly transferred to the Army infantry. A major step forward came on 28 December, when the War Department created the Alaskan Air Force, which replaced the localized two month old Alaska Defense Command. The Alaskan Air Force was now a prescribed unit, and Davis had more authority, as he reported administratively to DeWitt, and operationally to Buckner. Also, the AAF (the AAC had been redesignated as the Army Air Force, AAF, on 20 June) notified Davis

When the early B-18Bs arrived in Alaska they were in their natural aluminum finish, but later arrivals came painted in Olive Drab. (USAF)

that winterized B-26 medium bombers and P-40 fighters were on their way, along with five Douglas C-53 transports. By late December the 11th Fighter Squadron and 77th Bombardment Squadron were ordered to be deployed to Alaska. The 18th Fighter Squadron was scheduled to receive P-40Es to replace their P-36s; the 36th Bombardment Squadron would replace their few remaining B-18s with B-17Es; and the 73rd Bombardment Squadron would receive the B-26s. The 77th was a B-18 unit, but soon was equipped with the hot new Martin B-26A Marauder at McGowen Field, ID. The Marauder had a high landing speed and a tricycle landing gear which, with its relatively short range, made it unsuitable for Alaskan air operations, and it would be withdrawn from the theater in late 1942. The arrival of these new types was sorely needed, as the Alaskan Air Force aircraft reserves were further depleted when a P-36 engine malfunction caused a pilot to bail out, and when one engine on a B-18 quit on take off and totaled the aircraft, killing one crewman. The new aircraft were used for desperately needed training, as the newly arriving pilots had mostly trained in good weather climates, and their deployment was a failure until the old Alaska hands trained the new pilots.

The old B-18s were sent to Kodiak Island with the 36BS, commanded by Capt. Russell Cone, where they patrolled the Gulf of Alaska and the North Pacific aided by PBY-5As of VP-42. The B-18s flew the inner search pattern, while the PBYs, equipped with new search SCR-521 radar, flew the larger outer pattern. Later two B-17s and three LB-30s (B-24s) arrived, allowing the 36th to extend its search pattern. While at Kodiak Capt. Cone and Capt. Donald Dunlap flew several flights in April to Otter Bay to determine if the airfield could accommodate heavy aircraft for the upcoming campaign against the Japanese in the Aleutians. The take offs and landings were an adventure, as the field was covered with Marston steel matting that flexed or rolled up when an aircraft landed on it. The condition of the field was improved, and in mid-May a B-17 was able to land there, and finally the 54FS and 406BS were stationed there.

The old B-18s were sent to Kodiak Island with the 36BS, where they patrolled the Gulf of Alaska and the North Pacific, aided by PBY-5As of VP-42. The B-18s flew the inner search pattern, while the PBYs, equipped with new search radar, flew the larger outer pattern. Pictured is a B-18 parked next to a B-26, whose more advanced communications system is being utilized as a control tower. (USAF)

The 406th Bombardment Group had been activated on 15 January 1941 at Fort Douglas, UT, as the 16th Reconnaissance Squadron. The unit moved to McGowen Field, ID, in early 1941, equipped with B-18s for only a short time in late 1941, and in early 1942 was equipped with B-26s, keeping a few B-18s. In late January 1942 the squadron had their B-26s assigned to other units, and was equipped with Lockheed A-29s, which were modified Hudson IIIs originally intended for Lend Lease to Britain. The squadron was sent on anti-submarine patrols, and on 22 April redesignated as the 406BS. Maj. Douglas Courtney, CO of the 406BS at Paine Field, WA, was notified by Col. John Hart, 42BG Commander, at 1100 on 2 June 1942, and ordered to assemble his crews and aircraft immediately for a month's temporary duty in Alaska. The 406th had only a few aircraft available, as it had been released from anti-submarine patrols in A-29s on 4 May, and was to transition to B-25s. At the time the squadron had six A-29s, two B-18s, and two B-25s on loan for transition training. Hart informed Courtney that he was immediately dispatching ten more A-29s and two more B-18s to Paine Field. Four hours after his first orders Courtney sent Capt. Henry Mitchell with a flight of four B-18As and the most airworthy of the A-29s to Alaska. The route took the unit to Great Falls, MT, and across the Canadian border to Calgary, then Edmonton, Fort Nelson, White Horse, and into Alaska to Ladd Field. At Ladd Field they flew to Elmendorf and Kodiak, where they were greeted with disappointment, as Kodiak had been informed that the 406th was a B-25 unit. Once the B-18s arrived they were reluctantly put into service with the A-29s on anti-submarine patrols over the Gulf of Alaska and the Bering Sea until their B-25s arrived. What was to be a temporary 30 day duty extended to 17 months! The 406th also flew anti-submarine patrols over the Bering Sea, as Courtney sent two B-18s to Nome and Naknek. The detachment remained at Nome for a month from 30 June 1942, when it was relieved by the 404BS. A detachment of A-29s and B-18s led by Capt. James McCall was based at Naknek, in southwestern Alaska, from 20 June until 18 November 1942, when it was transferred to Thornborough Field at Cold Bay until 1 May 1943 flying patrols, maritime escort, and other special missions. Another detachment of A-29s and two B-18s under Capt. M.F. Itz was stationed at Yakatut from 21 June 1942 to 22 July 1943. These anti-submarine detachments were generally classified by their participants as not only miserable and monotonous, but also a waste of time, as the enemy submarine activity in the area was nil, despite the fact the Japanese submarines could have been successful, as there were many targets available.

Flying in the Alaskan weather was always an adventure, as there were days of thick fog interrupted by icy rains or snow, often driven by "williwaw" winds. These unexpected gale force winds swept the Aleutians and buffeted aircraft, and their damp air fouled carburetors and coated wings with ice. Most patrols were forced to be flown above the fog and clouds, which were punctuated by occasional volcano tops, but prevented spotting any enemy vessels below. Large local mineral deposits deflected magnetic compasses, so that even experienced pilots lost their way and ran out of fuel before returning to base. There was a lack of radio beam navigation facilities, so lost pilots could not "fly the beam," which was navigating by tuning their radios to directional beams from airfields of

This photo shows a crude attempt to paint a B-18 with white snow camouflage in spring 1942. (USAF)

a known location. The runways and taxiways were usually muddy in the spring, summer, and fall, and icy in the winter, and aircraft and personnel were chronically mired in mud. If these conditions made flying bad, they made living conditions even worse. Tents were constantly blowing down, and the bomber crews often slept in their aircraft to stay out of the weather. The food was mostly tinned—chili, corned beef, Spam, and powdered eggs—and served in damp, draughty mess tents.

Lt.Col. Davis was the senior officer in Alaska, but had no tenure, and with the formation of the 11th Air Force a general officer was mandatory, as commander. Lt.Col. Davis was too junior to be considered for another promotion, and he was replaced by Col. Lionel Dunlap on 17 February 1942; Davis became his Chief of Staff. On 8 March Dunlap was replaced by Col. William "Bruce" Butler, who became a Brig. General a short time later, and Davis was promoted to full Colonel, and assigned as Butler's Chief of Staff. The stern Butler faced the problem of resentment from Alaskan personnel, who felt that the likable and easy going Alaskan veterans, Davis and Eareckson, had been short changed in promotions, and would be better suited for Alaskan command. To complicate matters, Butler also faced the old nemesis of Alaskan air commanders, the shortage of pilots and aircraft.

As the months passed so did the B-18; the remaining few were used as squadron hacks and transports. The last B-18 reported lost in Alaska was on 16 February 1944, when #37-630 went missing. The B-18's final plight in Alaska was to slowly fade into oblivion, as records of their demise were not kept. Some were probably flown back to the States before they gave out altogether, while others were abandoned, as there were no spare parts, or were just not worth the effort to repair.

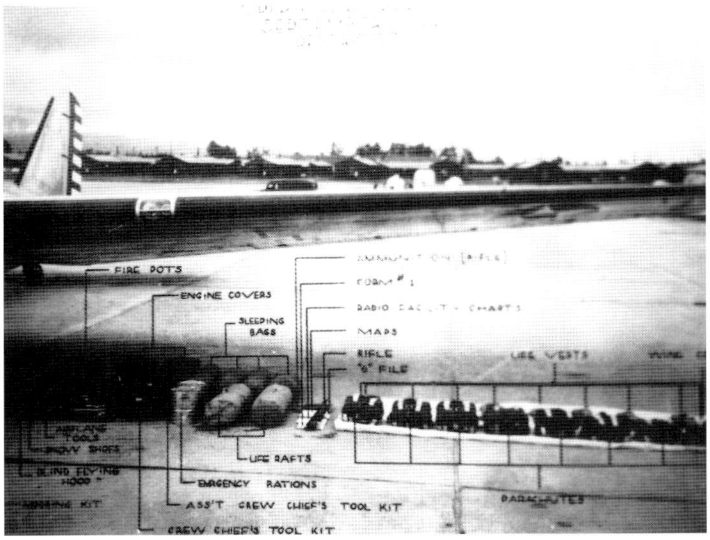

Flying in Alaska was a dangerous practice, and each aircraft was equipped with extensive emergency equipment. (USAF)

Brig.Gen. William "Bruce" Butler (left) was appointed as the CO of the new 11th Air Force over the popular and "Alaska experienced" Maj. William Eareckson (right), and Maj. Everett Davis. The stern Butler faced the problem of resentment from Alaskan personnel, who felt that the likable and easy going Alaskan veterans (Davis and Eareckson) had been short changed in promotions. To complicate matters, Butler also faced the old nemesis of Alaskan air commanders the shortage of pilots and aircraft. (USAF)

9

Anti-Submarine Warfare (ASW) The B-18 Finds its Niche

By mid-1943 the fortunes of Germany's U-Boats had turned, and in "Bloody May" alone the *Kriegsmarine* had lost 41 U-Boats, and had another 37 damaged due to the improvements in Allied weapons and tactics. This marked the first time that U-Boat losses exceeded merchant ship losses. In a memorandum issued in June 1943, Adm. Karl Doenitz, head of the *Kriegsmarine*, recognized this reality:

"The war at sea is at present characterized by a decrease in victories of our navy against enemy merchant shipping. The principal exponent of this type of warfare, the U-Boat, is limited in operational capacity by the ever growing strength of the enemy's anti-submarine defenses and in particular by the enemy air force, using yet unknown equipment and weapons."

Once America was drawn into the war, German U-Boats were prowling unchecked off the northeast American coast, and soon moved south into Caribbean waters. America needed a well-built, dependable aircraft to fly maritime reconnaissance, but the Flying Fortresses and Liberators were in high demand and short supply, and could not be spared for such a relatively mundane purpose. At the time the B-18 had not found a niche, as it was an outdated second line bomber waiting for obsolescence, but there were a large number of B-18 and B-18As available. A total of 122 of the B-18As were modified at Air Force modification depots and converted into the B-18B. Of the 122, 79 had MkIV magnetic anomaly detection (MAD) equipment installed in a tail boom. The B-18B was the first aircraft in the war to be equipped with SCR-517 centimetric air-to-surface vessel radar (ASV), with a scanner enclosed in a radome. Other equipment tested was the absolute altimeter, LORAN (<u>LO</u>ng <u>R</u>ange <u>A</u>id to <u>N</u>avigation) airborne microwave radar, radio sonic buoy, and improved depth charges. These aircraft and weapons were in the vanguard of Doenitz's "unknown air equipment and weapons" that plagued his U-Boats.

The German U-Boat

Type VIIA/B/C/D/F

At the start of the war in 1939, the German U-Boat Command (BdU) had only 56 commissioned U-Boats, of which 39 were operational, but of these only 22 were the 220 foot long and 20 foot wide, 750 ton (surfaced) ocean-going Type VIIB. This single hull type was capable of a maximum surface speed of 17 knots, and a maximum range of 8,500 miles at 10 knots, and was capable of diving to 300 to 600 feet. Submerged, the Type VII had a limited battery life that allowed it a range of 90 miles at four knots, and maximum underwater speed of eight knots for a short time. It was armed with five torpedo tubes, carried 14 torpedoes (a combination of G7e electric and some G7a steam torpedoes), and mounted an 88mm deck gun; it also mounted on the conning tower either a brace of small caliber machine guns, or a 20mm cannon. It had a crew of four officers and 40 to 44 men. Since the U-Boat had restricted submerged speed and range it spent most of its time on the surface, and its fundamen-

Adm. Karl Doenitz, head of the *Kriegsmarine*. (Author/B.C. West)

The oceangoing single hulled Type VII U-Boat was 220 feet long and weighed 750 tons (surfaced). It was armed with five torpedo tubes, and carried 14 torpedoes. This type was capable of a surface speed of 17 knots, with a maximum range of 8,500 miles at 10 knots, and could dive to 600 feet. Submerged, the Type VII had a limited battery life that allowed it a range of 90 miles at four knots, and maximum underwater speed of eight knots for a short time. It had a crew of four officers and 40 to 56 men. (USN)

tal defense from surface or air attacks was to be alert and dive to safety, which took just under a minute. In 1942, with the increase in Allied air attacks the Germans equipped their U-Boats with a special platform fitted aft of the conning tower to carry 20mm cannon capable of firing 150 rounds per minute, and two to four 7.9mm machine guns, which proved to be ineffective, except as a deterrent. The most successful Type VII was U-48, which sunk 53 merchantmen of 304,000 tons under five captains on 12 patrols.

Type IXA/B/C/C40/D-1/D-2/D-42

The 254 foot long by 22.25 feet wide, 1,050-1,150 ton (surfaced), double hulled, Type IX ocean-going U-Boat was armed with six 21 inch torpedo tubes and carried 22 torpedoes. It was armed with a 105mm deck gun and a 37mm and 20mm AA gun on the conning tower. It had a range of 12,000 to 13,800 miles at 10 knots (maximum 18 knots), and a submerged range of 68 miles at four knots (eight knots maximum). This type was crewed by four officers and 44 to 50 men. The IXs sank more tonnage per boat than any other type. The most successful Type IX was U-123, which sunk 44 ships of 225,000 tons on 13 patrols.

The oceangoing double hulled Type IX U-Boat was 254 feet long and weighed 1,050-1,150 tons (surfaced), was armed with six torpedo tubes, and carried 22 torpedoes. It had a range of 12,000 to 13,800 miles at 10 knots (maximum 18 knots), and a submerged range of 68 miles at four knots (eight knots maximum). This type was crewed by four officers and 44 men. (USN)

The Type VII U-Boat vs. the Type IX

The Type VII was the workhorse of the U-Boat fleet, and the largest single class built in submarine history. During the war the BdU was supplied with 709 VII U-Boats, of which 665 were C-types, while only 159 IX boats of all types were built (there were 1,171 U-Boats of all types commissioned during the war).

The Type IX U-Boats were considered to have relatively mediocre combat qualities when compared to the older Type VII design. The IX was characterized by its broad beam and large deck, which carried a large and more visible conning tower, and was nicknamed the "Seekuhe" (sea cow) by its crews. The IXs were so large and cumbersome that they could not dive in daylight operations off much of the eastern U.S. coast, as shallow water extended to the east over 100 miles, and ended at the continental shelf. The IXs had to move into the shallow waters at night, and then flee back to the deep waters off shelf before sunrise. The VII was over 300 tons lighter, and dimensionally smaller than the IX, giving it a lower silhouette, and making it more maneuverable on the surface and underwater; they were also more stable in maintaining depth under water, and thus less easily located by detection apparatus. The VII dove faster, reaching 43 feet in 30 seconds, while the IX required 35 seconds to reach the same depth. The five second difference meant that a B-24 would be 2,000 feet closer to the IX U-Boat on a depth charge run. The IX was more complicated, and thus more susceptible to depth charge attacks. Doenitz and the BdU favored the VII (and constructed more), as the IX required more raw materiel, time, and work force to construct, and it was estimated that two Type VIIs could be built for every Type IX.

Due to their great range, endurance, and increased torpedo and surface armament, the Type IXs gained success out of proportion of their numbers, particularly early in the war, as they were able to act as individual commerce raiders against defenseless merchantmen in distant areas such as the Caribbean, where they were operational in numbers nearly equal to the VIIs. Although the 141 IXC U-Boats comprised only 12% of the total U-Boat force, they accounted for nearly 37% of the total Allied shipping losses to U-Boats. However, the IXs suffered heavier losses proportionately, and in some months numerically than the VIIs, particularly after the Allies upgraded their convoy defenses, causing the IXs to submerge and become more vulnerable than the VIIs.

TYPE VIIC

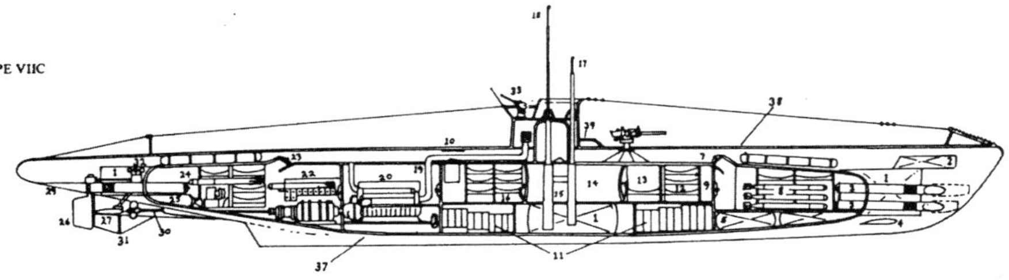

TYPE VIIC (1942 Configuration)

- Displacement - 769 tons surface/ 871 tons submerged
- Dimensions - 220¼ x 20¼ x 15¼ Feet
  67 x 6 x 4.8 Metres
- Armament - One 88mm, One 20mm AA,
  Five 21 inch torpedo tubes (4 forward 1 aft)
- Torpedoes - Twelve 21 inch
- Machinery - Two Shaft diesel/electric motors BHP/SHP 2800/750
- Speed - 17 Knots surface/ 7.6 Knots submerged
- Bunkers and Range - Oil fuel 113.5 tons - 9,400 nms at 10 kts
- Complement - 44

TYPE IXC

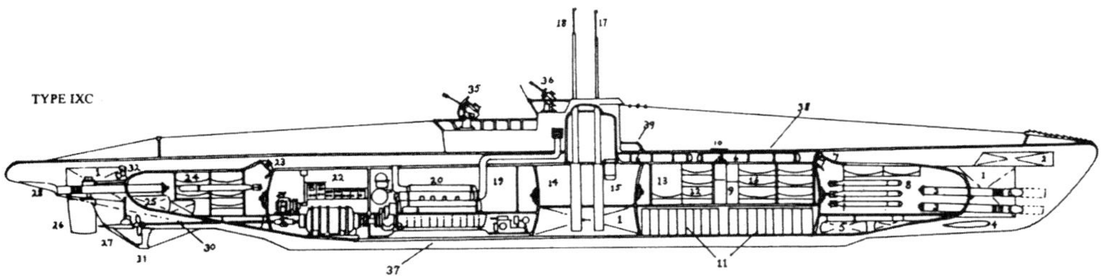

TYPE IXC (1943/45 Configuaration)

* Extended conning tower and enhanced AA armament

- Displacement - 1120 tons surface/ 1232 tons submerged
- Dimensions - 252 x 22¾ x 15½ Feet
  76.8 x 6.8 x 4.8 Metres
- Armament - One 37mm mounting, Two twin 20mm mountings
  * Note - Main deck gun removed
  Six 21 inch Torpedo Tubes (4 forward, 2 aft)
- Torpedoes - Twenty-two 21 inch
- Machinery - Two shaft diesel/electric motors BHP/SHP 4400/1000
- Speed - 18.3 kts surface/7.3 kts submerged
- Bunkers and Range - Oil fuel 208 tons, 16,300 miles at 10kts
- Complement - 49 - 57

KEY

(1) Main Ballast Tanks
(2) Stabilizing Tank
(3) Bow Torpedo Tubes
(4) Fore Hydrophone
(5) Forward Trim Tanks
(6) Spare Torpedo Containers
(7) Forward Torpedo Loading Hatch
(8) Forward Torpedo Room and Torpedo Ratings Mess Spare Torpedoes
(9) Galley
(10) Escape Hatch
(11) Batteries
(12) Officers Wardroom
(13) Captains cabin
(14) Control Room
(15) Control Room Hatch
(16) Petty Officers Mess
(17) Air and Sea Search Periscope
(18) Attack Periscope
(19) Diesel air intakes
(20) Main Diesel Engines
(21) Main Electric Motors
(22) Control Panel
(23) Aft Torpedo Loading Hatch
(24) Aft Torpedo Room Engineer Ratings Mess Spare Torpedo stowage
(25) Aft Trim Tank
(26) Twin Rudders
(27) Aft Hydrophone
(28) Aft Torpedo Tubes
(29) Aft Torpedo Tube
(30) Propeller Shafts
(31) Propeller and Rudder guards
(32) Steering gear
(33) Single 20 mm cannon
(34) 88mm Main Deck Gun
(35) 37mm Mounting
(36) Twin 20mm Mountings
(37) Pressure Hull
(38) Deck Casing
(39) Compass Housing

TYPE VII & TYPE IX U-Boats

ASW Weapons

LORAN and the Absolute Altimeter LORAN (LOng Range Aid to Navigation)

In October 1940 American scientist Dr. Alfred Loomis proposed the use of shore based transmitters to send out electronic pulses as the basis for a new navigation system. In early 1941 Dr. Melville Eastham began work on the project at MIT's Radiation Laboratory, and by September had completed a working example. The aircraft's LORAN set received radio signals transmitted from three known points, allowing the operator to locate his position to within four miles at a range of 1,200 to 1,500 miles from the transmitters. LORAN coverage extended over the Eastern Sea Frontier, Gulf Sea Frontier, and most of the Caribbean Sea and North Atlantic, allowing the effective assembly of anti-submarine forces.

The Philco AN/APQ-4 LORAN determined the geographic location of the aircraft by picking up LORAN signals from a known position on a grid. The LORAN was relatively simple, and could be used in bad weather until nearing target, when the more accurate airborne radar could make a fix. The relatively low frequency (1.950MHz) LORAN consisted of two units that together measured approximately 1 foot x 2 feet x 2.5 feet, and weighed about 80 pounds. One unit was the power supply, and the other contained the oscilloscope display tube, receiver, and timing circuit. The oscilloscope screen was about four inches in diameter, and would display both a stationmaster and associated slave signal. Shore-based transmitters operating in pairs, typically separated by 400 miles, sent out pulsed and precisely synchronized radio signals. The receiver identified these signals by their radio frequency and repetition rate, with the time differential between the two signals measured in microseconds. By consulting the LORAN charts for the area, the navigator identified from this pair of stations a line of position that located the aircraft. On these charts there were lines of constant time for every pair of transmitting stations. The operator needed two successive fixes to determine ground speed, drift, and ETA. The LORAN's nighttime range was more than its theoretical daytime range. An experienced operator could get a fix in three minutes, and on a 1,400 mile mission have a minimum error of 28 miles.

Absolute Altimeter

The Radio, or Absolute Altimeter, utilized a modification of microwave radar to determine the aircraft's altitude, replacing the less accurate barometric altimeter that responded to barometric changes, and could give incorrect readings in areas of irregular barometric pressures. An incorrect reading could be devastating in a blind landing approach, or could foil an otherwise well conducted low level attack. The Absolute Altimeter was accurate to within 10 feet at altitudes of less than 400 feet. It allowed the aircraft to fly safely as low as 50 feet in a low level attack, and became standard AAF anti-submarine equipment by May 1943.

Magnetic Anomaly Detector (MAD), Retrobombs, and the Sonobuoy

While Airborne Surface Vessel Detection (ASV) radar promised to detect surfaced U-Boats, a technique needed to be devised to find submerged U-Boats, and the solution seemed to point toward a Magnetic Airborne Detector (MAD). When a ship's steel hull is

Since magnetic fields, unlike radio waves, were unaffected by their passage through earth or water, the Magnetic Airborne Detector (MAD) was based on magnetometers used by mining companies. The MAD detected distortions in the earth's magnetic field that would indicate large underground mineral deposits, and were adapted to locate small objects, such as submarines. (USAF)

constructed it undergoes heating by welding and impact through riveting. Ferrous metal is comprised of clusters of iron molecules called "domains," each of which is a minute magnet with its own north-south magnetic field. Normally these domains are arranged randomly, but with heating and hammering they tend to align themselves, so that their south poles point north and their north poles point south, so that the accumulative effect is that the ship's hull acquires a permanent magnetic field that is quite substantial in comparison to the earth's magnetic field. Also, the steel in a ship's hull causes the earth's lines of force, called "flux," to move out of their normal positions, and be concentrated at the ship, which is called the "induced field." The ship's total magnetic field ("magnetic signature") at any point on the earth's oceans is the sum of its permanent and induced magnetic fields. The earth's natural magnetic field does not always run in straight north-south lines, but weaves in places to the east and west, and dips at different angles to the horizontal. These directional and dips in the area of a large mass of ferrous matter are measurable by a sensitive magnetometer. An airborne magnetometer was to detect a metal submarine in the earth's magnetic field. The depth at which a submarine can be detected is a function both of the size of the submarine, and how close to the water's surface the magnetometer is flown. Since magnetic fields, unlike radio waves, are unaffected by passage through earth or water, prewar investigation on a Magnetic Airborne Detector was based on magnetometers used by mining companies to detect distortions in the earth's magnetic field that would indicate large underground mineral deposits. However, contemporary devices could not locate small objects, such as submarines.

Close up of MAD aircraft tail antenna, which was placed as far as possible from the aircraft's magnetic field. (USAF)

Late in 1940, Dr. Victor Vacquier of the Gulf and Research Development Company introduced the "saturable-core" magnetometer for mineral prospecting, which was two to three times more sensitive than previous types. Early in 1941, the National Defense Research Committee (NDRC) recognized the potential of Vacquier's device, and began its development for anti-submarine warfare. The detection of a submarine presented the NDRC with a very challenging dilemma, as Vacquier's device had to measure the distortion of a comparatively minute ferrous object like a submarine in comparison with the earth's powerful magnetic field. The strength of the earth's magnetic field varies with the distance from the poles (e.g. latitude), but averages about 50,000 gammas, while the magnetic field of the World War II U-Boat was only 10 gammas at 400 feet. This strength decreased with the cube of the distance, so that at twice the distance (800 feet) the sub's magnetic field was only 1.25 gammas. In addition, there were other problems to contend with involving the aircraft and equipment. If the magnetic detector in the search aircraft were not kept in alignment with the earth's magnetic field to within only 0.1 degree, its sensitivity would drop off noticeably. Because the metals used in constructing the aircraft would interfere with the device, it had to be positioned in the furthest end of the tail or wing tip, and ferrous components had to be replaced with non-ferrous ones. Even so, the range of the NDRC device was so short that the search aircraft had to fly at 100 feet, and directly over a submarine submerged at 300 feet to locate it. This drawback meant that the device could not be used at night or in poor visibility for fear of colliding with the ocean's surface.

The next investigation of a MAD device was started in early 1942 by Westinghouse and the Airborne Instruments Laboratory. Trials began in the spring using Navy patrol aircraft and airships, and when German U-Boats appeared off the U.S. East Coast the MAD apparatus was used operationally for the first time, but only for a short time, as the Germans shifted their U-Boats to southern waters. Trials showed the range of the MAD to be very short, and the operator received signals of a contact only when he was directly over the submarine. If depth charges were released at that time, the forward speed of the search aircraft would cause the depth charges to fall forward of the submarine's location. Thus, it was obvious that a special bomb that would fall vertically, even though the aircraft was moving forward, needed to be developed. By mid-1942, Dr. Carl Anderson, the California Institute of Technology head of Aircraft Projects, developed a 35 pound Torpex filled, impact

A B-18 could carry sixteen 35 pound retrobombs, eight under each wing. When a MAD operator made a contact he fired the retrobombs, whose rockets propelled the bombs backwards off their launching rails, bringing them to a mid-air stop. After their solid rocket fuel was expended the bombs would drop vertically into the water over the submarine. (USAF)

fused "retrobomb" with a solid fuel rocket installed in its tail. Rocket propulsion had two characteristics that made it an ideal solution. Firing backwards, it would cancel out the forward motion of the aircraft, and having no recoil, relatively large missiles could be carried without damaging the aircraft structure. Field trials were conducted using a B-18 bomber, and Dr. Anderson described a test flight flown by a Douglas test pilot to the Goldstone test facility:

"One day one engine was sputtering and popping, but the test pilot said, 'Well let's go.' The other engine quit right after take off, and that was a harrowing experience. But they patched that up and we did get out to Goldstone with it. When we got there, I asked the pilot how the airplane flew. He said, 'Fine. It has a cruising speed, top speed, and stalling speed, all of 100 miles per hour."

The retrobomb went into service in mid-1943. When a MAD operator made a contact he fired the retrobombs, and the rockets propelled the bombs backwards off their launching rails, bringing them to a mid-air stop. After their solid rocket fuel was expended the bombs would drop vertically into the water over the submarine. A B-18 could carry 16 under wing retrobombs, eight under each wing on special launching rails, while a PBY could carry 24, 12 on each wing. They were launched in salvoes of eight—four from each wing—with a second salvo of eight—again, four from each wing—being launched automatically a half second later (followed by another eight in another half second on the PBY installation). The retrobomb launchers were divided into groups of three, with each group being set at a slightly different angle, so that the bombs from each salvo hit the water in a line about 100 feet long, and at right angles to the aircraft's line of flight. The half second intervals between salvoes gave 90 foot spacing between these lines. Before the attack was initiated the crew tracked the submerged submarine with smoke markers so that they could attack it lengthwise, hopefully resulting in at least two of the bombs exploding over the target. Of the 122 B-18Bs, 79 had MkIV magnetic anomaly detection (MAD) equipment installed in a tail boom.

The British Admiralty established a special committee for anti-submarine measures headed by Dr. Patrick Blackett. In May 1941 the committee proposed an expendable sonar buoy that could be dropped from ships to listen for trailing submarines. Tests showed the idea to be practicable, but due to limited funding the British abandoned the idea. Fortunately, RCA had supplied the sonar buoy prototypes and continued the tests with the Navy, that found when using MAD equipment the operator had problems distinguishing if the object detected was indeed a submerged submarine or another metallic object, such as a wreck. To remedy the problem, the Colombia University's Undersea Laboratory at New London, CT, developed a passive hydrophone buoy with a radio transmitter based on the RCA design designated as the AN/CRT-1 sonobuoy. The 14 pound cylindrical, 3.75 foot long, and 4 inch diameter device consisted of a small floating transmitter under which a hydrophone was suspended on a cable. The parachute at the top of the buoy slowed its fall, and on impact with the water the hydrophone's 24 foot long connecting cable was released, and the buoy's batteries were turned on. The cable had to be long enough so that the sounds of the waves splashing the canister would not interfere with any faint U-Boat propeller sounds. Underwater sounds were transmitted to the floating transmitter, and then to an operator in the aircraft patrolling overhead, who listened to the sonobuoy's HF transmissions on a special receiver. The operator had the challenging job of trying to decipher U-Boat sounds from those produced by the sea and its inhabitants.

The range of the hydrophone depended significantly on conditions, but the U-Boat's propeller had to be turning to produce cavitation, which is the sound of bubbles produced by a turning of propellers, and varied with depth (i.e. more depth meant lesser cavitation). Detection could be as high as 3.5 miles if the U-Boat were moving at seven knots submerged at 60 feet in calm water, or as little as 300 feet if the U-Boat were moving at three knots at a depth of 250 feet in a rough sea. The early sonobuoys were non-directional, and the operator was unable to determine the direction of the noise's origin. To obtain an approximate location of possible U-Boat noises it was necessary to lay a pattern of five sonobuoys. The life of the batteries was about four hours, and to prevent the enemy from recovering a buoy, a stopper was slowly dissolved by seawater, and the buoy would sink once the batteries ran out.

The first feasibility trials were held in March 1942 by the Navy, who launched the sonobuoy by boat, with its S-20 submarine acting as the target, and the blimp *K-5* carrying the receivers. During the trials the operator in the blimp could hear the submarine's propellers at ranges up to three miles from the buoy. In June 1942 *Project Sail* was initiated at Quonset Point for MAD system research and testing by the Naval Ordnance Laboratory and the NDRC. In July, the first so-called "fast drops" of sonobuoys were made by parachuting sonobuoys from a B-18 cruising at 120mph; after the success of the tests 200 units were ordered, and the AN/CRT-1 went into production in the fall of 1942. The first operational use of a sonobuoy was in August 1942, and the first U-Boat (U-568) sunk in conjunction with a sonobuoy occurred on 30 October 1942 by a 145 Squadron RCAF Hudson. Some early sonobuoys were used by a few Caribbean B-18s.

The MAD/Retrobomb/Sonobuoy combination had limited operational usefulness, as the U-Boats changed their tactics, and remained on the surface using an increased AA armament capability to fight it out, making it too dangerous for the attacking aircraft to fly directly over the bristling U-Boat to drop their retrobombs.

Airborne Surface Vessel Detection (ASV) Radar

Long Wave Radar ASV MkI

The British development of radar in the late 1930s is well known through the movies and documentaries on the Battle of Britain that show the massive "Chain Home" radar antennas, which gave the RAF "few" the forewarning that saved England. But the development of any airborne radar would be much more difficult. A radar transmitter requires an antenna to send out its signal, and a receiver needs an antenna to pick up that signal. The basic antenna is the dipole, which is reciprocal, being able to produce radar waves at a particular frequency, and also is effective in receiving them. The dipole antenna functions best if its length is half of the wavelength of the radar beam, so the objective in developing airborne radar was to develop a shorter wavelength beam. With the longer wavelength apparatus available to focus the beam in any direction, a large array of dipole antennas were required, which limited the number and length of the dipoles. The airborne radar needed to be able to automatically sweep and sample manually guided radar with an A-scope, and was the first step toward ASV (airborne surface vessel) and AI radars (airborne intercept), which were essentially similar. The next step was to develop 360 degree radar, which was difficult, as a dependable rotary coupling system needed to be designed so that the electrical connections between the antenna system and the radar electronics would rotate freely. Also, radar's accuracy depended on the focusing (e.g. "spotlighting") of the beam, as a narrow beam would pinpoint the direction of a distant target more accurately than a broad beam.

In 1936 E.G. "Taffy" Bowen formed a team at the Telecommunications Research Establishment (TRE) at Swanage to investigate airborne (AI) radar. Installing radar in aircraft was difficult due to the size and weight of the set and its antenna, its power supply, and operating in cold and vibration. Bowen and his group solved most of the problems over the next year, and developed early experimental sets of a very long 6.8 meters, which made the transmission of narrow beams unfeasible, as most of the return was from the ground and limited AI radar to the height of the aircraft from the ground. In early 1937 during flight testing Bowen flew over the ocean, and discovered that ships could be detected at short ranges, and ASV radar was conceived. By the summer of 1937, TRE had developed shorter wavelength ASV radar that was based on the Western Electric 316A valve, which developed 100 watts of power at a 1.25 meter wavelength. This radar was installed in an Avro Anson, and first flown on 17 August 1937 in weather so poor that standard reconnaissance was impossible. The Anson was able to track the aircraft carrier HMS *Courageous* and aircraft taking off from her decks, the battleship HMS *Rodney*, and the cruiser HMS *Southampton*. However, this radar was not ready operationally due to antenna difficulties, along with administrative and priority problems. As the European war approached in 1939 ASV and AI were given increased priority, and the wavelength was increased to 1.5 meters, and two antenna designs were developed. One antenna design utilized a dipole transmitting antenna with a reflector that sent a broad forward directed beam. Two receiving antennas that had overlapping polar diagrams were installed on the aircraft's wings. The target's direction was ascertained by comparing the signals from the two receivers, which were displayed on a cathode ray tube that had a center vertical baseline, one to the right, and the other to the left of the vertical line. If the target was to the left, then the left receiver gave the stronger signal. The vertical time base displayed the distance to the target with a 1,000 ton vessel detected at ten miles, and an elevated coast line at up to 40 miles.

In late 1939 a test was conducted using a Hudson of the Coastal Command 220 Squadron and RN submarine L 27. The submarine was detected broadside at three miles when the Hudson was flying at 1,000 feet, and later at six miles when flying at 6,000 feet. The encouraging results led the Coastal Command to equip 12 Hudson bombers of its 220, 224, and 233 Squadrons with the first ASV radar, designated as the ASV MkI, in January 1940. The radar had its fixed antenna located in the aircraft nose, and was only able to search forward. At 3,000 feet the MkI could detect a U-Boat at 5.5 miles, but at 4.5 miles the target was lost in reflections from the water's surface, called "sea clutter." Flying at 200 feet was found to be more effective, as the U-Boat could be detected at 3.5 miles, and the sea clutter would not obscure the target until a half mile, but flying at 200 feet proved to be difficult to maintain by a pilot for long periods. The forward looking MkIs were greatly improved when a second antenna system was developed called the Long Range ASV (LRASV), which was a sideways looking system. The transmitter was an 18 foot long array of ten dipoles installed in five pairs (later four) on top of the fuselage, while the two 12 foot receiving antennas were installed on the sides of the fuselage. Because of their length these antennae were nicknamed "stickleback" by the British. The nose or side antennas had to be used individually, but the side antennas could be quickly switched from side-to-side. Routinely the side antennas swept ten miles on each side, so a patrolling aircraft flying at 125 mph could theoretically scan 20 miles on each side, covering 2,500 square miles in an hour. This array gave a much better resolution and range (2.5 times better), and was able to detect a submarine at 10 to 15 miles.

MKI reliability was poor due to inferior manufacturing quality control of its components. It required skilled tuning by a highly trained operator, and was difficult to maintain, and in practice was considered to be inadequate for U-Boat detection, but was very good for bad weather navigation, and for rendezvousing with convoys for escort duty. By 1940 some 200 sets were manufactured and placed in Coastal Command Hudsons and Sunderlands, but another radar system using a shorter wave length was needed.

Long Wave Radar ASV MkII

The MkI was replaced by the more reliable and better performing shorter wavelength ASV MkII, developed by the spring of 1940 by Pye Radio Ltd. and E.K. Cole Ltd. The MkII was very similar to the AI Mk4, and also operated at 176MH. Whereas early ASV sets required separate receivers and transmitters that used multiple sets of separate antennae, the MkII had an Aerial Coupler that used a motor driven switch to allow the transmitter and receiver to send and receive radar beams through a single set of antennas. The MkII had a more powerful transmitter than the MkI that could focus a narrow beam either through the nose, or the two new side scanning antennas placed one on the port, and the other on the starboard. However, a variation used a large antenna array placed on top of the fuselage. This array was used to sweep shipping lanes, or as a receiver beacon at ranges up to 90 miles. Radar returns were received through a new and more powerful receiver and indicator

The MkI ASV radar was replaced by the more reliable, and a better performing shorter wavelength ASV MkII developed by spring 1940. The MkII receiver used the early Range Azimuth screen (A-scope), which was an oblong screen with a vertical base line running down the middle, and all ground returns came up as blips from the bottom. (Pima)

units. When the side scanners detected a suspected target the aircraft would turn at 90 degrees toward the target and switch on the nose radar and close. In tests flights at 1,000 feet the MkII could detect a surfaced submarine at 10 miles through its side antennas, or at seven miles through its nose antennas. The first MkII success was on 30 November 1940 when a Whitley MkIV bomber damaged U-71 in the Bay of Biscay, and half a year later more sets and more experience had increased attacks on U-Boats by 20%, and night patrols were initiated.

The MkII receiver used the early Range Azimuth screen (A-scope), which was an oblong screen with a vertical base line running down the middle, and all ground returns came up as blips from the bottom, either to the right or left of the base line, depending if the signal were to the port or starboard, and the distance from the bottom of the trace was equivalent to the range. Three distance ranges were selectable on the indicator, and the suitable scales were imprinted on the cursor. There were two blips on the screen: the lower and larger was the simulated sea echo at the height of the aircraft. The upper blip was the simulated target, and the radar operator had to line up the target blip on the base line and determine the distance to the target. The A-scope returns made it difficult for the operator to determine the U-Boat's location in relation to the aircraft and distorted images below, so it was almost useless for range and bearing navigation in U-Boat detection or navigation on anything but a coastline. When a blip was seen the radar operator gave the pilot course corrections so that he could approach the target head on; the radar return would move from either side of the base line, and the operator would give directions such as: "right, right, left, left." The echo would grow larger on the screen as the target was approached, and would move down the screen, and the range was measured by a graticule attached to the front of the screen. The pilot was constantly maneuvering, which meant that the aircraft would be banking over the target and dropping all the depth charges in salvo instead of in string, a much less accurate method.

The noise at the bottom of the screen was sea clutter, and the target echo would merge into it and be lost. The MkII was far less susceptible to sea clutter, and it did occur at a half mile, at which time the U-Boat could be visibly observed. At night, unless there was bright moonlight, a surfaced U-Boat once detected by radar would become lost in the radar's sea clutter return at one half mile. The initial answer was to drop a string of flares to light up the U-Boat, which hopefully should be directly under one of the flares. However, by the time the bomber returned to attack, the U-Boat had plenty of time to dive. The next fix was to drop a buoy into the sea containing time delayed flares near the U-Boat. As the bomber circled and approached for its attack run the flares would be shot into the air, hopefully silhouetting the U-Boat. RAF HQ administrative officer Humphrey Leigh suggested that a better solution would be to mount a powerful forward pointing searchlight on the aircraft that could be switched onto the U-Boat as it disappeared into the radar sea clutter. Contractors Savage & Parsons had problems installing the searchlight due to its size and power source. In the meantime, the Air Ministry also considered the Turbinlite searchlight, which had initially been developed to intercept enemy bombers at night, but had failed testing. Savage & Parsons continued to work on Leigh's searchlight, and developed a powerful new 24 inch searchlight that could swivel about 20 degrees downwards or sideways that was initially powered by a generator, but later by a battery pack. On 4 May 1941, Leigh personally piloted a radar/Leigh Light (L/L) equipped Wellington to illuminate the Royal Navy submarine H-31, which had been lost in the radar's sea clutter at a half mile. On 21 December 1941 a MkII radar equipped Swordfish made the first successful night attack on a U-Boat using the Leigh Light. By June 1942 ASV/L/L equipped patrol aircraft were swarming the U-Boats crossing the Bay of Biscay to reach the Atlantic, and were so successful that by August Doenitz decided that it would be more advantageous for his U-Boats to fight it out on the surface.

Short Wave Microwave ASV Radar

Cavity Megnetron

High microwave frequencies could produce more focused beams with greater resolution, and at the end of the decade there was no source that was powerful enough to generate short waves. The British Admiralty established a committee to study radar operating at ten centimeter microwaves. A team at Oxford's Clarendon Laboratory was assigned to develop a microwave receiver, and another team at the University of Birmingham was to develop the microwave transmitter. To test their designs, microwaves had to be generated for the circuits to detect. At the time there were two methods to generate microwaves: the magnetron, and the new American invention, the klystron. Drs. John Randall and Henry Boot combined the features of both and made the initial breakthrough in centimetric radar when they built their 3,000 Mc (10cm) "magnetron oscillator," or Cavity Megnetron. They made their first microwave transmission on 21 February 1940, which had a maximum power output three times that of any other device. The device was supplied to Cambridge's Cavendish Laboratory, headed by Dr. Philip Dee. Dee's team built a 9.1 centimeter radar system that was referred to as ten centimeter for expediency. On 12 August 1940 the radar was used to track an aircraft, and on the next day it tracked a technician carrying a metal sheet on a bicycle without any of the previous radar crippling ground clutter.

The advent of the new centimetric radar enabled the search aircraft to locate a U-Boat, and then the powerful Leigh searchlight could be switched on to aid the attack when the target became lost in the radar's surface clutter. (USN)

British Centimetric ASV Radar

The British developed ASV MkIII centimetric radar, but its story is a checkered one. The original team that developed British airborne radar was disbanded, as Taffy Bowen was in America, and personal conflicts arose, causing a new group to be assembled to develop centimetric radar for nightfighters. Centimetric radar had the benefit of having a small parabolic reflector to focus an accurate beam that produced better range and resolution, and eliminated strong ground returns that were inevitable with the broad beams of the long wave MkI and MkII sets. When Drs. Randal and Boot developed their magnetron, the first sealed unit suitable for airborne use was produced by the GEC in June 1940. While centimetric radar development concentrated on AI use in fall 1940, the Royal Navy conducted ground based experiments under Captains B.R. Willett and C.E. Horton. They were able to track the submarine HMS *Usk* over seven miles on 11 November 1940. After testing and refining the equipment and antennas, the first operational British centimetric radar was the Type 271 S-band shipboard set, which was put into production within months, and underwent sea trials in March 1941. The rudimentary set had manual direction, and separate large box-like transmit and receive antennas, and led to a series of other naval and ground radars. By March 1941 the Type 271 was put into operation, only because it was the straightforward option for both utility, and the placement of its large antennas in the large areas available. On 16 November 1941 the Type-271 equipped Royal Navy corvette HMS *Orchis* sank U-433 near Gibraltar, and by mid-1942 236 Royal Navy ships were fitted with the Type 271.

However, the priority was for S-band AI radar for nightfighters, and ASV radar for anti-submarine aircraft. In April 1941, tests on the AIVII were conducted against the RN submarines HMS *Sea Lion*, and in August against the HMS *Sokol* that demonstrated that the radar was effective, but the development of centimetric ASV was lagging, and the first prototype ASV radar was not test flown until December 1941. The production version was being finalized by the summer of 1942 by Ferranti, but on 30 September 1942 the company was ordered to halt work, a directive that astounded Coastal

Command. The reason was that simultaneously Bomber Command was developing H2S ground mapping radar, which was very similar to ASV radar, under Philip Dee and Bernard Lovell at TRE. The first test of H2S—which was very successful, as it picked up the outline of a town 35 miles away—was conducted on 1 November 1941, and the first production model was ready by mid-March 1942. H2S attracted the attention of Churchill, and centimetric ASV was halted in favor of the H2S, which was to equip four engine bombers flying at 20,000 feet. Coastal Command leaders were piqued that H2S was selected over ASV radar, and were unwilling to continue to work with the Bomber Command concept for ASV purposes. Finally Coastal Command was forced into an attitude adjustment, and reluctantly began to redesign the Bomber Command high altitude H2S concept for ASV functions at 2,000 feet or less. It would not to be until 1 March 1943 when a Wellington of 172 Squadron flew its first patrol over the Bay of Biscay. The small 28 inch scanner was mounted in the nose, giving a 60 degree scan that was received on a Plan Position Indicator Screen that represented the area below as a map on a TV-like screen (to be described later).

America Develops British Centimetric ASV Radar

The RAF ordered 4,000 ASV MkII sets, but production problems due to the demand for other types of radar, particularly bomber H2S targeting radars, slowed its delivery, and the MkII was not available in numbers until early fall 1940, when 45 sets were delivered to Coastal Command. Thus, in mid-1940 Britain found that its scientific, engineering, and manufacturing resources were being stretched thin, and it was apparent that the development of centimetric radar would require the time, money, and engineering capabilities of the United States to develop. Churchill wisely decided to share Britain's secret technological developments, including the cavity magnetron, with American scientists. In August 1940 the prominent British scientist Sir Henry Tizard, with eminent physicist Dr. John Crockett as his deputy, Edward "Taffy" Bowen as the British radar representative, and four other scientists made up the so-called "Tizard Mission," which was sent by Churchill to meet with American Dr. Vannevar Bush. Bush had recently created the National Defense Research Committee (NRDC), which was "to coordinate technology development between civilian scientists and military officials, and to conduct preliminary technical research studies using such funds as were available." The initial British technical information was examined by Alfred Loomis, who was a very successful investment banker, and an avid lay scientist and researcher, and in this position Loomis was put in control of the NDRC's "Microwave Committee." In September Loomis hosted a gathering for members of the Tizard Mission, who unveiled the cavity magnetron, and at once the perceptive Loomis realized its potential in saving U.S. scientists "two years of development." Bush and Loomis understood that for the prompt development of cavity magnetron technology it would have to be kept out of military control, and a civilian research laboratory would have to be established using NDRC funding. Both Bush and Loomis had strong bonds to the Massachusetts Institute of Technology (MIT), and set up the Microwave Laboratory there. Soon the Microwave Laboratory was referred to as the Radiation Laboratory, or "Rad Lab." At the time American long wave radar was generally on a par with its British counterpart due to the work of the Navy's NRL and Army Signal Corps Laboratories.

In October 1940, Loomis and Bush began to enlist scientists and researchers for the Radiation Lab, and Edward Bowen of the Tizard Mission proved to be an important recruiter, enthusiastically promoting the new British technology. The knowledgeable and genial Bowen has been credited with establishing the amiable Anglo-American scientific collaboration during the war. Among the top scientists signing on with the Radiation Laboratory were physicist Dr. Isidor Rabi and a number of his top students, and Dr. Ernest Lawrence of the University of California, Berkeley, who also was accompanied by a number of top students and researchers, including the brilliant Luis Alvarez. By mid-November 1940 the Radiation Lab was ongoing, and a month later 30 physicists were conducting research under the management of Lee DuBridge, another Lawrence protégé, and under the supervision of the NDRC and Loomis.

The Radiation Laboratory had a 6 January 1941 deadline to demonstrate its centimetric radar, and on 4 January the large prototype, consisting of a very large transmitter and receiver, was set up on a MIT roof. Radar beams were bounced off surrounding buildings in an unimpressive demonstration of the impractical unit. A 1 February deadline was set for an airborne demonstration, which meant the size of the set had to be significantly reduced. To achieve this goal the same antenna had to be used for transmission and reception of radar beams. Although the scientists developed a rudimentary duplexer that allowed a single antenna to transmit and receive beams, at the deadline there was no functional airborne radar ready for testing. The MIT researchers increased their efforts, and on 10 March a B-18 was fitted with a prototype centimetric set that was used for initial calibration tests, and on 27 March the first tests using the B-18 set successfully picked up surface shipping. The prototype set was sent to the Research Construction Company division of the Radiation Laboratory for construction of sets to be supplied to the Air Corps and Navy for refinement for operational use, as neither service had any operational ASV radar of its own. On 10 March the adapted British AI centimetric radar had successfully detected a flying aircraft.

Meanwhile, since the centimetric radar was in the slow developmental stage, U.S. Secretary of War Henry Stimson desperately wanted examples of the secret, and reportedly successful, long wave British MkII ASV radar, and ordered the Chief Signal Corps Officer to purchase sets, with the Army and Navy to split the cost on a 60-40 basis at $6,000 per set. In late February 1941 the MkII radar was shipped under guard from Canada, and reached the Army Signal Corps Aircraft Radio Laboratory (ARL) at Wright Field. However, these sets were not provided with the necessary antenna, as each had to be fabricated for the specific aircraft that was to carry the MkII. Also there were no spare parts supplied for maintenance, and these omissions delayed the American ASV radar program. By mid-April the Aircraft Radio Laboratory was ready to demonstrate its progress using a laboratory test version of the set that was installed in a B-18 at Wright Field. The B-18, with ARL technicians on board, flew to Lake Erie, successfully detecting ore freighters, and the tests seemed to be successful enough for the Signal Corps to procure 30 service test models from Western Electric. The AAC trumped the Signal Corps order with a production order for 500 sets on the Air Corps Test Board's supposition that further tests would be as successful. However, there was concern among the scientific community about the Test Board's conclusions about the tests, which had been conducted by skilled technicians, and particularly about the standardization of the British design to American production specifications. The British had great faith in their design, and were using it operationally, and the Americans had nothing else, and ordered a further 6,000 sets from Research Enterprises Ltd.—a Canadian Company—and another contract was licensed to Philco Radio Company. The production standardization and quantity manufacture did prove to be arduous as predicted, as the Metric measurements, external antenna layout for U.S. aircraft, wiring, British built components that had to be modified for American use, and defective oscilloscopes all caused delays in the program. Research Enterprises had promised initial delivery by December 1941, but by February 1942, the 1st Bomber Command had outfitted only four B-18s with the "Americanized" MkIIs, which were designated as the SCR-521. With the U-Boats devastating merchant shipping off American coasts, and the Navy unable to cope with the problem, the War Department was eager to have the Army Air Force and Signal Corps utilize the SCR-521 sets on a large scale operational basis as soon as possible. An investigation reported that the AAF did not have sufficient long range B-24s and B-17s available, and would have to manage with the B-18 medium bomber for the new anti-submarine campaign. Compounding the problem was the continued lack of necessary radar antennas and personnel to install them. The manufacture of antennas was delayed by the shortage of steel tubing and insulators, while the ARL had only one crew capable of installing the antennas, which required a day per aircraft. The investigation concluded that by mid-April the AAF would have 150 sets available, but only 33 were installed.

Although the long wave SCR-521 could dependably locate a surfaced U-Boat at six miles with a trained operator, the operational results were disappointing, as many operational contacts were false, usually being floating debris or high waves; nonetheless, the set saw extensive American ASW service through default, as nothing else was available. Later the Germans captured a MkII set from a downed RAF Wellington in Tunisia during the North African campaign, and soon the Germans developed the long range *Metox* (FuMB-1) radar detector. *Metox*, named after its French manufacturer, was quickly installed on U-Boats, and by September 1942 almost all U-Boats sailing from French ports into the Atlantic were equipped with it. The wooden, cross-shaped antenna, that had to be taken inside before diving, was the distinguishing feature of the unit, and was called the "Biscay Cross" by U-Boat crews. The *Metox* receiver was able to detect radar wavelengths between 1.25 and 2.5 meters, which was within the range of the 1.5 meter MkII radar beams of an approaching Allied patrol aircraft. It could detect these airborne radar signals at a distance up to 30 miles, which was well within the six to ten mile range of the SCR-521, and thus rendered this radar ineffective, as the U-Boat would identify an approaching attacker and dive long before it arrived.

With the disappointing operational performance, and then negating of long wave MkII ASV radar by *Metox*, the necessity for operational centimetric ASV radar became a priority. The strategic priority for centimeter radar sets before 1941 was for detection of aircraft by aircraft (AI radar), but after the Battle of Britain, Hitler's U-Boats, not aircraft, became the priority, as hundreds of merchant vessels were being sunk almost without Allied air intervention. By mid-1941, both the British at TRE and the Americans at MIT were developing ten centimeter AI radar, called AI-10. Studies of the two receiver and transmitter units found that the British AI receiver was better, while the American AI transmitter was better, and it was decided to use the best in a new system, which increased the range three fold. The new AI-10 sets were tested by ARL, and though

The AI-10 SCR-520 air interception radar set was the strategic priority in 1941, but when the U-Boats menaced America the SCR-520s were converted into ASV-10s—anti-submarine sets that were designated as the SCR-517A ("A" denoting a conversion set). (USAF)

very bulky at 600 pounds, were found to be very successful, and were designated as the SCR-520. The Signal Corps ordered several hundred sets to be built by Western Electric, and the first pre-production set was available in December 1941 for air-to-air testing. However, in early 1942 the priority had shifted from the aircraft threat of 1941 to the very real U-Boats operating off the U.S. coast. Now Gen. Arnold had SCR-520 sets available for air interception, but he needed an ASV radar unit better than the 6,000 long wave SCR-521s he also had on order from Canada and Philco. Fortunately, both the ARL and Radiation Laboratory had made great strides in ten centimeter radar research, and were able to convert the air-to-air SCR-520 AI sets into anti-submarine ASV-10s that were designated as the SCR-517A ("A" denoting a conversion set). The SCR-517A units employed a compact dish antenna that could be mounted entirely inside the nose of a bomber, and could be focused into a narrow beam that radiated from a rotating scanner able to cover the surface in a full 360 degree circle under the aircraft. The set presented much more distinct reflections, and experienced less interference from sea clutter. Ten SCR-520 sets were quickly modified and installed in B-18s at Wright Field for further testing as the SCR-517A. On 27 March 1942 a SCR-517A equipped B-18 flew over Block Island Sound to test its set against the submarine USS *Mackerel*. During this successful demonstration radar echoes were being received as far away as 19 miles. By the beginning of April four SCR-517A B-18s were stationed at Westover, MA, and six at Langley Field, VA, for further field testing by the First Sea Search Attack Group (1SSAG). The new radar was to be tested operationally along with MAD equipment against U-Boats. In the first three months of operational field testing, the Group's history reports that 11 contacts were made, resulting in six attacks where depth charges were dropped, with three reported as "possibly successful." On 1 April 1942 a B-18 of the Attack Group detected a surfaced U-Boat 11 miles away by radar, tracked it, and made a "definitely successful" attack. This claim and the first three have been denied, as postwar *Kriegsmarine* records do not show a U-Boat sunk in those areas on those dates. Because of the increased U-Boat activity in the south Caribbean, the Attack Group operated from Trinidad between 24 September and 16 October 1942 in an attempt to more extensively test the new radar operationally. The Group conducted patrols at a search rate of 50 square miles around the aircraft, and 12 contacts were reported, but only four resulted in a depth charge attack. The final report concluded that no U-Boats were sunk, but the Germans were unable to detect the centimetric radar, as they had been able to with the long wave SCR-521 sets.

Converting the SCR-520 sets to the SCR-517A configuration caused problems that arose from changing short range air interception (AI) radar to long range air to surface radar (ASV). It would not be until into the fall when these conversion problems would be solved, so that scratch built SCR-517s (designated with no "A") could be scheduled for manufacture. The major conversion difficulties were in changing the speed of rotation of the spinner, and the angle of rotation of the reflector, and work progressed very slowly, while the U-Boats continued to ravage Allied merchant shipping. Western Electric was contracted to supply ten sets in February 1942, and 20 for each of the next two months, but that schedule was not met, and was reset to 60 SCR-517 sets for the Navy in June, and 40 in July; 28 SCR-520 sets for the AAF in June, and 35 in August; and 60 SCR-520 sets for the British in July, 100 in August, and 40 in September. Obviously the Americans were beholding to the English for their development and sharing of the cavity magnetron. As the demands for ASV (SCR-517) and AI (SCR-520) radar shifted these allotments were changed several times during the summer of 1942, and production numbers and delivery dates suffered. To make the situation more difficult, the Navy decided that it needed 100 of the SCR-517 sets to install on its small subchaser vessels. To escape this subchaser allocation responsibly the Signal Corps said that the Air Force was to decide the allotment. Surprisingly, the AAF agreed to let the Navy have 100 SCR-520s to be converted to SCR-517As, provided that the AAF first receive its scheduled delivery of 100 SCR-517s and an additional 25-30 SCR-520 sets for AI service tests. Despite the reshuffling and agreements, neither the Army, Navy, nor British were receiving very many SCR-517As, nor -520s, because the spinner manufactured by General Electric was in short supply, and Philco had to be awarded a contract to manufacture additional spinners. At the end of June the AAF only had 11 aircraft equipped with the SCR-517A, but these did not have spinners, and so far the Signal Corps' 1942 target of 1,142 sets to be manufactured had reached only 203! In the meantime, improvements on the SCR-517A were being made, with the B (heavier due to the addition of beacon and IFF features) and the C (lightweight) versions being introduced, causing further delays. The lack of certain raw materials, such as mica, and components, such as switches, condensers, and potentiometers slowed production. Since Arnold and the AAF had ordered thousands (but fortunately only hundreds were manufactured) of the inferior long wave SCR-521 sets, they were stuck with them, and by the end of summer 1942 over 90 B-18s were equipped with either the SCR-521 or the SCR-517A, and many with a Leigh searchlight. Western Electric had to labor to convert the 102 available AI-10 (SCR-520) into the ASV-10 (SCR-517A), and it would not be until late summer that the conversions were completed, and SCR-517As could be built from scratch as SCR-517 (no A). In September, Arnold felt he could safely justify replacing the inferior SCR-521s with the SCR-517A version; fortunately, he was soon vindicated when the SCR-521 sets were compromised by German *Metox* detection. The SCR-520s would continue in production as AI equipment, with the possibility of conversion to -517As. By the end of September 1942, the Army and Navy had each received 100 SCR-517A sets, but 48 of the new SCR-517C sets had gone to the Army. Sixty-two of the Army's SCR-517A sets were installed in B-18Bs, while the remainder went as spares and for parts, or to maintenance and operational training schools. The failure of the delivery schedules caused problems, as aircraft remained out of ASW operations while parked in depots awaiting ASV sets for installation. The AAF was under the gun of Secretary of War Stimson and his advisor, Dr. Henry Bowles, who were impatient for the AAF bombers to get on with the war against the U-Boats. Ten SCR-517A equipped B-18s of the 1SSAG under Col. William Dolan had been operating out of Langley Field, VA, and Jacksonville, FL, and Gen. Arnold reported to Stimson that he was "enthusiastic over the possibilities of this equipment."

The success of Dolan's ASV B-18s led to the establishment of the Sea Search Attack Development Unit (SADU), which was to be the operational testing unit for new types of ASW weapons (the SADU and 1SSAG will be discussed in depth later).

So by the fall of 1942 Secretary Stimson's avocation of ASV radar for patrol aircraft was finally coming to fruition, and had proved to be the most effective means of combating the U-Boat menace. But the days of the stopgap ASV equipped B-18A were numbered. In September 1942 the first two B-24s with their longer range, especially with auxiliary fuel tanks, were equipped with the SCR-517, and in December the 1SSAG had its first squadron of SCR-517 B-24s to complement its aging B-18s.

The SCR-517A radar apparatus included a rotating three foot diameter concave dish type antenna mounted behind a bulbous egg shaped dome, which replaced the shark nosed Plexiglas windows covering the bombardier's compartment in the B-18A. The spinner was the rapidly rotating and highly accurate antenna mechanism that allowed a narrow radar microwave beam to scan a wide area. The antenna was connected to the antenna junction box (AB-874) by two braided connections. The junction box was connected by three braided connections to the radar control unit (MD-7/ARC-5), which was located just adjacent to the dome enclosure on the port side just below the window. The two knob control box was situated on top of a 28 volt DC unit. The left knob controlled: ON> STB> NORM> CNR> MAP> TEST, while the right knob controlled antenna tilt to +10 degrees UP and -15 degrees DOWN. The BC-1040 radar transmitter resembled two black stove pipes, and was located in the nose compartment on the starboard fuselage wall. The Signal Corps BC-1101-A Receiver (indicator) was placed on top of the Western Electric BC-1043-B Synchronizer in the radio operator's position in the cabin.

SCR-517A Radar Gallery

Radar antenna and junction box located in the nose of a B-18B. (Pima)

Radar and antenna tilt controls. (Pima)

Radar transmitter unit. (Pima)

Radar receiver (above), with PPI screen and synchronizer (below). (Pima)

Training determined that the chief operational deficiency of the early SCR-520 and -517A sets was that they used the early Range Azimuth screen (A-scope) of the MkII sets, rather than the Plan Position Indicator (PPI) screen. The PPI scope was divided into compass headings, with a "lubber line" extending from the center of the circle (the true heading of the aircraft) toward the target, along with a rotating time base that gave a blip out from the center. The PPI screen presented the terrain below much like that on a TV screen, and so was much easier to use. The PPI could be adjusted in range to enlarge the image of the area below on the scope, and to bring out the light and shadows on the scope that could represent a U-Boat. As the bomber closed on the target, the blurred images on the scope slowly moved nearer the center of the scope, which constantly represented the position of the bomber. Unlike the A-scope, the PPI scope could plot the position of the target it was homing, and track it until the bomber was directly over the target. Using the PPI, the radar operator could direct the pilot on an attack track, so that he could drop the depth charges at intervals that would have a greater chance of doing damage. In night patrols, the radar operator would not know the nature of the target (native fishing boat, merchantman, or U-Boat) until the Leigh searchlight was switched on at a mile or less. The pilot had to make a quick target ID to decide if he should drop the depth charges immediately; if he was too late, or uncertain of the ID, he had to switch off the searchlight and turn at once to circle quickly to make an attack, but there was a good chance of the U-Boat escaping in a crash dive.

Early tests determined that under ideal conditions ASV radar could detect individual U-Boat-sized targets at 25 miles, and a group of ships at even greater distances. The range of ASV radar was then acknowledged to be 25 miles in every direction, and capable of sweeping an area of 50 square miles around the aircraft. The 1SSAG used these range assumptions as the basis to train combat crews in the tactical use of the new equipment for the ASW patrol concept in the Caribbean. However, operational conditions were not always ideal, and results mostly depended on the skill of the operator.

In operational training missions flown by the 2nd Anti-submarine Squadron—the first ASW unit in the UK—the new ASV radar proved to be difficult to use, and an initial report found a 55% failure rate due to poor maintenance. 2nd Squadron CO Lt.Col. Jack Roberts did not believe the high maintenance failure rate, and defended his maintenance personnel by ordering another study. But this new study found the sets operated satisfactorily only 63% of the time, with 42% due to electrical equipment failures (mainly defective vacuum tubes, 65%), 33% due to "undetermined" causes, and only 5% due to maintenance deficiencies. However, it appears that the majority of the 33% "undetermined" category actually fell into the maintenance deficiency category.

In August 1942 the U-Boat war had reached a perilous phase, and the Air Force began an ASAP program to convert 122 B-18As to the B model with the ASV radar conversion configuration. The B-18As were ferried to San Antonio for the conversion, which removed the glass in the upper bombardier's position and replaced it with a bulbous ASV radome, as well as the installation of a rudimentary Metallic Anomaly Detector (MAD) boom located aft of the tail. Radar equipped B-18Bs did not appear until October 1942, when the four old B-18s of the 35BS, which were nearly worn out from flying exhaustive patrols off the east coast, were replaced by four radar equipped B-18Bs that were immediately sent to the Caribbean. Here the squadron was to suffer many teething problems with the newly installed radar. By October 1942, three ASV B-18Bs were being delivered per week to the Caribbean. By December 1942, there were 45 B-18Bs and nine old B-18s in the Antilles, and only three B-18s and two B-18As in the CZ. Some 43 B-18Bs served with the 6AF and Antilles Air Command, and probably even more with the assorted ASRONs of the Anti-submarine Command.

ASV radar was obviously the foremost and continual obsession of the ASW combatants, as it was the essential factor for the success of the campaign. Training of operators and technicians was a major problem, as was the shipment of parts, particularly because the equipment was constantly being improved and changed. By early 1943, well tuned radar equipment controlled by experienced and skilled operators could identify landmarks at 100 miles, and even buoys at 35 to 40 miles. A surfaced U-Boat could be distinguished at more than 50 miles, and the conning tower of a U-Boat running with its decks awash at 15 to 30 miles. But ASV radar continued to have its limitations, as the equipment was not reliable, and required constant and expert maintenance.

Depth Charges (DC)

In their early attacks on U-Boats, the RAF and Coastal Command found that their 500 pound anti-submarine aerial bombs originally developed in World War I were often more dangerous to the user than to the enemy, as they had to be dropped at low speeds and altitudes, and an inaccurate explosion could bring down the attacker. In 1940, the British developed an improvised aerial depth charge based on the well-proven standard Mark VII 450 pound drum-shaped naval depth charge. It was fitted with a conical fairing on one end, and fins on the other to stabilize it as it fell from the aircraft, but the weight and shape of these depth charges restricted their use. This device used the reliable hydrostatic pistol, which would detonate at a preset water pressure (depth), and prevented it from detonating in the air if it bounced off the ocean's surface. These new aerial depth charges proved to be ineffective, and improvements were needed, as during almost two years of war (September 1939-June 1941) Coastal Command calculated that a paltry one percent of attacked U-Boats had been sunk. Attacking aircrews were disgruntled when U-Boats were seen to be bracketed by depth charges; only to emerge from the spray of the explosions unscathed.

In early 1941 the British had created the Operational Research Section of the Coastal Command, headed by Dr. Patrick Blackett and his assistant, Dr. E.J. Williams. One of their first projects was to analyze attacks on U-Boats by studying attack records and photographs. By the summer of 1941, the Research Section had determined that an aircraft's best chance of causing significant damage to a U-Boat was to depth charge it while it was on the surface, or at the latest when it was in the first 15 seconds of its dive. There were two methods to improve a depth charge that had to be of a given size and weight so that it could be accommodated inside an aircraft. The first was to use a more powerful explosive, and the second was to have the depth charge detonate closer to the U-Boat. The early Coastal Command depth charges were filled with Amatol, but by the end of April 1942, Mark XI depth charges were filled with Torpex, a mixture of RDX, TNT, and aluminum, which was 30 to 50% more powerful than Amatol-filled types, and became the British depth charge explosive of choice. The preset hydrostatic pistols detonated at 100 to 150 feet, which was the hypothetical average depth the U-Boat would reach during a crash dive initiated at the hypothetical average distance at which the attacking aircraft was first seen by the U-Boat. Since the aircraft dropped the depth charges on the eddy of diving bubbles, the U-Boat would actually be too far ahead of this hypothetical position to be damaged. The only U-Boats likely to be attacked successfully were those caught on the surface, or just initiating their dive, but these U-Boats escaped because the depth charges were set too deeply at 100-150 feet. The apparent remedy was to preset the hydrostatic pistols at a more realistic depth using a depth charge that sank slowly, so it would explode at the desired depth. The ideal depth for detonation was determined to be 25 feet, but the existing hydrostatic pistols had been designed for shipboard use, where the minimum safe detonation depth was 50 feet. A new shallow firing device needed to be developed, and the Mark VIII detonator was introduced in Spring 1942. However, it had a minimum depth setting of 34 feet, which was not shallow enough to contend with a surfaced U-Boat, considering that the Torpex-filled depth charge's lethal range was 19 feet. Also, a dropped depth charge had a tendency to form a coating of bubbles upon impact that delayed the action of water pressure on the pistol. By July 1942 the Mark XI depth charge had been adapted to use the Mark XIII Star pistol, a break-away tail, and concave nose spoiler that allowed the depth charge to detonate at 15 to 25 feet.

In August 1943, the American aerial depth charge arsenal consisted of 14 Marks that were derived from two basic types: the 325 pound Mark 17, and the 650 pound Mark 29. The Mark 17 design yielded the Marks 41, 44, 47, 53, 54, and 75, while the Mark 29 design yielded the Marks 35, 37, 38, 48, 49, and 71. The first operational depth charge was the Mark 17, which became available in quantity in April 1942. The round nosed Mark 17 was 15 inches in diameter and 52.5 inches long, and weighed 325 pounds, including

234 pounds of TNT (the Mark 44 was a Torpex-filled version), which gave it a lethal range of approximately 17 feet from the U-Boat's pressure hull. The 1SSAG helped to test this depth charge, which initially had a tendency to skip or ricochet due to its rounded nose. A flat nosed attachment was added to cure this problem in the TNT-filled Mark 41 and the later Mark 47, which was filled with Torpex. The TNT-filled Mark 53, armed with a hydrostatic fuse, was introduced late in the war, and was followed by the similar Torpex-filled Mark 54, which remained in service for 30 years (the Mark 75 was a thicker cased limited version). The Mark 29, introduced in May 1942, was the other main U.S. depth charge. It weighed 650 pounds, was filled with approximately 425-450 pounds of TNT, and measured 18 inches in diameter and 67 inches in length. This Mark was plagued by a weak tail, unstable underwater course, and its round nose also caused surface skipping. This surface skipping could detonate the explosive charge and endanger the attacking aircraft due to its larger charge and resulting explosion, but the problem was also remedied by the attachment of a flattened nose. The TNT-filled Marks 35 (no data), 37 (new tail), and 38 (shorter at 61 inches, with more explosive at 425 pounds of TNT) were versions of the basic Mark 29, which remained as the main 650 pound type depth charge. The Marks 48 (enlarged, 18.6 x 68 inch Mark 29 weighing 850 pounds), 49 (a Mark 38 filled with 472 pounds of Torpex), and 71 (a modified 525 pound M64GP bomb version armed with hydrostatic fuse) were later 650 pound type depth charges.

When the First Bomber Command initially investigated the bombing of submarines, it believed that a bombsight was not necessary for optimal attacks from 50 to 100 feet. But tests showed that the average range of error using this dead reckoning type of attack was 175 feet. When the standard round nosed depth charges were found to be erratic, both the Army and Nay pushed to develop a low level ASW bombsight. By the summer of 1943 several efficient bombsights were developed, and with the more accurate flat nosed depth charges produced improved results.

Because the British were at war against the U-Boat since September 1939, the Coastal Command had completed extensive studies on formulating an anti-submarine doctrine in relation to depth charges. In July 1942 it had prescribed that its policy was "to concentrate efforts on sinking those U-Boats which are still on or near the surface." A "stick" of six dropped depth charges spaced at 36 foot intervals yielded a hypothetical "lethal area" of approximately 20 feet wide by 220 feet in length. The Type VIII U-Boat was 220 feet long, and the Type IX was 254 feet long. The lethal area of each depth charge was the diameter of a circle (20 feet) around its explosion, and the lethal area of a stick of depth charges would vary with the interval (spread) of the charges; the greater the interval, the less overlapping of explosive charges. This depth charge interval was to allow for the average range error that occurred when a pilot was attempting to maneuver his aircraft at 150 mph or more while descending to 50 feet across a moving target. To further complicate matters, while aiming the plane and preparing to release the depth charges, the pilot and bombardier had to compensate for the forward movement of the U-Boat, and for the trajectory of the depth charges as they fell through the air and then into the sea, each having a different density that affected their speed. With experience, the initial 36 foot interval was later lengthened to 60 feet, which increased the lethal area to 20 by 340 feet, and gave the pilot more much needed leeway in adjusting for range error. There are two types of error when making an attack: range error and line error. Typically range error (150 to 180 feet) was two to three times that of line error (60 to 90 feet). Coastal Command attack policy also recommended that the route of attack should be the shortest possible, even if the angle of attack would be more difficult, as it was much better to attack while the U-Boat was still on the surface. Depth charges were to be dropped from 50 feet, not only for a more accurate attack, but also to preclude the depth charges from breaking up when hitting the surface when dropped from higher altitudes. Depth charges dropped from 50 feet took less than two seconds to hit the water's surface, and then two to three seconds to reach 25 feet and explode. During this time it was determined that the forward movement of the depth charges was about 40 feet from their point of release using the conning tower as the center of the drop by the attacking aircraft. The speed of a crash diving U-Boat was about ten feet per second. A U-Boat that submerged entirely, with its conning tower leaving a swirl on the surface, would have moved forward about 150 feet beyond the head of the swirl in 15 seconds, which was the maximum period of submergence for an effective attack. First the pilot would have to initiate an immediate attack, hoping to catch the U-Boat on the surface, or in the preliminary act of a crash dive, and then he had to make an estimate of the U-Boat's movement when calculating the time to drop the depth charges. There was no accurate low level bombsight available, and attacks on U-Boats were more of an art than science, and constant practice was required to become proficient.

The AAFAC Anti-submarine SOP Manual of 29 July 1943 began by noting that attacks on a U-Boat were very uncommon, but reassured the aircrews that their efforts were successful, nonetheless, because their presence overhead forced the U-Boats to stay submerged, and reduced their operational effectiveness. The American anti-submarine attack directives issued for patrolling aircraft followed the British methods. Ideally, the patrol aircraft was to patrol at 1,000 feet, as flying above that altitude was ineffective, because the aircraft could not dive quickly enough to the 50 foot attack level before the U-Boat submerged and escaped. Surprise was essential to catch the U-Boat in time to drop a stick of depth charges near enough to be lethal, and an attack via the most direct route was recommended, even if it was not the most efficient. Attacks on U-Boats that had submerged more than 15 seconds were deemed useless. With experience and new types of depth charges the American procedure of spacing dropped depth charges differed from the British method. The American 325 pound-type depth charge was dropped at 50 foot intervals, while the more powerful 650 pound type was dropped at 70 foot intervals. The Americans felt that patrols over five hours were unproductive (the Coastal Command conducted much longer patrols), and that 30 minutes was the radar watch duration limit (vs. 45 minutes in the Coastal Command).

Initially, American ASW patrol aircraft were inadequate. The Navy's PBY Catalina could fly 800 miles, but had a limited bomb capacity, and flew so slowly that the U-Boat could easily crash dive before the lumbering amphibian could reach it. The AAC's aircraft of expediency was the B-18 medium bomber, which had only barely adequate range, speed, and bomb capacity, but was all that was available until the B-24 could be produced in sufficient numbers to meet all its operational obligations. The four engine Liberator in its long range configuration (LR) could patrol to 1,800 miles with ten to twelve depth charges, and its very long range version (VLR) could do 2,400 miles with six or eight. Also, its larger crew (ten) allowed for watch rotation, which meant shorter and more efficient visual searches.

Tactical Response to the U-Boat

In responding to the U-Boat offensive, the U.S. developed and implemented tactics that took advantage of the U-Boat's need to surface daily, usually at night, when it was necessary for the U-Boat to recharge its batteries (usually requiring four hours for a full charge), ventilate the boat, and to allow the crew some fresh air time. The U-Boat only had a limited submerged battery speed of four knots, and a surface diesel speed of 15-18 knots. Because their surfaced speed was five knots faster than most merchant vessels, U-Boats had to maneuver on the surface to intercept a convoy, and then fired their torpedoes or deck guns at a target,and only submerged to escape pursuing surface vessels, which soon had to return to escort the convoy. After its escape the U-Boat would surface and race ahead to make another attack on the convoy. The use of round-the-clock aerial patrols caused the U-Boats to dive frequently and remain submerged for long intervals, and prevented them from catching up to the convoys. When the air patrols extended into the night, the two to four hour battery recharging procedure became dangerous for the U-Boat from nine days before and five days after the full moon. When patrol aircraft were fitted with ASV radar recharging became even more dangerous, and had to be accomplished even further offshore.

When an aircraft was using radar to detect a surfaced U-Boat that was using its *Metox* radar warning set to detect aircraft radar; the two opposing radar operators were confronted with different liabilities for detections and false alarms. If the aircraft radar operator made a contact he could ask the pilot to investigate without misgivings, because if the contact was a false alarm the aircraft would continue its patrol. However, if the U-Boat *Metox* signaled an alarm its operator had to risk an air attack, weighing the option of a crash dive and stopping recharging of the batteries to the option of a false alarm, and remaining on the surface to recharge the batteries and continue on a quicker and more comfortable voyage. When centimetric radar became available, negating *Metox*, the U-Boats spent much more time submerged. Three general types of anti-submarine procedures were used:

1) Routine aerial patrol of areas in which U-Boats were thought to be present.
2) Aerial escort of convoys within range of air coverage.
3) Intensive and unrelenting patrol of an area that had a reported U-Boat sighting. The AAF termed this operation "killer hunt," while the Navy used the term "hunter-killer."

Early in the war the AAF utilized the aerial patrol to contain and obstruct U-Boat operations in areas where they were known to be operating. During this phase air patrols had to be flown using accurate navigation, and dependable communications that were often neither accurate nor dependable. The air crews had to be able to quickly and accurately identify surfaced vessels so as not to attack friendly vessels. It was to be assumed any submerged vessel was a U-Boat, as Allied submarines did not operate in the Caribbean. In early 1942 Capt. C.D. Meadowcroft of the 90BS out of Zandery Field, Surinam, was flying a patrol to search for a reported U-Boat when the crew spotted a dark shape moving slowly beneath the surface. Meadowcroft turned and dove into the attack, dropping four depth charges in a perfect pattern. As the crew watched for a result they saw an oil slick come to the surface, but closer inspection identified it as the mangled remains of a whale. There were at least three confirmed "whale kills" in the Caribbean, but how many other whale oil slicks seen after a depth charge attack were reported as a possible U-Boat sinking? In reality, the large majority of patrols were routine, and thousands of hours were flown looking over an empty ocean, especially after the summer of 1942, when the Germans withdrew most of their U-Boats from American waters.

Aerial convoy escort, while universally loathed by aircrews because of its defensive nature, became essential to keep the U-Boats at bay, especially after the establishment of the coastal convoy system, which initially ran from Chesapeake Bay to Key West, and before long extended into the Caribbean and Latin America. Later, the Air Force favored the offensive killer-hunt search operations using their newly developed anti-submarine tactics and equipment. These patrols extended hundreds of miles from the coast in an attempt to keep reported U-Boats submerged and on the defensive, away from the convoys. Accurate navigation, dependable communications, and constant vigilance by either human eye or radar was again required, as well as the ability to make quick and decisive attacks when the rare U-Boat was encountered. With experience and time navigation, communications, and radar equipment and their use all improved. If the patrolling Army aircraft was unsuccessful in its attack it would radio the position of the U-Boat, not only to other AAF units, but also to the Navy, which would post warships and aircraft to the area to maintain contact, and continue its unremitting hunter-killer operation. The offensive killer-hunt and hunter-killer operations required large numbers of aircraft and shipping, which had to be taken from defensive convoy escort and patrols, and so it was not used extensively until mid-1943, when patrolling aircraft off escort carriers very successfully employed this tactic in conjunction with the Ultra code breaking U-Boat intelligence. The Navy killer-hunt and the Army Air Force hunter-killer systems required close collaboration of the two armed forces, which was difficult due to the unfortunate inter-service rivalry that had existed early in the war, and continued on to a lessening degree as the war progressed. The Navy had control over the American anti-submarine organization, and thus the use of the AAF's land-based patrol aircraft. To Army aircrews' chagrin, the Navy had employed AAF aircraft on everlasting and fruitless search and convoy patrols off the eastern U.S. coast, and then over the Caribbean. The important story of the Army/Navy contest for administrative control will be discussed at length.

Typical Anti-submarine Patrol

After a predawn breakfast a briefing was held by the Operations Officer, who discussed navigation charts, radio codes, visual signals, recognition signals and their expiration, Allied shipping data, and reports of U-Boat activity. There were map and information boards located on the side of the room with the positions of surface vessels, and the possible positions of U-Boats, the positions of Army and Navy aircraft flying missions, and those about to fly missions. The Met Officer discussed weather, cloud conditions, and weather signals. Radio silence was to be maintained, except for unusual circumstances, and then it was to be broken only in code.

The B-18 ASW patrol crew consisted of a pilot, co-pilot, bombardier, radar operator (if equipped), navigator (often absent, as there was a shortage), and crew chief, who also acted as the gunner. Each of the crew was dressed in coveralls, Mae West life vest, and carried a pistol, knife, and a canteen. The parachutes and emergency supplies were packed in the seat. The aircraft was equipped for a ditching with two rubber life rafts packed with emergency rations, an emergency radio, and automatic SOS transmitters, and one of the rafts had a five gallon wooden keg attached. The bomber was also outfitted with a Tommy Gun, hatchet, and smoke bombs, which were useful to mark the position of a diving U-Boat, or as an aid in navigation to check drift readings. If a crew were forced into a jungle landing or bailout their chances of survival were slim. There is a story of a pilot who landed in the jungle only 15 minutes from Zandery Field, Dutch Guiana, and it took a search party four days to reach him through the impenetrable jungle. Also, the natives on Trinidad disliked both the British and Americans, and were uncooperative in giving aid.

After the pre-flight check the engines were started, and the control tower flashed a green light to clear the aircraft to taxi to the end of the runway. The tower again flashed a green light signifying clearance for take off, and the pudgy bomber lumbered down the wet runway, which was constantly drenched by tropical rains. At 80mph the Bolo would lift off and begin to climb until it reached normal patrol cruising altitude of 1,000 feet, and a cruising speed of 135mph. At sunrise the pilot would trim the Sperry autopilot, turn on the automatic flight controls, and he and the crew began to visually search the ocean surface for U-Boats. In aircraft without radar these searches were futile, as U-Boat captains would usually submerge at dawn to conning tower or periscope depth, where detection was difficult in the white caps that covered the U-Boat's diminished wake. While few U-Boats were detected during these daily patrols, the threat of a possible attack kept the U-Boats submerged, and less likely to make daylight attacks. Once the B-18s were equipped with ASV radar the search was easier and more productive, as the B-18 could detect the U-Boat before being detected itself, and the U-Boats then weren't safe day or night.

The Caribbean was notorious for its poor weather, especially during the rainy season, and the air crews became accustomed to being wet, or at least damp, while on the ground, and during a mission the pilots were chronically wet, because the aging B-18's windshield seals and side windows leaked relentlessly, not only in the rain, but also in cloud banks. Many patrols were flown under the solid cloud cover, and were often hampered by rain squalls. The rains were heaviest during the morning searches, but afternoon thunder storms were frequent when the pilots were on the return leg back to base. In this poor weather U-Boat captains would frequently surface, as visibility was restricted, and lookouts and listening devices would give them enough warning of a slowly approaching bomber, so that the U-Boat could dive before the bomber was overhead, where the U-Boat would become visible. When a U-Boat was spotted the pilot immediately put the bomber into a full throttle power dive, but usually it arrived over the bubbling wake of a U-Boat's crash dive, at which point the depth charges were often released through frustration, with little hope of result. A tactic called "baiting" was used when a U-Boat escaped. The attacking aircraft would leave the area for an hour or so and then return, hopeful that the enemy may have resurfaced and would be vulnerable to attack.

The ASW patrol crew consisted of a pilot, co-pilot, bombardier, radio operator, and crew chief, who also acted as the gunner. This crew is from the 1st Observation Squadron, which flew patrols of the Atlantic side of Panama. (USAF)

After three hours the patrol had flown for about 400 miles, and it was time to change the heading to take the aircraft back toward a familiar landfall near its base. The boredom and constant strain of staring at the ocean's surface was the most difficult part of a patrol. If something unusual was spotted the pilot was informed, and he would turn to investigate. Mostly the sightings were oil drums and garbage, as merchant crews in those days were notorious for using the ocean as a dump. While searching for U-Boats, all sightings of Allied shipping were recorded, along with debris, such as rafts, oil slicks, and flotsam that were indications of a successful U-Boat attack. The prevailing easterly Caribbean winds often carried debris across the entire Caribbean Sea from the many shipping losses in the Puerto Rico and Lesser Antilles area. Again, ASV radar relieved the crew of much of the stress and tedium of visual U-Boat search.

The final leg of the return trip was the most anticipated and hazardous part of the patrol. Making landfall at a predetermined point, even with a navigator, was problematical, and the afternoon heavy rain squalls and low cloud layers made visual identification of landmarks difficult, and radio navigational aids of the day were not reliable. To complicate matters, base anti-aircraft gunners, especially at the start of the war, were instructed to fire at any aircraft not approaching from the correct direction. Amazingly, the 45BS did not lose one aircraft to navigational error in over a 1,000 patrols, and lost only one aircraft overall, after it ditched due to the loss of one engine. This crew was saved, and the bomber was towed to Colon, Panama, and cannibalized; another example of the uncanny floatability of the B-18.

After the patrol, the crew relieved themselves and grabbed some coffee, and was debriefed by the Squadron Operations Officer and Intelligence Officer. The debriefing dialog and Flight Log were immediately sent to Navy Intelligence, who used the information to update Allied shipping data, and rarely update U-Boat sightings.

For the typical night patrol, the crew spent the morning sleeping, woke, relaxed, then ate an early dinner, and assembled at 1700 in the Operations Room for briefing similar to that of the daytime patrol. The mission began just before darkness at about 1800, and soon settled into the familiar routine of flying a triangular point-to-point-and-return course; trying to find a U-Boat before it could attack shipping, or at least keep it submerged. Early war night offensive air operations were effective only when the moon provided some modicum of visibility, so on nights of poor visibility the patrols usually were relegated to convoy coverage. One advantage in patrolling tropical waters was that they were highly phosphorescent, and surfaced U-Boats chanced having their sparkling wakes spotted. For four to six hours the crew looked down on the water and mostly saw nothing, and when they did, it was usually a native vessel, rarely a whale, or just as rarely a U-Boat. Once ASV radar became standard offensive night patrols were routinely flown, but the norm continued to be hours, days, and weeks of fruitless searches, which frustrated the crews, and made them doubt their contribution to winning the war. The unexpressed satisfaction of these seemingly unproductive missions was that perhaps they kept a U-Boat or two submerged and away from the convoys. These hours on patrol gave many pilots and crews valuable experience when they later transferred to other bombers. Many of the pilots and crews of the 45BS, in particular, and other ASW squadrons would later train in and fly the B-29 that would definitely contribute to winning the war.

The American ASW Crisis of December 1941: B-18 Finds its Niche

At the time of Pearl Harbor, the war in Europe and the Battle of the Atlantic—the life and death struggle between the convoys supplying England and the German U-Boat—had been going on for over two years. Although the British had endured the *Luftwaffe* aerial onslaught during the Battle of Britain, causing the Germans to abandon *Sea Lion* (their invasion of Britain), the future, nevertheless, remained ominous. In June 1941 the Nazis turned east to attack Russia, and made swift, huge gains in territory, annihilating all Soviet opposition in its path. The U-Boat blockade of the United Kingdom was sinking more shipping than was being constructed, and the island nation faced starvation of not only food, but military supplies and equipment, and it was possible that Great Britain would not have the means to continue the war. When war came, America was ill prepared to confront the German U-Boats with either trained men or equipment, much less a comprehensive anti-submarine strategy. In November 1941 the Coast Guard was transferred to the Navy,

Regardless of the type of patrol—day or night—surprise and a speedy attack were crucial, as once a U-Boat was sighted, either visually or by radar, a successful attack had to occur in no more than 15 to 30 seconds after a U-Boat began to submerge. The pilot could use cloud cover or the sun behind him to achieve surprise after a sighting. The attack was to be made at 15 to 45 degrees, as low as possible, preferably at 50 feet, and then the pilot had to drop the depth charges within 20 feet of the U-Boat's pressure hull. (USAF)

and in December Roosevelt took a major step by appointing Adm. Ernest King as C-in-C of the U.S. Fleet. King, while capable, was considered curt, undiplomatic, and stubborn; so much so, his demeanor prompted Eisenhower to write in his diary that "if someone shot King it might help win the war." After the U.S. entered the war King was also named the Chief of Naval Operations (CNO), which gave him both the authority and the resources to administer U.S. anti-submarine warfare in the Battle of the Atlantic. Included in his command was the American Theater, which incorporated the North and South American Continents and the oceans around these continents to the mid-Atlantic and mid-Pacific, not including Greenland and Alaska.

Early in the Battle of the Atlantic, the British were quick to appreciate the necessity for close sea and air coordination between their anti-submarine forces, not only at the administrative, but also at the operational levels of command. Historically, the American Navy and Army had disagreed over their respective roles in air operations. Immediately after the Pearl Harbor attack a crisis arose in America's anti-submarine capabilities that can be traced to the 1935 Joint Action of the Army and Navy Agreement. This Agreement gave the Navy responsibility for "all inshore and offshore patrol for the purpose of protecting and defending the coastal frontiers," whereas the Army "held primary responsibility for the defense of the coast itself." The Agreement charged the Navy with responsibility for sea patrols and the protection of shipping, but did not answer the question of the Navy's command of all air operations in coastal defense, or if it should only control those air operations in support of the fleet. The Navy assumed it had "preeminence" in coastal defense and, of course, the Army saw this as a threat in which the Navy could gain control of Army air forces. Therefore, the Navy and Army continued to have confused and overlapping responsibilities that would hinder ASW operations for far too long. By 1941 this interservice debate had not been resolved, but for all intents joint air action was under the jurisdiction of the North Atlantic Naval Coastal Frontier (NANCF), and so effectively the Navy was responsible for coastal patrol and shipping, and thus anti-submarine defenses. Meanwhile, the AAC was essentially relegated to a support role, and the training that was essential to anti-submarine functions was neglected, as the AAC became more concerned with its enormous pre-war expansion and its primary mission, strategic bombing. For administrative purposes the NANCF would soon be incorporated into the Eastern Sea Frontier.

By July 1941, the Navy had divided its existing naval districts into Sea Frontiers:

Eastern Sea Frontier extended from the Canadian border to northern Florida.

Gulf Sea Frontier extended from the northern Florida Atlantic coast east to the northern Bahamas, south to the eastern half of Cuba, and also included the Gulf of Mexico as far south as the Guatemalan/Mexican border.

Caribbean Sea Frontier included the remainder of the Caribbean and the northeast coast of South America, to become the largest anti-submarine command in the world, measuring 1,000 miles north-to-south and 2,500 miles from northwest to southeast, and totaling 2.5 million square miles. It was initially divided into three Sectors: Puerto Rico, Panama, and Trinidad

The future U-Boat threat was not the only reason for the establishment of the Sea Frontiers, as there was a large German population of over a million in both Argentina and Brazil that had a substantial political and economic influence in South and Central America. A portion of the exiled French navy, including an aircraft carrier, had been interned on pro-Nazi Vichy island of Martinique, and a large number of Lufthansa airliners that could possibly be used as bombers against the Panama Canal were confined on a number of South American airports.

The Navy and AAC were directed to pool their insufficient assets for ASW patrols, and on 8 December 1941 the Navy requested that the Air Corps' 1st Air Support Command (1ASC) and the 1st Bomber Command (1BC) of the 1st Air Force (1AF) begin patrols on the Eastern Sea Frontier. The bases of these 1AF units were Westover Field, Chicopee Falls, MA; Langley Field, VA; and Mitchel Field, Hempsted, Long Island, NY. The 1ASC sent its observation and pursuit aircraft to patrol out to 40 miles offshore from Portland, ME, south to Wilmington, NC. At the time the 1ASC had

(USN)

no more than ten aircraft in service to conduct these patrols. The 1BC was equipped mainly with B-18 bombers, and fewer B-25s that flew to 300 miles offshore, while its very few B-17s flew patrol up to 600 miles offshore. Initially, it was considered a good day when three aircraft each from Westover and Mitchel Fields could fly patrols. The first patrols were flown by unarmed planes, or planes with bombs, rather than depth charges, and there were no night flights until the installation of some rudimentary ASV radar sets in March. The aircrews were not sufficiently trained in navigation or ship recognition, which was a problem, as the search was done visually. On 29 December 1941 a bomber crew mistakenly identified a Navy destroyer near Newport, RI, as a U-Boat, and dropped four bombs on it; fortunately, the crew lacked the fundamentals of U-Boat attack tactics, and the bombs exploded harmlessly. On 31 December 1941 the AAC and Navy formed the Joint Control and Information Center in New York City, which would track the movements of merchant shipping, plot and track U-Boat contacts, and determine the locations of all surface and air anti-submarine patrols.

It soon became apparent that an effective American anti-submarine campaign would necessitate more aircraft than the Navy could provide, and that Navy seaplane or carrier-based aircraft were not as effective as ASW aircraft, nor as AAC land-based aircraft with their better speed, range, and armament. R.Adm. John Towers, Chief of the Navy's Bureau of Aeronautics, requested that 200 B-24s and 900 B-25s and B-26s from future production be transferred from the AAC allotments to the Navy for anti-submarine duties. Of course, this bold request antagonized the AAC, which was protecting its planned strategic bombing campaigns, into declaring that there could be no heavy or medium bombers available to the Navy.

In January 1942, with the Pacific Fleet destroyed at Pearl Harbor, Adm. Chester Nimitz, C-in-C in the Pacific, was given the greater priority for American naval forces, as the Japanese were considered a direct threat to America, rather than Nazi Germany. German pocket battleships and cruisers had been eliminated as a threat by the Royal Navy, allowing Adm. King to dispatch portions of the Atlantic Fleet to the Pacific, as only a small number of U-Boats were operating off the eastern American shore. The Eastern Sea Frontier under Adm. Adolphus Andrews was responsible for the defense of 1,500 miles of the U.S. East Coast from the Canadian border to the border of the Carolinas. Andrews' meager defense force consisted of 20 vessels: the largest was a 165 foot Coast Guard cutter, and none was faster than a surfaced U-Boat. Andrews' air force was made up of 103 aircraft that were generally obsolete. Most of the Navy's Catalinas had been directed to the Pacific coast to search for Japanese surface vessels, and Andrews had little hope of receiving new Catalinas coming off the Douglas production lines, as they were either being sent to the Pacific, or to Britain to aid the RAF Coastal Command. The possessive AAC Bomber Command was reluctant to divert aircraft from expected strategic bombing duties to an anti-submarine function, and the AAC aircraft most suited for ASW duties (the B-17, B-24, and B-25) were not available in large numbers, and the B-18 was chosen as the ASW bomber by default.

Because the Navy lacked anti-submarine aircraft the Air Corps inherited ASW air duties, and in March 1942 Gen. Arnold approached Adm. King with the logical proposal to establish an AAC organization that would control all air anti-submarine operations. Of course, King and the Navy did not accept this brash proposal, because it would give the AAC a long established Navy responsibly, and even place naval vessels under Army control. To placate Arnold, on 26 March King centralized the ACC's anti-submarine responsibility by transferring the 1st Bomber Command to the Eastern Sea Frontier. This move made sense, as the 1BC controlled most of the Air Corp's aircraft involved in anti-submarine operations, and it was the only command capable of administering such an undertaking. However, the 1BC wished to maintain its standing as a bombardment force, so to carry out its new responsibilities as an anti-submarine force it had to divide its tactical and training en-

Adm. Adolphus Andrews administered the Eastern Sea Frontier, which was responsible for the defense of 1,500 miles of the U.S. East Coast, from the Canadian border to the border of the Carolinas. Andrews' meager defense force consisted of 20 vessels; the largest was a 165 foot Coast Guard cutter, and none was faster than a surfaced U-Boat. Andrews' air force was made up of 103 aircraft that were generally obsolete. (USN)

deavors into two parts: one for high altitude bombardment, and the other for low level anti-submarine warfare. Over time the 1BC retrained its personnel and modified its equipment for anti-submarine warfare, but to the detriment of both bombardment and ASW responsibilities. But this centralization of AAC resources was by-and-large neutralized by the Navy's allocation of AAC anti-submarine air squadrons to the control of Navy sea sector commanders, and the Navy's refusal to permit aircraft assigned to one sector to operate in another sector where the U-Boat threat at the time was more serious. When aircraft transfers were allowed they usually came too late. To add to administrative problems was that, while under the control of the Sea Frontiers, all AAF units were also administered along the AAF chain of command: through the First or Third Air Force that, in turn, had to operate through their respective defense commands before reaching AAF Headquarters. The AAC considered the Navy command structure and policy too rigid, and its strategic philosophy too defense oriented, concentrating on the convoy system. King answered this AAC criticism of the Navy with a criticism of the Army's doctrine of a broad air offensive against U-Boats that it could not accomplish due to their lack of sufficient aircraft. Later, King's belief in the convoy system was validated when the Navy finally initiated its defensive coastal convoy system, protected by available Army aircraft. The first convoy sailed south from Hampton Roads, VA, on 14 May 1942. The convoy system was ultimately very successful, and if it could have been instituted earlier, even on a smaller scale, it could have saved more merchant ships from destruction.

So as America entered the war B-25s, B-17s, and B-24s were in short supply and, being superior to the B-18, were assigned to other more urgent duties than anti-submarine warfare. The few Mitchells, Fortresses, and Liberators that were available went to reinforce the Philippines and Hawaii, and then for the defense of Java and Australia, while others were used in the escalating training programs that required the use of combat aircraft in their final stages. By the end of January 1942 the 1BC had 119 aircraft available for anti-submarine duty to patrol large areas off the East Coast, but only 46 could be considered as "in commission," and of these only nine were long range B-17s; the remainder were mostly B-18s and a few B-25s. Since there were relatively large numbers of B-18s available, and because the bomber had never really found its niche, it was assigned ASW duties as the only available option.

When America entered the war its aircraft numbers were insufficient, and there was a serious lack of organization, anti-submarine equipment, and training, but ultimately the blame for America's inadequate anti-submarine defense lies in insufficient pre-war planning. Fortunately, Hitler and the BdU also suffered flawed planning, and there were only a limited number of long range U-Boats available to attack America.

West Coast ASW Operations December 1941-February 1943

After the Pearl Harbor attack, it was thought the greatest danger from submarine attack would be from the Japanese along the U.S. West Coast. However, at the time the Japanese had only 20 submarines capable of voyaging from the Homeland to America, and Japanese naval strategy limited submarine attacks principally to enemy naval vessels, with merchant vessels being relegated to secondary targets. During December 1941, nine Japanese submarines patrolled off the American West Coast and encountered no warships, and were credited with sinking ten merchant vessels and three tankers. Between February and October 1942, four other Japanese submarines patrolled off the West Coast and sank seven ships, and shelled three onshore installations, causing minor damage. In October 1942 the Japanese withdrew their long range submarines from the West Coast until late 1944, when one returned to sink two ships.

The 2nd Air Force and 4th Air Force were responsible for West Coast air training and defense, and on 28 November 1941, AAC Headquarters ordered these Air Forces to support the Navy in ASW air patrol. Although the interservice rivalry was not nearly as intense on the West Coast, initially, the lack of experience and the diverse organizational and operational procedures of the two services prevented cooperation in establishing offshore patrols to cover the primary areas of patrol without duplication. In late December 1941, interservice liaison was improved by the establishment of a common information center in San Francisco. However, the different methods of patrol caused problems. The Navy patrolled in a search pattern shaped like a fan, with each aircraft flying out from a central position on diverging courses, while the AAC flew a parallel track search pattern, with each patrolling aircraft flying parallel within sight of the aircraft on either side. Soon, the AAC pilots were ordered to fly the Navy's fan search patterns, with the AAC ASW missions ranging out to 600 miles, and the Navy flying close offshore. The search area ranged from Seattle in the north and south to the Baja, California, coastline.

In December 1942 the AAC had only 45 modern fighter aircraft, 35 B-18 medium bombers, and ten B-17 long range bombers on hand on the entire West Coast. Immediately after Pearl Harbor the AAC detained Philippine-bound B-17s and their crews, and with other available aircraft were formed into the so-called "Sierra Bombardment Group." The Sierra Group contributed to offshore patrols until February 1942, when it became apparent that the immediate Japanese threat had subsided, and the scheduled transfer of aircraft to the Philippines and Australia was resumed. The Sierra Group was a temporary ASW fix, using personnel untrained in ASW patrols relying on eyesight and conventional bombs, instead of radar and depth charges, to detect and attack Japanese submarines, which were often found to be nothing but floating debris. After the Sierra Group was disbanded, the AAC continued to fly ASW patrols, mainly with B-18s supplemented by some B-25s, a few B-24s, and fewer B-17s. These improvised ASW patrols continued until February 1943, when the Navy's air and surface forces had developed sufficiently to deal with the Japanese submarine menace that never materialized.

Other ACC Groups involved in West Coast ASW Patrols:

12th Bombardment Group (Light) was constituted on 20 November 1940, and activated on 15 January 1941 at McChord Field, Tacoma, WA. The Group was formed from a cadre from the 2nd Bombardment Group, and from the 8th Pursuit Group at Langley and the HQ organizations of the 2nd Wing and GHQ Air Force. The 12th trained in B-18s, B-23s, and PT-17s, and flew coastal patrols after Pearl Harbor.

41st Bombardment Group (Medium) was constituted on 20 November 1940, and activated at March Field, CA, on 15 January 1941 under Capt. Lawrence Douthit, and then by Lt.Col. Archibald Smith on 1 June 1941. It trained in B-18s and A-29s, and later transitioned to B-25s. The Group flew ASW patrols off the West Coast until October 1943, and then deployed to Hawaii.

47th Bombardment Group (Light)
The 47th was constituted on 20 November 1940, and activated on 15 January 1941, with cadres for Headquarters and three tactical squadrons (the 84th, 85th and 86th), and were manned by experienced personnel from the 17th Bombardment Group from McChord Field. The 20th Reconnaissance Squadron (later the 97th Bombardment Squadron) was manned by personnel from the 89th Reconnaissance Squadron. In July 1941 the new Group and its squadrons left McChord for the new base at Hammer Field, Fresno, CA. Here the 47th was joined by several hundred more experienced personnel to form an experienced, cohesive unit. While at Hammer Field the Group trained in its B-18s, and to break the monotony, flew many search missions over the Southwest desert and Mexico looking for downed ferry pilots who were flying new aircraft east from the Southern California factories. They also participated in infantry exercises at Fort Ord. Immediately after the Pearl Harbor attack the Group was assigned a number of new B-18s, and deployed to Hamilton Field, where they were loaded with bombs and sent out on sweeps of the Pacific Ocean off the California coast looking for Japanese submarines or surface vessels. The patrols continued until the end of December, when the 85th Squadron out of Sacramento Airport took over the patrols of the area. Bombs were in such short supply that at the end of the month the bombers still carried the same bombs that were loaded into them on 8 December. At the turn of the year, the Group began to receive small consignments of long range B-24s and LB-30s that would eventually replace their aging B-18s.

Introduction to the U-Boat Campaign
When Hitler declared war on the United States, most of the *Kriegsmarine's* U-Boats were supporting the Axis campaign in North Africa, which Commander-in-Chief of the U-Boats Adm. Karl Doenitz (BdU) considered to detract from the U-Boat's primary task of curtailing the transport of food and war materiel to Great Britain via the North Atlantic, and now the western Atlantic.

The U-Boat Campaign in the West was directed by Doenitz and the BdU in three separate theaters at three different periods of the war. The first phase was the so-called *Operation Paukenschlag* (*Operation Drumbeat*) that began in January 1942 off the U.S. east coast, and continued in diminishing intensity until mid-July 1942, when it was continued as a minor nuisance campaign until the end of the war. In February 1942 *Operation Neuland* began as a separate campaign in the Caribbean, and continued through 1942, accounting for 35% of worldwide Allied shipping losses. During the first half of 1943 the Germans discontinued major U-Boat operations in the Caribbean, with only three boats present in January. In July 1943 the BdU initiated another substantial Caribbean offensive that was decisively overwhelmed by U.S. naval vessels and naval and army patrol aircraft. After this setback Doenitz operated his U-Boats in small numbers to tie down disproportionate numbers of anti-submarine vessels and aircraft that by that time the U.S. had in increasing numbers, and could be easily spared. The third U-Boat offensive in the west was then directed southward against Brazilian coastal shipping, and continued intermittently until the end of the war.

The U-Boat Campaign began on 3 September 1939, when U-30 sank the liner SS *Athenia* off the Irish coast. The campaign that evolved into the Battle of the Atlantic was limited by the number and type of U-Boats available, the location of their Baltic bases, and logistics. Nonetheless, the U-Boats threatened to place Britain under economic siege, as it had during World War I. Again the English resorted to the convoy system, but had too few purpose-built escort vessels, and by the end of 1940 over a thousand merchant ships had been sunk, and the vessels sunk to vessels constructed ratio was at a critical stage. With the fall of France in June 1940, the U-Boats could be based on the French Atlantic coast, and more were available to be formed into *Die Rudeltaktik* "Wolf Packs" of eight to 20 U-Boats. In the Wolf Pack concept, the first U-Boat to

Hamilton Field, CA (pictured), near San Francisco, and March Field, near Los Angeles, were the major West Coast B-18 bases. (USAF)

make contact with a convoy would inform BdU at Lorient, France, of the convoy's speed, course, and make up. The other U-Boats would be instructed to gather and intercept the convoy. After initial teething problems the Wolf Pack concept became so successful between May and December 1940 that it was called the "Happy Time."

East Coast ASW Operations December 1941-June 1942
Operation Drumbeat (Paukenschlag)

Japanese submarines and aircraft carriers would prove to pose no threat to the American West Coast, but the German U-Boat became a serious threat to Allied shipping off the American East Coast. For the British, it was imperative that Allied merchant vessels import goods and materials to the British Isles, and conversely the Germans urgently wanted to prevent this from happening. At the start of the war in September 1939 the Germans had only 56 operational U-Boats, of which only 22 Type VIIs had sufficient cruising range to be operational in the North Atlantic. As America entered the war in the beginning of 1942—a critical stage in the Battle of the Atlantic—the *Kriegsmarine* only had 98 oceangoing U-Boats, and four months later there would only be an additional 26 available. To conserve his U-Boat fleet, Adm. Doenitz planned a strategy that would allow them to attack Allied shipping at the least defended areas to gain the greatest destruction at the least risk of loss. Plans to deploy the U-Boats off the U.S. coast were made soon after Pearl Harbor. At a naval conference with Hitler on 12 December, Grand Adm. Raeder, head of the *Kriegsmarine*, asserted that America would be too preoccupied with the Japanese, and the U-Boats could operate against the East Coast shipping lanes with little risk. But Doenitz had no U-Boats immediately available to patrol American waters, and it was not until 23-27 December that five long range U-Boats of the Second U-Boat Flotilla were able to sail from the French port of Lorient to carry out *Operation Paukenschlag*. The U-Boats, modified to carry an additional 20 tons of fuel, sailed from France, and took two weeks to reach American waters, where they could patrol individually for two weeks, and then take two weeks to return.

Courageous and professional, Capt. Reinhard Hardegen was the stereotypical U-Boat ace. During his stay in East Coast waters during *Operation Drumbeat*, Hardegen claimed eight ships totaling 53,360 tons of the 25 ships sunk totaling 136,661 tons he sunk during his five career patrols. (Author/ B.C. West)

Beginning in March 1942, U-Boat patrol time could be extended up to nine weeks by being refueled and resupplied by specially modified submarines named "Milch Kaus" (Milk Cows). These 1,700 ton U-Tankers were armed with one antiaircraft gun of 37mm and two 20mm guns, but carried no torpedoes. The U-Tanker carried 700 tons of diesel fuel, of which up to 600 tons could be transferred to operational U-Boats. This amount of diesel could supply 12 Type VII U-Boats or five IX boats. The Milch Cow also carried stores, spare parts, ammunition, and a physician. At first refueling was hampered by inexperience and bad weather, and this resulted in the dangerous massing of a number of U-Boats in the vicinity, and vulnerable to discovery and attack. Inexperience was overcome, but bad weather was not.

U-Boat patrol time could be extended to nine weeks by being refueled and resupplied by specially modified submarines named "Milch Kaus" (Milk Cows). A second refueling could extend the patrol time another 2 1/2 weeks. (USN)

The first U-Boat success in the new American Theater was by Type IX U-123, skippered by U-Boat ace Capt. Reinhard Hardegen, who sunk the 9,000 ton British passenger steamer *Cyclops* 300 miles off Cape Cod, MA, on 12 January. Three days later he sank the Panamanian tanker *Noress* with three torpedoes in a surface attack that led to the New York *Times* headline: "Tanker Torpedoed 60 Miles off Long Island." *Operation Drumbeat* had commenced. Over the next 17 days, 13 vessels were sunk off the U.S. East Coast by German U-Boats. During his stay in East Coast waters Hardegen claimed eight ships totaling 53,360 tons of the total 25 ships of 136,661 tons he sunk during his five career war patrols.

The offshore geography of the U.S. east coast dictated U-Boat tactics, as much of it (except the Outer Banks off North Carolina and the southern Florida coast, both of which became U-Boat hot spots) had shallow water extending up to 100 miles to the east before ending at the continental shelf. The VII U-Boats, and more so the large and cumbersome IX U-Boats, could operate on the surface in the shallow coastal waters in daylight operations, as they were unable to crash dive, and had to move there at night on the surface, and then flee back toward the deeper waters off the shelf before sunrise, lying submerged at about 100 feet during the day waiting for nightfall. After dark the U-Boat would surface and move toward shipping lanes to make an attack with two to four torpedoes and/or deck guns. There were times early in the campaign that U-Boat commanders sometimes attacked from the surface during the day, as they became aware of the vulnerability of merchant shipping and the incapacity of Allied ASW defenses to protect them. Between mid-March and mid-April 1942 was the most prolific period of U-Boat successes off the U.S. coast, when 2.2 ships were sunk per day, with the large majority of the sinkings occurring off Cape Hatteras, where the U-Boats had the advantage of deeper waters. This period of practically no resistance was called the "Second Happy Time" by the U-Boat commanders.

American Response to the *Drumbeat* U-Boats

Because of inadequate prewar ASW planning and preparations, the problem American ASW faced in early 1942 was the need to establish an effective ASW doctrine. The unfortunate delay in organizing antisubmarine defenses, and even the use of rudimentary precautionary measures to protect vessels led to a disastrous loss of shipping. Between January and the beginning of July 1942, U-Boats sank 171 vessels off the East Coast, including many tankers, and most of the total three million tons of shipping lost worldwide during that time was lost off America. U-Boat captains found merchant shipping cruising in peacetime shipping lanes, often individually instead of in convoys, and to make matters worse, at night these ships were silhouetted against the lights of brightly lit coastal cities, oblivious of the effectiveness of a blackout.

The rare initial attacks on U-Boats failed because surface and air attacks were individual and random, due to the lack of adequate forces, and thus were not coordinated and sustained. The attackers lacked proper weapons and tactical training, and the experience to be successful. Poor communications and lack of navigational skills prevented follow up attacks by other air or surface units. Also, the U-Boats had the advantage of attacking individual vessels, as Adm. King did not have the resources to institute the convoy system, since escort vessels had been transferred to the Pacific, and had been sent to the British as part of the Lend Lease agreement. The BdU had passed along careless radio intercepts from merchant vessels that led the U-Boats to lucrative hunting grounds. By the end of January the first wave of U-Boats had sunk 41 vessels without loss to themselves.

To patrol the enormous area off the 3,000 mile eastern U.S. coastline required trained personnel and specialized surface vessels that the Navy did not possess, despite having two years from the outbreak of the European war to develop an anti-submarine force, much less a doctrine. Despite the success of escorted convoys by the Royal Navy, after Pearl Harbor Adm. King had no escort vessels available, and his policy was that independent vessels were better off than a number of vessels exposed in convoys that were either not escorted, or weakly escorted. At the time there was the irrational premise that the method to defeat the U-Boats was to build merchant ships quicker than the U-Boats could sink them, and Roosevelt and the War Department placed a higher priority on merchantman construction than building convoy escort warships. Destroyers were in short supply, and those available were World War I era types, and were needed for North Atlantic convoy escort to aid the British. The Navy had only 20 escort vessels, and had to depend on smaller vessels, including civilian yachts, for offshore anti-submarine duties. The Navy's General Board chose the Hamilton Class Coast Guard Cutter as their definitive escort ship, but as late as October 1942 there were only five available, and even then the Navy had few qualified officers and men available to captain and crew escort vessels. By February and March 1942 the destruction of merchant shipping in the North Atlantic and off America had risen to such a magnitude that King was forced to consider implementing the convoy system, even though the Navy had scraped together only 122 escort vessels of the 600 that were required. Many of these escorts were assigned to North Atlantic convoy duty, and the remainder were assigned to small convoys sailing the waters off the Americas. So it was that many single unescorted merchantmen continued to ply the U.S. coasts from Maine to Key West, and the carnage would continue.

After Roosevelt rejected Adm. King's request for four engine B-24 and B-17 land-based bombers for the Navy, the understaffed 1BC, under Brig.Gen. Arnold Krogstad, was the East Coast's first line of defense. The 1BC needed to be reorganized, as it had lost all but one of its bomb groups (the 2BG) in December and January. Many of its experienced pilots had been transferred to form the core of new units being formed in the rapidly expanding Air Force. In January and February 1942 it integrated the largely untrained 13th and 45th Bomb Groups, and the 3rd and 92nd Reconnaissance Squadrons. What Krogstad now commanded was a few experienced pilots and officers directing a large number of enthusiastic, but largely inexperienced air crews that required extensive training. So the 1BC was confronted with providing standard bombardment training, but also specialized ASW training, which had to be learned operationally, primarily through trial and error, as aircraft had to be put into the air for anti-submarine patrols. The ASW aircraft were neither equipped with ASV radar, nor depth charges, and there was no ASW doctrine on finding, attacking or, much less, sinking a U-Boat. At this point there was no chance that the 1st Bomber Command could be an effective anti-submarine force.

In mid-December 1941 Krogstad established his Headquarters Staff in the Federal Building in Manhattan, which was on the same floor as Adm. Adolphus Andrews' HQ. In order to preserve the separation of services Krogstad set up an independent operations center that had no links to the Eastern Sea Frontier, but conducted operations in conjunction with North Atlantic Naval Coastal Frontier (NANCF), which was incorporated into the Eastern Sea Frontier. By the end of January this liaison was extended to the Sixth Naval District in Charleston, SC, and the B-18s of the 66th Observation Group of the 1st Air Support Command were sent to patrol that area. Meanwhile, it was evident that the Eastern Sea Frontier, with its headquarters in New York City, was unable to administer the defense of the entire Western Atlantic. On 6 February 1942 the Gulf Sea Frontier, commanded by Capt. Russell Crenshaw, was created to defend the southern coast from Jacksonville, FL, to Texas. The Seventh and Eighth Naval Districts, headquartered in Key West and New Orleans, respectively, were the components of the Gulf Sea Frontier.

This improvement in administration did not lead to a concomitant improvement against the U-Boats, which virtually cruised unscathed in the first two months of 1942. Nearly 8,000 hours of air patrols were flown, and only four unsuccessful attacks were made on U-Boats. During March an improvement was seen as, although the U-Boats increased in numbers and inflicted increased damage, 1BC was able to send up 16,000 hours of patrols. These patrols, too, resulted in relatively few attacks, but did force the U-Boats to remain submerged for increasing periods, and restricted their freedom of movement. Only four B-18s carried the first ASV radar sets, and there was a shortage of spare parts and qualified maintenance personnel for these unreliable; the arrival of new equipment was months in the future.

Early in the war the British had broken the German U-Boat codes by using a captured German Enigma code machine. The broken codes, known as ULTRA, gave the British vital information on the disposition of U-Boat Wolf Packs, thus they routed their convoys away from them. The British passed U-Boat intelligence to the Navy, which initially failed to pass it along to the AAC in a timely manner, and its usefulness was lost; because of this intelligence breakdown attacks on U-Boats were unsuccessful. However, in on 1 February 1942 the Germans changed their Enigma machines on the Atlantic U-Boat net with new, more complex machines, and it was not until 13 December 1942 that another machine was captured, and the code could again be deciphered.

Army/Navy Administration Problems Remain to Be Solved

On the High Command level, the basic problem with the U-Boats savaging merchant shipping was that it interfered with the future conduct and strategy of the war. The Navy continued to be unable to provide the necessary number and type of aircraft required for ASW, and while the AAF did have the required aircraft; it was torn between utilizing them for its commitment to the development of their strategic bombing doctrine, which was an essential part of grand Allied strategy, and the need to counter the U-Boat menace that was also threatening the grand Allied strategy. When decisive leadership was needed President Roosevelt either did not appreciate the extent of the U-Boat threat due to the shock of Pearl Harbor and the rapid successes of the Japanese in the Pacific, or perhaps he chose to remove himself from the increasingly hostile Army-Navy enmity.

The AAF had provided anti-submarine aircraft, but more were required, and it appeared that the provisions of the 1936 Joint Action of the Army and Navy would be compromised even more. On 9 March Gen. Hap Arnold proposed a solution in a letter to Adm. Ernest King. Arnold advocated the "establishment of a Coastal Command, within the Army Air Corps, which will have for its purpose operations similar to the Coastal Command, Royal Air Force, operating when necessary under the control of the proper naval authorities." In the RAF Coastal Command system the British Admiralty issued a broad objective for ASW, and allowed the Coastal Command to independently direct its units within that agenda. British sea and air officers worked together from the same intelligence, and had developed a close and effective organization. Under Navy control the AAF experienced almost complete Navy management of its assigned ASW units, down to the lower echelons of command by the Sea Frontiers and naval districts. The advantages of an American Coastal Command would be that it was an essentially independent unit specifically trained and equipped for ASW training, and having the ability to return to normal bombardment functions once the U-Boat threat diminished without being under continued naval control.

Not surprisingly, Arnold's Coastal Command proposal was not acted upon, but on 16 March 1942, the AAF and USN were presented with an "agreement" dictated by the Joint Chiefs of Staff that defined the relationship of the 1st Bomber Command and 1st Anti-submarine Command to the naval commands. Since most of the Air Force ASW units were under the operational control of the 1BC and 1ASC, it was thought that centralizing ASW air operations with the 1BC would resolve the Army/Navy control impasse, but not to the end wished by the AAF. Previously the AAC commands had operated under the direct control of the naval Commander of the Eastern Sea Frontier. But now the Joint Chiefs of Staff also gave clear jurisdiction over all naval forces and all AAC air units that were "engaged in operations over the sea for the protection of shipping and against enemy seaborne activities" to the various sea frontier commanders, who were all naval officers. To this end Lt.Gen. Hugh Drum, CG of the Eastern Air Command, further relinquished all units of the 1BC, 1ASC, and Civil Air Patrol to the Commander of the Eastern Sea Frontier. Excluded were three bombardment groups and four observation squadrons that were to be used as operational training units for sea search personnel. Nothing had changed for the AAF in the administration of the anti-submarine campaign, except that there was more micromanagement and concentration of local command, which would expedite operations. Nonetheless, the Navy still ruled.

Under the Joint Chiefs of Staff's Army-Navy "mutual accord" agreement of 16 March, AAF units continued to be assigned to the

Navy Sea Frontier commanders, who regarded these assignments as essentially permanent, with the tendency of each Sea Frontier commander not to allow AAF units to operate in another Sea Frontier except in dire emergency. Again, the AAF ASW operations were hindered. For the AAF to participate successfully in anti-submarine warfare a centralized command was needed, but not under a single Navy commander. AAF leaders desired a separate AAF command, responsible to AAF Headquarters, that was trained and equipped for anti-submarine warfare. This concept was particularly important due to the fundamental difference in the strategic approach to anti-submarine warfare that persisted between the USN and AAF in their utilization of land-based ASW aviation. The Navy wanted to employ ASW (i.e. AAF) aircraft defensively for convoy protection, while the AAF wanted to use their aircraft as an offensive mobile striking force to seek out U-Boats where they were operating. On 21 June Adm. King articulated the official Navy policy in a lengthy letter to Gen. Marshall: "...(convoy) escort is not just one way of handling the submarine menace; it is the only way that gives any promise of success." He went on to attack the use of the AAF's hunter-killer offensive doctrine of air patrols that have "time and time again proved futile" to the exclusion of defensive convoy patrols. King continued, stating that the only efficient anti-submarine method was to attack "continuously and relentlessly those U-Boats that have been drawn to the convoys." In his conclusion King made five recommendations. Regarding aerial ASW he asked for a build up "as soon as practicable" of a force of 1,000 ASV equipped Army bombers to patrol the projected 7,000 miles of convoy routes extending along the Eastern, Gulf, Caribbean, and Panama Sea Frontiers. He went on to note that "airpower was not to be a temporary measure pending augmentation of naval surface forces," but rather "a permanent arrangement to protect our shipping properly." The "1,000 ASV-equipped" and "as soon as practicable" clauses in King's letter, while correctly considering the massive numbers of radar equipped aircraft required for convoy escort, "as soon" was certainly not at hand or, in fact, anywhere "practicable" in the near future. Neither were the radar equipped, well trained naval surface vessels available for anywhere near widespread convoy protection.

However, it should be remembered before criticizing King that the U.S. Navy was engaged in a two ocean war, with limited resources (particularly destroyers and patrol aircraft) for convoy protection. King also was in disagreement with the British over 1942 naval strategy. With limited resources King had to choose between protecting either troopship convoys or cargo vessels and tankers, even when pressed by Britain's plea for its survival through the delivery of food, oil, and war materiel by cargo vessels and tankers. If Roosevelt had not sent the British and Canadians the 60 four stack WW-I era destroyers in 1940-41 these warships could have been used as convoy escorts off the East Coast, the Gulf of Mexico, and the Caribbean, and prevented the loss of numerous merchant men and hundreds of seamen. While many historians place the blame for American ASW failure on King and his uncompromising stand, the Army Air Forces should also bear some blame for not more easily relinquishing its mandated prewar control of its land-based aircraft to Navy control for ASW duties for whatever reasons, political or doctrinal. This self interest resulted in an early American ASW response that was totally ineffective and costly.

The AAF maintained that, while convoy protection "was important to the immediate task of protecting shipping and for that reason deserving a high priority, (it) could never dispose of the U-Boat menace but must be supplemented by a vigorous offensive campaign in which the strategic movement of the submarine fleet could be promptly countered by a corresponding shift in the weight of air attack." To meet this objective, the question of the AAF creating a RAF Coastal Command type again arose, and was supported by RAF Coastal Command experience. In August, visiting RAF Air Marshall P.B. Joubert concurred with this AAF view when he stated that "while a certain amount of close escort of convoys, particularly when threatened, is a necessary feature of air operations, the main method of defeating the U-Boat is to seek and strike. The portion of air available should always be engaged in the direct attack on U-Boats and the smallest possible number in direct protection of shipping."

The U-Boat situation became so critical that the War Department demanded action, and on 20 May 1942 the Assistant Chief of Staff, Operations Division, requested that the leaders of the AAF and the Eastern Defense Command take action to improve the efficiency of the First Air Force and the 1st Bomber Command, with the 1BC being reorganized "to fulfill the special requirements of anti-submarine and allied air operations, in consonance the Army responsibility in operating in support of, or in lieu of naval forces for protecting shipping." This statement clearly deemed the AAF as no longer a stop gap, emergency anti-submarine measure, but a vital force that had developed ASW techniques, and laid the groundwork for it to continue a more significant effort in the future. Although the 20 May Directive gave the Navy continued jurisdiction, it added that the AAF "must be prepared to submit recommendations and to take every action to make anti-submarine warfare fully effective." The AAF finally had a foot in the administrative door.

Under the 20 May Directive, Deputy Chief of Staff Maj.Gen.Joseph McNarney proposed that the 1BC was to be organized as a "unit for anti-submarine and related operations" on the East and Gulf Coasts, with air bases to be established at "strategic locations in order to take maximum advantage of the mobility of land-based aircraft." Gen. Arnold was to provide every available B-18 in the U.S., was to equip them with centimetric ASV radar as soon as sets became available, and was to install the necessary depth charge and bomb racks for anti-submarine warfare. As soon as these ASV-equipped B-18s were available they were to be "welded into units particularly suited for hunting down and destroying enemy submarines by methods developed by our experimental units (e.g. the 1st Sea-Search Attack Group (1SSAG), author), which have been operating off Cape Hatteras." The reorganization stressed mobility, operating from a network of new bases on the East and Gulf coasts. When a unit was moved outside the Eastern Defense Command it would operate under the control of its new Sea Frontier, but continue to be assigned to the 1BC, and be regarded as a temporary detachment.

When the Gulf Sea Frontier was organized in February 1942 it was given only nominal surface and air forces. The AAF contributed a paltry 14 observation aircraft and two old B-18s. In early May, a detachment of 20 B-18s was urgently dispatched south from the Eastern Sea Frontier. Between 8-10 May a squadron of A-29 light bombers was sent to Jacksonville, FL; six B-25 medium bombers to Miami, FL; and on 20-21 May, a detachment of B-25s was sent to Havana, Cuba, to patrol the Yucatan Channel. In late May the Seventh Naval District/Gulf Sea Frontier had a new commander, R.Adm. James Kauffman, who had been transferred from Iceland, where he was instrumental in forming the successful North Atlantic convoy routes. His first step was to move his HQ from remote Key West to Miami, which provided better communications with Gulf Sea Frontier naval and air units, thus permitting increased and more coordinated anti-submarine searches and attacks. On 26 May Maj.Gen. Follett Bradley, CG of the First Air Force, established the Gulf Task Force (GTF) as part of the 1BC at Charleston, SC. Transferred to the GTF was a detachment of 20 B-18s that had previously been sent south, along with the 66th and 97th Observation Squadrons, and all local Civil Air Patrol units. The GTF operated in a similar manner as the 1BC did under the Eastern Sea Frontier. As was typical, the GTF found it did not have sufficient aircraft to patrol its assigned area, and Gen. Arnold dispatched several training units of the Third Air Force to conduct ASW patrols as part of their training missions under the direction of the GTF. In June the GTF moved its joint operations center from Charleston to Miami, FL, to join the Gulf Sea Frontier HQ. With this expansion the 1BC's radius of action did not extend into the Caribbean, where the Antilles Air Task Force (to be discussed later) operated as an independent AAF unit under the control of the Caribbean Sea Frontier. By September 1942 the 1BC was operating from ten bases extending from Westover, MA, to Galveston, TX. During the second quarter of 1942, while flying slightly fewer monthly patrol hours than the record number in March, there were 54 (seven times more) attacks, of which seven resulted in damage, but no U-Boats destroyed by aircraft alone.

On 11 June 1942 U-157, a new Type IX U-Boat (the C model) under Wolf Henne was patrolling the Old Bahamas Channel, and sank the American molasses tanker SS *Hagan*, causing Gulf Sea Frontier commander James Kauffmann to order all available forces to "hunt this submarine to exhaustion and destroy it." Early the next morning U-157 was spotted by a GTF ASV equipped B-18 that closed to two miles with the U-Boat still on the surface. The pilot quickly attacked, but the bomb bay doors malfunctioned at 900 feet, and the Bolo made a quick diving turn back into the attack as the doors opened. Four depth charges were dropped at 300 feet as the U-Boat was nearly submerged, shaken but undamaged. Later that morning Henne's U-Boat was spotted by a Pam Am airliner that called in its position, but a mix of three B-18s and A-29s searched and were unable to find it. In the meantime, nine vessels of the Key West Killer Group and six from the Miami Killer Group were sent into the Florida Straits between Key West and Havana. For the next two days the aircraft of the GTF and 15 Navy vessels tenaciously hunted for the U-Boat. During the night of 12-13 June an ASV equipped B-18 made contact with the U-Boat as it was crash diving. The B-18 radioed his position, and Kauffmann immediately sent the Key West Group to the area. On the 13th at 1600, the 165 foot Coast Guard cutter *Thetis* picked up a strong sonar contact and dropped ten depth charges that brought up an oil slick and air bubbles. Four other vessels arrived and dropped an additional 22 depth charges on the slick. Afterwards the GTF took partial credit for the sinking, but Navy man Kauffmann officially gave full credit to the *Thetis*, and nothing was mentioned in the dispatches crediting the GTF B-18s.

By late June 1942, shipping losses to U-Boats continued to such an extent that Gen. Marshall stated in a memo to Adm. King that "another month or two (of such losses) would so cripple our means of transport that we will be unable to bring sufficient men and planes against the enemy in critical theaters to exercise a determining influence on the war." Adm. King responded to Marshall's concern by defending his commitment to the convoy system, and by disparaging the AAF's ASW air offensive position, emphasizing that patrol operations had not only proved to be ineffectual, but required more aircraft than were available. He strongly recommended the expansion and strengthening of the convoy system, and that the AAF should provide a minimum of 500 aircraft, including 200 to the Caribbean and Panama Sea Frontiers, to protect the critical oil and bauxite shipping from South America.

American ASW Tactics and Administration Improve

Despite the continuing administrative and communications inflexibility, progress was being made in ASW techniques through the continuous experience gained by hours of patrolling. By the beginning of April 1942 Brig.Gen. Westside Larson had succeeded Krogstad, who had made progress in ASW operations and tactics as 1BC CG. Still personnel and equipment shortages continued, and radio and telephone communications needed improvement. Army-Navy non-cooperation continued to hinder operations, but despite the interservice rivalry, the organization, equipping, and training of the Eastern Sea Frontier forces to protect vessels sailing in convoys along the East Coast was slowly accomplished. By June 1942 the Frontier's improved air and surface patrols forced German U-Boat commanders to operate in the safer waters of the Gulf of Mexico and the Caribbean. In the second quarter of 1942 only seven of 54 air attacks resulted in U-Boat damage, but in the third quarter, with more experience, eight of 25 attacks resulted in damage, one sinking, and more submerged time for the U-Boats.

Although a claim for sinking a U-Boat was good for almost a daily news article, the first verified successful aircraft attack on a U-Boat did not occur until 7 July 1942. On that day Lt. Harry Kane and a crew of four flew a Lockheed A-39 Hudson of the 396th Bombardment Squadron (M), which had been based originally in Alameda, CA, but was operating out of MCAS Cherry Point, NC. Kane was flying at 1,500 feet in broken clouds on a routine six hour daylight patrol between Cape Hatteras and Charleston, SC, when at 1412 he spotted a U-Boat (U-701) at seven miles to port, and immediately turned and dove to the attack. At 50 feet Kane dropped three 325 pound Mk-17 depth charges armed with new detonators set to explode at 25 feet on the swirl of the diving U-Boat. The well timed explosions flooded the U-Boat aft of the conning tower, but 37 of the 44 crew were able to escape. Subsequently, over the next several days the seven German crewmen, including captain Horst Degen, survived to be rescued by a Navy blimp (K-8).

On 19 July Doenitz withdrew the last two of his U-Boats operating off the Cape Hatteras area, and committed them to the Gulf and Caribbean. At the end of July 1942 Doenitz officially cancelled the U-Boat campaign off the eastern American coast after five U-Boats were sunk—two by aircraft—but not before 109 merchantmen of 610,000 tons had been sunk. From July 1942 until the end of the war only ten merchantmen were sunk for six U-Boats lost off the American East Coast. Doenitz had been criticized for not having his U-Boats intercept ships sailing in the North Atlantic carrying supplies directly to Britain, but he argued that it did not matter where a ship was sunk, as every ship sunk had to be replaced, and ultimately the Battle of the Atlantic was a fight between ships sunk and new construction.

AAC Bomb Squadrons of the Eastern Sea Frontier

The squadrons of the 2nd Bombardment Group out of Langley Field, VA, were active in ASW patrols. On 7 December 1941 the 96th Bombardment Squadron had an assortment of B-17Bs and Cs, and B-18As, and the following day all (six) B-17s except one were deployed for March Field, CA, to intercept a possible sea of air Japanese attack. The squadron's B-18s were active flying ASW patrols from Langley, and from deployment bases in the Carolinas and Florida. The B-18s were loaded with 100 pound bombs, and patrolled along the coast to 200 miles out to sea. The large search area extended from Langley as far south as Opa Locka, FL, just north of Miami, and stretched the resources of the squadron.

At the time of Pearl Harbor, the 20th Bombardment Squadron was equipped with B-25s, and was one of the best trained and equipped medium bomber units in the AAC. But by April 1942, most of the Squadron's B-25s had largely been transferred to various newly formed medium bomb units, and it was left with an assortment of old aircraft, mainly the B-18A. Infrequently, the Squadron would receive a B-17 D or E that would join the Bolos in ASW patrols, as the AAC was the only ASW presence in the Atlantic, since the Navy was fully committed to the Pacific. The Squadron's B-17s patrolled 600 miles offshore, while its B-18s and B-25s conducted shorter range patrols of Chesapeake Bay, and large portions of the Atlantic Coast on searches for U-Boats, U-Boat milk cows, and German commerce raiders, but no enemy shipping was found. By July the unit began to receive the B-24D equipped with ASV radar that was used in combination with MAD equipment, and the B-18 was phased out.

The 41st Reconnaissance Squadron (Long Range) was constituted on 22 December 1939, and activated into the 2nd Wing on 1 February 1940 at Langley. The group was furnished with B-18s and B-18As, B-10s, and the lone XB-15. The Squadron Commander was Maj. Caleb Haynes, aided by the capable, but not yet distinguished, Capt. Curtis LeMay. In August 1941 the Squadron was deployed to Newfoundland and attached to the Newfoundland Base Command, where it flew ASW patrols in B-18s and B-17s. On 22 April 1942 it was redesignated as the 429th Bomb Squadron.

Marshall Takes Steps to Resolve the Administrative Dispute

A land-based AAF anti-submarine air force using ASV equipped bombers was imminent, and would require the mobility that centralized control would give it to be effective. To resolve the anti-submarine control impasse the passing of letters and memorandums would no longer suffice, and decisive official action was needed. It

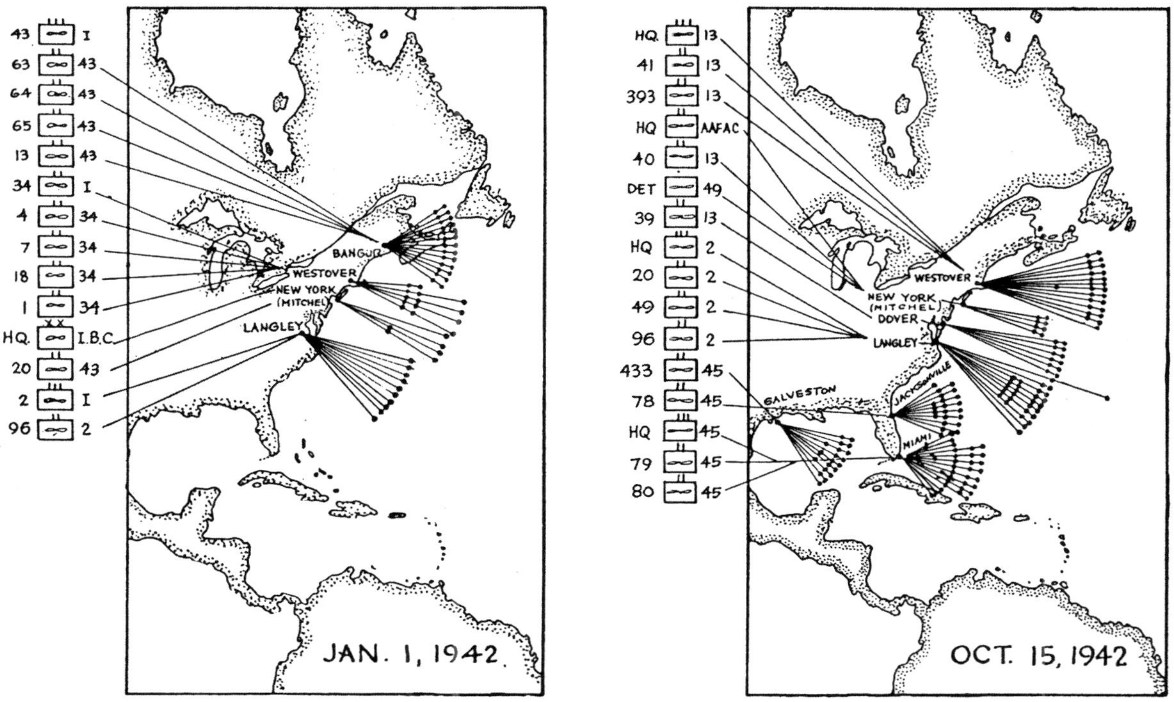

(USN) AAF ANTISUBMARINE OPERATIONS

would not be until September 1942 that official action on the often debated ASW command structure began. On 14 September Gen. Marshall stated: "Experience with the First Bomber Command in anti-submarine operations since March indicates that the effective employment of air forces against submarines demands rapid communications, mobility, and freedom from restrictions inherent in command systems based upon area responsibility." Marshall proposed the establishment of the First Antisubmarine Army Air Command, which was to include the ASW units of the 1BC. Control of this new unit was to be centralized in the War Department, so that it could be quickly sent to areas of U-Boat activity. Initially, it was to operate off the Atlantic coast, the Gulf of Mexico, and Caribbean, and then to expand into other areas as aircraft became available. Again, Marshall seems to have acquiesced to the powerful Adm. King when he continued: "Naturally, (it) will be under the operational control of the Sea Frontier involved," meaning continual naval control, but with more latitude of action as dictated by the War Department. Operational autonomy would be increased by subsequent orders that clarified the unit's mission. Marshall urged the Navy's "closest cooperation," and also insisted on "liaison between our immediate headquarters," especially in the matter of transmission of intelligence, which the Navy collected and distributed.

On 22 September 1942 Marshall's Deputy Chief of Staff, Maj.Gen. Joseph McNarney, directed Gen. Arnold to organize the new anti-submarine command using the 1BC as the cadre. The unit, designated as the Army Air Forces Anti-submarine Command (AAFAC), was activated on 15 October 1942 under Brig.Gen. Westside Larson, who was responsible to the Director of Military Requirements through the Director of Bombardment. The unit was directed to "attack hostile submarines where ever they may be operating," a charge that allowed the unit great autonomy. Now, with the authorization of the Operations Division of the General Staff, units could be transferred outside the U.S. on detached service. Matters of policy, planning, and the development of new weapons and equipment continued to be the province of the AAF. (The operations of the AAFAC will be discussed later.)

Although Adm. King was forced to comply to his superior, Marshall, he did so only reluctantly, maintaining "that the preferable method to be the allocation of air units to the Sea Frontiers, such allocation to change from time to time and from Frontier to Frontier as the exigencies of war dictatedÖand would continue to exercise control over Army planes through the commanders of the various Sea Frontiers." So Marshall's resolution, while giving the Army better organization and operational latitude, did not define the limits of operational ASW control by the Navy. The settlement allowed the Army and Navy, each with long-range land based air forces, to develop and operate more or less independently without costly duplication and, more importantly, without lapses. This ongoing struggle for operational control will be discussed further in the context of its consequences on ASW operations in the various Sea Frontiers.

The SADU and the Sea-Search Attack Squadrons

On 7 July 1942, Secretary of War Henry Stimson sent a memorandum to Secretary of the Navy Frank Knox, pointing out that the current ineffective system of command and communications would compromise the mobility of AAF ASV radar equipped patrol aircraft once they became available in numbers. To aid this mobility, Stimson urged a single Sea Frontier extending from Maine to Mexico, including the Atlantic and Gulf of Mexico under the control of a naval officer. Knox, like King, did not want change, but wanted the augmentation of the existing forces "which now seem to be working effectively." Stimson had selected English radar expert Dr. Edward Bowles of the Tizard Mission to develop the use of U.S. land-based ASV equipped bombers "within a flexible system of command." Stimson authorized 175 ASV equipped aircraft, initially mostly B-18s, to outfit the 1BC as "an adequate force if freed from the restrictions inherent in inflexible command systems based upon area responsibility."

The basis for an ASV equipped bomber force was in place when, at the end of May 1942, Gen. Arnold directed the director of Technical Services to establish the Sea-Search Attack Development Unit (SADU) at Langley Field. So that the SADU could function without encountering red tape, Arnold assumed direct control through the Technical Services Director. Operationally, the SADU, as the 1st Sea-Search Attack Group (1SSAG), was under the direction of the 1st Bomber Command, which had been activated on 8 June 1942. On the same date the 1SSAG was allocated a tactical unit, the 2nd Sea-Search Attack Squadron (2SSAS), that was equipped with ten B-18s. The 1SSAG was under the command of the able Col. William Dolan from its activation until its disbandment on 10 April 1944, and worked closely with the National Defense Research Council (NDRC) and the Radiation Laboratory at MIT. The Group's work was to be pioneering the development and testing of anti-submarine equipment and tactics using existing equipment under development. This experimental unit was to be hands on, taking part in actual sea-searches, and training crews in ASW tactics. From its activation to 15 July 1943, the Group reported 43 total U-Boat sightings: 22 by MAD equipment, 18 by ASV radar, and 3 visual sightings.

On 17 June 1942 the 1st Sea-Search Attack Squadron (1SSAS), stationed at Langley Field, was chosen to investigate the use of ASV and MAD anti-submarine equipment. As part of the test it was sent to Key West, FL, from 17-23 August 1942 to fly actual anti-submarine patrols to operationally test their radar equipment. After a rather inactive Florida stint the Squadron was sent to Edinburgh Field, Trinidad, from 20 September to 21 October, where U-Boats were more active. During the month the Squadron's ASV-equipped B-18 would fly unending test patrols off Trinidad, and made 20 U-Boat contacts and attacks. The attack reports concluded that the U-Boats equipped with *Metox* were unable to detect the ten centimeter ASV. Although the contacts led to no U-Boats being sunk, trial and error established the techniques for successful ASV employment later. On 10 December 1942 the 3rd Sea-Search Attack Squadron was activated to test and modify the B-24 for ASW work.

ASW Operations in the Gulf Sea Frontier January-September 1942

In April, Brig.Gen. Westside Larson had realized his 1BC needed air bases to be established south of Langley Field to extend the command's operating area, and increase its response time and mobility. This concern was soon realized when the U-Boats quickly shifted their operations to the Gulf of Mexico and the Caribbean. This change in U-Boat strategy again found the Americans unprepared, as the number of merchantman sunk in the Gulf Sea Frontier rose from two in April to 41 in May.

As the first wave of *Drumbeat* U-Boats returned to France, the second wave of six U-Boats was ready to leave for new hunting grounds off Florida. After the great success of the first wave, Doenitz proposed that a second wave of 12 U-Boats cross the Atlantic, half of which were to be the new longer range Type IXs, carrying five more torpedoes and more fuel than the Type VIIs. Doenitz' proposal was turned down, as the Norwegian and Mediterranean U-Boat campaigns were given priority by Hitler and Raeder. In February 1942 there were only two U-Boats operating off Florida: U-128 off central Florida, and U-504 to the south. The first ship sunk in the Gulf Sea Frontier was the American tanker *Pan Massachusetts*, on 19 February off Fort Pierce, FL. During the next week Reinhard Hardegen, who returned in U-123, and another U-Boat sank four more tankers, and then Hardegen sank six more vessels between 8-13 April. During the next two weeks five U-Boats sank 12 more ships, including five tankers. In May U-Boats moved their operations further west into the Gulf of Mexico in a Cuba, Key West, and New Orleans triangle. They sank 41 ships totaling about 220,000 gross tons, with 55% of the tonnage being valuable gasoline tankers. U-Boat U-506, captained by Lt. Erich Wurdemann, sank ten vessels totaling over 60,000 tons. During June 1942, in the Gulf of Mexico and the Caribbean and its approaches, U-Boats had destroyed more shipping than they had destroyed in any one month in all theaters of war combined. From May to September they sank 58 ships totaling 300,000 tons. Between 7 June and 21 July, due to few patrolling aircraft, and ships sailing without convoy protection in the Gulf of Mexico, seven U-Boats sank 28 ships of over 130,000 tons, with three U-Boats (U-67, U-158, and U-129) accounting for 103,000 tons.

When the Gulf Sea Frontier was organized in February 1942 it was given only nominal surface and air forces. But soon a B-18 framed by a palm would become a common sight in the Caribbean. (USAF)

After fighting the U-Boats for half a year, trial and error began to be transformed into experience, and more successful patrols were made by the 1BC. The GTF worked under the operational control of the Commander of the Gulf Sea Frontier to supply all of the AAF aircraft that flew ASW patrols in the area, and continued to operate until November 1942. In the three months (July through September 1942), eight of 24 attacks were believed to have caused damage to several U-Boats, with one sinking, that by the 396BS Hudson piloted by Lt. Harry Kane.

At the end of July 1942, Kauffman organized an effective convoy system that was protected by naval blimps and escort vessels together with AAF bombers, mainly the B-18s of the 29th Bombardment Group (Heavy). The 29th was constituted on 22 December 1939 at Langley Field, and activated on 1 February 1940. The initial cadre was transferred from the 49th Squadron, and 27 officers were assigned from the 2nd Bombardment Group, including its first commander, Maj. Vincent Meloy, the former commander of the 20th. On 21 May 1940 the Group moved to MacDill Field, FL, where Maj. Charles Lawrence became commander on 5 January 1941. The Group flew ASW patrols until June 1942, when it moved to Gowen Field, ID, where it transitioned to B-24s. A detachment of radar equipped B-18s was sent to Key West, and soon the U-Boats faced the same dilemma they had in the Eastern Sea Frontier—too much opposition. After 58 vessels had been sunk for the cost of two U-Boats (U-157 in mid-June and U-166 on 16 August), the Germans realized that this ratio could no longer be sustained in the face of increased American anti-submarine presence and escorted convoy. Doenitz withdrew his U-Boats from that now dangerous area, sending some back to the North Atlantic, and others to the more promising Caribbean area to the south.

From May 1942 the BdU had constantly shifted the emphasis of its U-Boat offensive southward, and by September had all but abandoned the Eastern and Gulf Sea Frontiers, with no merchantmen being sunk after 4 September. U-Boat captains were no longer willing to operate in areas of systematic, albeit previously undamaging, aerial patrols that reduced their tonnage sunk scores. The 1BC had flown 59,248 operational hours between January and October 1942, culminating in only 200 sightings and 81 attacks, which resulted in one U-Boat sunk, six seriously damaged, and seven sustaining damage. But these statistics do not reflect the many unproductive hours the U-Boats spent submerged fearing detection by the ungainly pot bellied B-18s and the Navy's makeshift ASW Hudsons.

Despite the Army-Navy ASW success in driving off the U-Boats from American shores the U-Boat anxiety remained, and from September 1942 until mid-1944 the U.S. would commit too many aircraft to counter too few U-Boats in the Gulf Sea Frontier. The U-Boats were by no means defeated, as Doenitz transferred his U-Boats to the productive oil and bauxite routes off Trinidad and Aruba to the south. This move was so successful that in November 1942 U-Boats would destroy more shipping than in any other month since June.

ASW Preparations and Operations in the Panama Sector

ASW Preparations in the Panama Sector Prior to Pearl Har-

bor

In 1929, airpower maverick Gen. William Mitchell wrote an article for *Popular Aviation* entitled "Panama is Defenseless," in which he warned that the Panama Canal could easily be put out of action by a few strategically dropped bombs. By 1939 some of the weaknesses Mitchell had delineated had been remedied, but the major fix was to be the establishment of an effective air defense system based on air patrols, radar interception, and a ring of outlaying air bases; all of which were absent.

"54 Group Program: First Aviation Objective"

The "54 Group Program" was intended to provide an adequate defense for the continental United States and its "outlying territories and the Caribbean area," and to contribute 1,000 tactical aircraft for potential use, in cooperation with the Navy, in "establishing and maintaining air control in South America." In prewar strategy, the building of outlaying bases was considered the Navy's bailiwick, but prospective Air Corps bases had been investigated in the Antilles to guard Panama's Atlantic approaches. The Pacific Panama approaches had no similar islands to build air bases, but in 1938 the Galapagos Islands, located in the Pacific off Ecuador, had been surveyed for possible airfields, seaplane bases, and advanced warning bases.

Albrook Field, located on the Pacific side of Panama near the Gulf of Panama, was completed in mid-April 1939 when its two (200 x 2,500 foot and 75 x 1,000 foot) runways were finally paved. Albrook had only one hangar that was used by the 44RS, but a new hangar for its pursuit squadrons was completed at the end of 1938. The weather was better on Panama's Pacific side, both over the sea and land masses west to the central mountains. Albrook was built to protect the Pacific approaches to Panama, while the Atlantic approaches were protected by France Field, which had been completed in 1917 as America entered World War I. The Atlantic approaches to Panama were further protected by the construction of Air Base One, on the far northwest end of Puerto Rico near Point Borinquen. This field was occupied by the 27th Reconnaissance Squadron's nine B-18s, the first (37-621) of which arrived on 3 December 1939.

The first three B-18s—37-31 and 37-32 (at France Field), and 37-33 (at Albrook Field)—arrived in the Canal Zone in early January 1939, where they shared the field with 28 of the original 43 Martin B-10s remaining in the Canal Zone. On 26 January ten of the old Martin bombers were ferried back to the U.S., and the ferry crews returned with ten new B-18s: six (36-299, -302, -319, -320, 37-10, and -27) assigned to the 6BG at France Field, and four (36-323, -324, 37-13, and -24) to the 44RS at Albrook Field. These ten B-18s were followed by 11 more in April 1939 (four to the 44RS and seven to the 6BG), and 11 more in June 1939. Despite these reinforcements, Canal Zone commanders requested more reconnaissance and transport aircraft to expand the planned new bases in Panama and the Caribbean. By August 1939 Albrook and France Fields based 73 combat aircraft, including: 35 Douglas B-18s; 24 Boeing P-26s; and 14 Northrop A-17s. Gen. Arnold and his planners considered these aircraft "good enough" to hold off any attack until reinforcements could arrive. At the time there was a shortage of pilots, as there were only 15 pilots available to fly the 35 B-18s, and there also was a shortage of trained bombardiers and no navigators. This personnel shortage extended to the Pursuit and Attack Squadrons, which were in as bad, and even worse state. In September 1939 30 more Douglas B-18s had arrived, and later that month 22 Curtiss P-36As also arrived—this time with pilots and crews. By the end of 1939 there was an influx of new flying personnel to the Bombardment and Reconnaissance Squadrons. The 74th Attack Squadron was converted to a Bombardment Squadron on 1 November when it exchanged its A-17s for B-18s.

Blackouts, Exercises, and Alerts

On 10 October 1939, just after the start of the war in Europe, a 15 minute test blackout of the CZ was ordered to be conducted between 2245 and 2300. The blackout was highly publicized for a week, so that the citizens, and especially automobiles, could comply. PCD Commander Maj.Gen. David Stone, and Republic Presi-

Albrook Field, located on the Pacific side of Panama, was built to protect Panama's Pacific approaches. It was completed in mid-April 1939 when its two runways were finally paved. (USAF)

Borinquen Field, located on the far northwest end of Puerto Rico, near Point Borinquen, was completed in late 1939 to protect the Atlantic approaches to Panama. Borinquen would become a major base, and by 1945 construction costs would exceed $51 million. (USAF via AFHRC)

dent Juan Demostenes observed the exercise from a hill while four B-18s of the 44RS, with new 19BW Commander Brig.Gen. Herbert Dargue aboard, checked the results from the sky. The blackout was described as "not all it could have been," as Panamanian officials and populace had little enthusiasm for it, since many city lights could be seen, along with automobile headlights.

When war came to Europe in September 1939 the Canal Zone was put on alert because of possible Nazi and Vichy French activities in the Caribbean. On 17 December 1940 the Hawaiian and Panama Canal Departments were alerted to the possibility of naval or air surprise attacks and sabotage. The Army prepared for the defense of the Canal by barring commercial airlines controlled and operated by Axis nationals, and replacing them with U.S. (Pan American Airways) or local airlines. Airfields and air facilities were to be developed to accommodate U.S. military aircraft, and preparations were to be made to allow U.S. military air operations to begin immediately in case of an actual enemy attack or imminent threat of an attack.

During 1939, 6BG B-18s participated in numerous search and patrol exercises using conditions of "actual warfare being simulated insofar as possible." Using information from the Cristobal Port Captain's office concerning the positions of shipping approaching and departing Panama, B-18s of the 7RS were ordered to find a specific ship. Initially, these searches were to extend 150 to 200 miles offshore, and were extended as the crews became more proficient. The 7RS B-18s maintained radio silence, except for intermittent position reports, and sent an encrypted message when the objective was intercepted, including the time, course, speed, and position of the ship. The message was decrypted and sent to the commander of the 25th or 74th Bomb Squadrons, who was to dispatch a specially equipped B-18 to drop an oil canister in the discovery area that would create a slick on the water, simulating an enemy vessel. Armed B-18s were then dispatched, and would fly to the reported position to attack the slick with real ordnance and machine gun fire.

In late spring 1941 a secret test was made to assess the vulnerability of Canal locks to a bombing attack. The Pedro Miguel locks, located near Balboa, were simulated using bed sheets to duplicate their size. The replica bed sheet locks were set up at the bombing and gunnery range at Rio Hato Field on the Pacific coast of Panama, near the town of Aquadulce. Over a three day period six B-18s, three each from the 25BS and 7RS, made almost non-stop bombing runs at different altitudes, using different size bombs (1,000, 500, and 100 pound). While officers observed from a hilltop about three miles away, the bomb bursts were recorded by a cameraman in an O-19 that would fly into the target area at 500 feet as the bombs were dropped. After the first day the courageous O-19 pilot found that his aircraft was shredded by shrapnel, and backed off for the remaining two days of photography. While the good bombing results were encouraging for the B-18 crews, they did not bode well for the vulnerability of the locks, as the first 1,000 pound bomb hit was determined to theoretically put the locks out of commission for "a while."

Panama Canal Department Air Force (PCDAF)

Lt.Gen. Daniel Van Voorhis was appointed as the commander of the Panama Canal Department (PCD) on 8 January 1940, and was a particularly poor choice, as the 62 year old ex-cavalry officer disregarded the importance of the Air Corps, and hindered the growth of the airpower in Panama until he was replaced on 19 September 1941. During his tenure the Panama Canal Department Air Force (PCDAF) was activated by Gen. Arnold on 20 November 1940, with Voorhis assuming dual command of the PCD and PCDAF. The Panama Canal Department Air Force consisted of the new 12th Pursuit Wing (16th and 37th Pursuit Groups) and the new 19th Bombardment Wing (6th and 9th Bombardment Groups). Fortunately, Voorhis' year and a half of apathetic air command would be offset on 8 May 1941 by the appointment of the dynamic and contemporary Maj.Gen. Frank Andrews to head the PCDAF, which was expanded and redesignated as the Caribbean Air Force.

Inspections and Goodwill Flights

The importance of the Caribbean and CZ was emphasized in early February 1940, when it was visited by Army Chief of Staff Gen. George C. Marshall and Brig.Gen. George Brett, who would later become the CO of the fledgling 6th Air Force. The Generals completed comprehensive aerial inspections flying B-18s of the CZ, Venezuela, and Puerto Rico. In mid-February President Roosevelt visited the Canal area aboard the USS *Tuscaloosa*. FDR reviewed Air Corps troops and the B-18s displayed at Albrook and France Fields in Panama. These visits were important, as they raised the flagging spirits of the troops stationed in these forgotten areas, whose facilities at the time were considered to be one step above primitive.

The Atlantic approaches to the Canal were favored by a string of islands extending from Cuba southeast to Venezuela, on which air fields could be constructed to base long range bombers that could guard the approaches to Panama. The deployment of aircraft and units, and the air defense of the Caribbean, extended in a triangle extending from MacDill Field, FL, to Puerto Rico and Panama. In September 1940, the U.S. sent 50 World War I destroyers to the British Royal Navy in exchange for the right to build air bases on the Bahamas, Jamaica, and further south to St. Lucia, Antigua, Trinidad, and British Guiana. Once these air fields were operational, the B-18s of bombardment and reconnaissance squadrons of the 19BW of the PCDAF flew monthly "goodwill" flights to these new fields. Also, the Bolos were assigned to fly to Central and South American Republics, particularly Peru and Ecuador, usually for some official occasion. These flights were made by a dozen or more B-18s, often accompanied by a few P-36s, and included impressive formation flights over cities along the route. Once in the host country there were public displays of the aircraft, and local officials were treated to demonstration flights. These goodwill flights were mainly used as show of strength displays that also reconnoitered various airfields that could be of possible use to the Axis. In addition to their goodwill function, the 19BW used them as training flights to familiarize the crews with the various airfields, weather conditions, and navigational difficulties associated with flying in the Caribbean, and as a chance to test their new long-range radio communications system to the 19BW HQ (call WZA), Albrook Field (call VO7), and France Field (call WYP).

In March 1940 Brig.Gen. Herbert Dargue, CO of the 19th Bomb Wing, scheduled a goodwill visit of 12 B-18s to Guayaquil, Ecuador, and Lima, Peru, to demonstrate the U.S. air presence in Central America, and to intimidate these governments, whose airlines were controlled by German and Italian nationals. Ecuador governed the Galapagos Islands, which are located about 1,000 miles to the southwest of the Pacific approaches to the Panama Canal, and American interest in building a base there continued, even though a Japanese attack from that sector was unlikely. To counter Axis presence in Latin America goodwill flights were increased, especially after the fall of France in mid-1940, when the threat to the CZ, Caribbean, and South America intensified with the French Vichy presence on the islands of Martinique and Guadalupe in the Antilles, and French Guyana on the northeastern South American coast. On 16 December 1940 three B-18As of the 9BG, one from each squadron (1BS, 5BS, and 99BS), left from France Field on a non-stop "diplomatic" flight to Quito, Ecuador. Their mission was to deliver a cargo of fresh vegetables and other indulgences to the U.S. Legation that was still trying to influence the Ecuadorian government in securing the Galapagos airbase. The 5RS was particularly active in flying goodwill flights with its three B-18s. Between 21 June and 4 July 1941 it flew to Lima, Peru, stopping at Trinidad, Panama, and Ecuador, and for 12 days in November 1941 a 5RS B-18 transported a number of Hollywood film stars, stopping at various Caribbean resort islands. During the same month a 5RS B-18 represented the AAC at the inauguration of the new Pan Am Airways facility at Port-au-Prince, Haiti.

1941 Caribbean Maneuvers

In January and February 1941 two large joint maneuvers were conducted to test the PCD defenses. In the first tests B-18s were to act as a defending force, and in the second as an attacking enemy force attempting to penetrate Canal defenses. In the first test, Navy patrol planes on reconnaissance from Coco Solo NAS spotted an "enemy carrier" (actually a small Navy vessel) in the Bay of Panama about 100 miles from the Pacific entrance to the Canal. The Navy radioed the "carrier's" position to the Army Control Center at Quarry Heights, which immediately dispatched defending B-18s of the 19BW located at Albrook, France, and Rio Hato Fields, and the Bolos located the "carrier" and sank it. As the enemy, the B-18s attacked the Pacific side Canal defenses, and were able to approach without being detected and strike the air fields without the interceptors of the 16th and 37th Pursuit Groups getting off the ground. The tests determined that long-range and standing patrols were necessary to discover attackers to give interceptors enough lead time to get off the ground.

Pre-war Alarms and the Strengthening of Defenses

On 10 February 1941 the Caribbean Defense Command (CDC) was

During November 1941 a 5RS B-18 represented the AAC at the inauguration of the new Pan Am Airways facility at Port-au-Prince, Haiti. (Pima)

created by uniting the Panama Canal Department Air Force (PCDAF) and Puerto Rico Air Force, which continued under the inept command of Lt.Gen. Daniel Van Voorhis. But the PCDAF was placed under the command of Maj.Gen. Frank Andrews. Gen. Marshall felt that too many aircraft were going to accumulate in certain areas of the Caribbean, and that aircraft in Panama, in particular, should be available for operations throughout the theater. Panama Canal Department CO Voorhis disagreed, stating that these aircraft should remain in Panama, as they were intended for the defense of the Canal. The issue was settled on 8 May 1941, when the CDC was formally redesignated as the Caribbean Air Force (CAF), to indicate the extent and area of operations of the organization at the time. Encompassing the aircraft in the PCD, the Puerto Rico Department, and the Trinidad Base Command, the CAF was the largest organization in the entire Air Corps at that time, and for the first time in American history hemispheric defensive operations included bases in foreign countries.

By June 1941 CAF air strength increased greatly with: 27 B-18s, 17 B-18As, 11 B-17s 13 A-20s, 13 A-17s, 92 P-40s, 24 P-39s, 17 P-36s, and 9 P-26s. By late 1941 the air force at Puerto Rico consisted of 6,000 men of the 13th Composite Wing, which was comprised of 21 B-18s and 92 fighters, mostly P-39s and P-40s, and a few P-36s. As more aircraft arrived in Panama, there was a need for a bomber field with a longer runway. Construction was begun in 1939 on a base at Bruja Point, and was completed in 1941 as Howard Field. As a result of the U.S. taking "protective custody" of the British possessions of Jamaica, Antigua, St. Lucia, Trinidad, and British Guiana, a major reorganization of U.S. air command was necessary. Although located on the British colony of Jamaica, Vernam Field, located 35 miles due west of Kingston, was one of the first projects on the Army Air Force's pre-war expansion agenda. Vernam was to be used as a critical staging area for training, and for the operation of a proposed Composite Wing, and was completed in mid-June 1941 when a detachment of the 27RS was the first to occupy it in April 1942. As U-Boat operations shifted to the south Caribbean Vernam was relegated to an insignificant status, and the 35BS was the last unit to operate from the base, leaving in October 1943.

The constant training flights could and did take their toll. On 9 June 1941 pilot 1Lt. Robert Walton and co-pilot 2Lt. Otto Ernst, as well as the three man crew of 74BS B-18 (37-580), were flying a navigation mission along the southwestern coastline of Panama when they were ordered to return to Albrook Field to assist in a search for a three man O-47 observation aircraft reported down in southwest Panama. The Bolo refueled and took off in the early afternoon, heading west to search a sector about 60 miles from Albrook that was located between the Pacific Coast and the central mountain range. During the search the pilots encountered a large dark cloud bank, and Walton tried to climb, but hit the top of a mountain near El Valle, and crashed in flames and exploded. Only the flight engineer, Sgt. Aldo Napolitano, survived.

Rumors of a possible Japanese and/or German attack on the Panama Canal were prevalent beginning in July 1941. Anti-submarine and torpedo nets were placed in front of the locks on both the Atlantic and Pacific sides of the Canal. The CAF was put on official and unofficial high alert, and the 19BW put at least one B-18 on standby or on patrol within 30 minutes emergency recall at several bases. On 22 August 1941 the Tenth Naval District issued an alert that two U-Boats and a tender were possibly heading into the Caribbean, and stated that the Navy was flying reconnaissance, and directing surface vessels to the area. Maj.Gen. J.L. Collins, CO of the Puerto Rican Department, was ordered to prepare as many aircraft "as might be deemed necessary" for action. Collins ordered nine new B-18s fueled and loaded with 300 pound bombs in anticipation of a Navy request to intercept these vessels, but the bombers remained on the ground, as the threat was just that.

On 18 September 1941 the Bomber Command, Caribbean Air Force was activated, but on 25 October became the 6th Bomber Command, with HQ at Albrook Field. Meanwhile, Puerto Rico Area of the Bomber Command, Caribbean Air Force, was activated on 27 October. On 27 October 1941 a number of CAF units were deployed to remote Caribbean airfields:

To Atkinson Field British Guiana:	44RS
To Beane Field, St. Lucia:	5BS

Atkinson Field, British Guiana. (USAF via AFHRC)

Wal er Field, Trinidad. (USAF via AFHRC)

To Coolidge Field, Antigua:	35BS(H)
To Piarco Field, Trinidad:	B-18 detachment of the 9BW
To Waller Field, Trinidad:	1BS and HQ & HQS, 9BG

B-18 Deployment: December/January 1941

Albrook Field, CZ:	1 B-18>HQS/12PW
Atkinson Field, British Guiana:	3 B-18s & 2 B-18As>44RS
	6 B-18As>99BS
Beane Field, St. Lucia:	4 B-18QAs & 1 B-18>5BS
Benedict Field, St. Croix:	3 B-18s>12BS
Borinquen Field, PR:	2 B-18s & 2 B-18As>10BS
	1 B-18 & 3 B-18As>27RS
	2 B-18s & 7 B-18As>40BG
	2 B-18As>5RS
Coolidge Field, Antigua:	4 B-18s>35BS
David Field Panama:	4 B-18s & 1 B-18A>35BS
Guatemala City, Guatemala:	6 B-18s & 1 B-18A>74BS
Rio Hato, Panama:	4 B-18s & 2 B-18As>3BS
Waller Field, Trinidad:	1 B-18 & 1 B-18A>HQS 9BG
	5 B-18As>1BS
Zandery Field, Surinam:	6 B-18As>99BS

B-17 Deployment 1941-1942

3-7 June 1941: 9 B-17Bs arrive:	4 to 6BG France Field
	(one lost 2 August 1941)
	3 to 9BG Rio Hato
	1 to 7RS France Field
	1 to 44RS Albrook Field
August 1941:	1 B-17B arrives and sent to 6BG France Field
13 November 1941:	5 remain in Panama (none airworthy)
	1 to 5BS Beane Field, St. Lucia
	1 to 44RS Atkinson Field, British Guiana
	(one wrecked 5 May 1942)
	2 to 1BS Waller Field, Trinidad
1 December 1941:	9 B-17Bs to 7RS France Field
Mid-December:	9 B-17Es arrive at CZ and
	were soon joined by 6 more
16 February 1942:	8 B17Bs, 4 B-17Es to 74BS Guatemala City,
	Guatemala (one wrecked 3 March 1942)
	4 B-17Es to 25BS Salinas, Ecuador

After mid-February 1942 the B-17s were sent mostly individually from base to base, and by early 1943 long-range patrols were having their affect on the B-17s, and a steady attrition was taking place.

The Panama Canal Zone Post-Pearl Harbor

In December 1941 the PDC uncovered critical shortcomings in its warning system. There were only two SCR-271 radar sets in operation at each end of the Canal, with three more on the way by the end of the year. Installation and operation of the equipment was hindered by lack of trained radar personnel, and the SCR-271s could not establish the altitude of approaching aircraft; the SCR-271 on the Pacific side had a large blind area over the Gulf of Panama. Also, surface vessel detection (AVS) airborne radar had not arrived in the theater to equip the B-18s.

On 8 December 1941 Maj.Gen. Davenport Johnson, CO of the Caribbean Air Force, released a message to his command announcing that Japan had attacked Hawaii. Almost immediately a number of reports of Japanese aircraft carriers off the Mexican coast and Japanese radio transmissions near the Galapagos were reported. The anxious CAF sent out hastily organized B-18 patrols to check out these spurious reports. The first wartime aircraft loss in the Caribbean was on 11 December 1941, when a patrolling 25th Bomb Group B-18A flown by 1Lt. M.N. Miller was "lost in action." The type of action was not stated, but was not due to U-Boat action, as there were no U-Boats in the Caribbean until February 1942.

Immediately after the Pearl Harbor attack, as was their responsibility, the Navy sent out PBYs from Panama into three sectors extending 600 miles out into the Pacific, supplemented by a number of patrol vessels. On 11 December the CDC received a message from HQ that there was "conclusive evidence" that there was a Japanese force operating in the Pacific west of Panama. At this point the CAF had not conducted any regular offshore patrols of the Pacific, as that mission had been assigned to the Navy. The situation made the Navy realize that it was unable to supply enough aircraft to patrol the Pacific sector, and requested that Bomber Command help out. But when called upon to supply patrol aircraft, Brig.Gen. Davenport Johnson replied that his command was also short of aircraft, and requested that 50 "modern" bombardment aircraft be sent to Panama, preferably B-17s, but B-18s would be adequate. In the meantime the Navy continued its PBY patrols, and the Air Force sent out B-18s on patrols within the limits of their range until more viable plans could be formulated for patrols of the Pacific approaches under the newly formed Air Task Force.

It was not until Christmas that the CAF produced an *Estimate of the Situation* that stated that, while the new Air Task Force had consolidated the Air Force and Navy air elements to conduct air reconnaissance of the Atlantic and Pacific sectors of Panama Naval Coastal Frontier; a plan was needed. On 4 January 1942 Bomber Command proposed a realistic plan based on the number of operational aircraft available from various bases. Every commander wanted the B-17, but only ten were available in the whole Caribbean area: six at Guatemala City, Guatemala, and four Guatemalan-based B-17s on loan to the AAC base at Talara, Peru.

This plan dispatched:
From Guatemala City: The Reconnaissance Task Force was to send six B-17s flying two daily reconnaissance flights (one at dawn, and the other four hours later) towards the Galapagos.
From Talara, Peru: four B-17s flying one daily recon flight toward the Galapagos.
From David Field, Panama: eight B-18s flying a daily 440 mile patrol.
From France Field, Panama: five B-18s would continue to fly their customary Pacific patrols to the extent of their range on 986 gallons of fuel and only one 300 pound bomb, while those on the Atlantic side were to carry a full fuel load and "the maximum bomb load under these conditions."

A-20s of the 59BS were to patrol off Panama Pacific coast.
Navy PAT Wing 3 PBYs were to conduct long range patrols off Panama.

Soon the Air Task Force found it was unable to conduct both the Atlantic and Pacific patrols with the forces available, so at the end of January the 40BG and 27RS were transferred from Puerto Rico to Panama. At this time the composition of the CAF was 373 aircraft of all types: 74 bombers (17 B-18s, 42 B-18As, six B-17Bs, eight B-17Es and one B-10); 225 pursuit aircraft; 25 attack aircraft; 20 Navy PBYs; and assorted observation and training aircraft. The CAF was only at 20% of its authorized strength, and many of its squadrons were squadrons in name only, with only a few operational aircraft available.

During the period there were numerous aircraft and base changes in the theater. By 12 February 1942 the patrol plan was:

From Guatemala City: four daily B-17 patrols of 350 miles wide by 900 miles long south of the Guatemala-El Salvador coast.
From Salinas, Ecuador: two daily B-17 patrols to the Galapagos, 100-200 miles wide.
From David Field: two daily B-18 patrols covering a triangular area 350 miles southwest of David.
From the Canal Zone: B-18 patrols flown over two routes: one along the north coast of Colombia to Barranquilla, and then out to sea back to the CZ. The second flight was from Rio Hato to Saint Lucia and Providence Islands, then southwest to the coast line and back to Rio Hato.

During the first few months after Pearl Harbor, despite the increased U-Boat activity, the 6AF continued to believe that the main threat to the Panama Canal was from Japanese carriers from the Pacific, and the 6AF increased deployment of its patrols in this direction. The patrols had to extend out to 900 miles to be effective, so if carriers were discovered a deterrent force could be organized and dispatched. On 20 February an "unknown green-colored object" that created a heavy wake as it dove was reported by a patrol plane in the Gulf of Panama near the approaches to the Canal. A force of B-18s was sent from bases in Panama, but the object could not be located, and the mission went into the records as one of the few against a submarine on the Pacific side of the Canal.

The 6th Air Force is Created from Disorder

Beginning 30 January 1942, Brig.Gen. Harry Malony and his Staff Officers, including Col. Ralph Wooten, spent 11 days conducting an extensive inspection of most bases in the Caribbean Defense Command. Wooten was charged with examining the air aspect of the Command, and was directed to draw up a report. The Wooten Report was an outspoken account of the multiple problems faced by the command. Wooten noted that command of the AAC units within the CDC was under Maj.Gen. Davenport Johnson, headquartered at Albrook Field, Panama, who directed the Bomber, Interceptor, and Service Commands, which included the Bomber, Pursuit, and Service units at all CDC bases. Each base had a Base Command that administered the base and its tactical units. But at each base the Bomber, Pursuit, and Service unit commanders operated virtually independently, and there was no AAC command under Johnson that was sanctioned to control all units in any of the Sectors or bases within these sectors. Johnson felt he could administer the CDC from Panama, but the heads of the Puerto Rico and Trinidad Sectors indicated that they needed an air officer to be put in charge of the air units in their Sectors. While the local command situation in Trinidad was unclear, the command structure of Puerto Rico Sector was particularly convoluted: Maj.Gen. James Collins commanded the Puerto Rico Department; Adm. John Hoover commanded both the Naval forces and the AAC units, not only in Puerto Rico, but also Trinidad; and Col. E.J. House commanded the 6th Interceptor Command, and also AAC units stationed at Antigua and St. Croix. The Bomber and Service Commands at Puerto Rico and Trinidad operated under the Bomber and Service Commands Headquarters at Albrock Field, Panama. As a solution Wooten recommended that the Caribbean be divided into three geographical defense sectors:

Panama Sector, including areas to the north, west, and south of the CZ, that were to protect Panama from attacks from the Pacific and the western Caribbean.
Puerto Rico, including the western Greater Antilles and the islands of the Lesser Antilles.
Trinidad Sector, including the islands of the Lesser Antilles and the bases at British Guiana and Surinam.

Wooten realized that the vast distances and the poor communications of the CDC prevented the Theater Commander from administering direct tactical control over units in the proposed three Sectors. Wooten then suggested that the Theater Commander continue to administer the units in the theater, but that each Sector was to have one Sector Commander who was to control the air operations of his Sector. Wooten's recommendations were to become basic to the administration of the 6th Air Force once it was formed, and it seems that the Army Air Forces would finally gain control of their aircraft from the Navy, at least in the Panama Sector.

Meanwhile, on 16 February 1942 HQ, Caribbean Air Force formed the Antilles Air Task Force (AATF) under the command of the Commanding General of the 6th Interceptor Command. The AATF is discussed later as part of ASW operations in the Caribbean. By April 1942 the U-Boat threat to Panama had increased, and the Commander of the Air Task Force at Albrook Field issued a directive that a standing alert force of two B-18s, six A-17s, and six P-39Ds be situated on the Atlantic and Pacific sides of the Canal. The force was to be fully fueled and armed, and on alert during daytime to attack any reported enemy submarine, with one B-18 on standby for night missions. The P-39 patrols were extended to 30 miles offshore (from three miles earlier). The few Navy aircraft in the CZ instituted anti-submarine patrols as far east as Curacao. Because of the increased U-Boat threat in the Caribbean, it became evident that another AAC force would have to be created to meet this menace.

On 6 March 1942 the Caribbean Air Force was renamed the 6th Air Force, which was assigned responsibility for the air defense of the Canal Zone. Initially Johnson sent the majority of his patrols to the Pacific side, as the fear of a Japanese carrier-borne attack on the Canal amazingly continued four months after Pearl Harbor. Despite intelligence to the contrary, rumors of a Japanese carrier task off the Mexican Pacific coast flourished, and caused the War Department to order reconnaissance patrols of the area be increased. The 6th Air Force gained additional importance, as the Navy only had 25 old PBYs of PAT Wing 3 and a rag tag fleet of two old destroyers, five sub chasers, six submarines, and a few miscellaneous vessels available in the Panama sector. The 53rd Pursuit Group and its P-40s were transferred from MacDill Field, FL, to Panama, remaining until November 1942, when it finally became apparent that the possibility of a Japanese carrier attack was no longer viable after the Japanese defeat during the Battle of Midway in June. From June to November 1942, the 45th Bomb Squadron of the 40th Bomb Group flew B-18s from the grass runway of France Field, which was located across the bay from Colon, Panama. The Squadron flew seven anti-submarine patrols a day to protect shipping in the southwest sector of the Caribbean. When the unit arrived there were only two navigators available, and for the first two months the pilot, co-pilot, and bombardier dealt with the navigation on flights without navigators.

In November 1942 the U-Boats moved to the South Atlantic, where land based aircraft could not reach them, the convoy system was not yet in effect, and Panama became a wartime backwater.

Trinidad and Puerto Rico Sectors: Situation after Pearl Harbor

After Pearl Harbor, the defense of these two sectors was much easier than that of Panama, as they were connected by chains of islands, and there was no critical and vulnerable target to defend (i.e. the Canal). By mid-February 1942 the air operations in the Puerto Rico and Trinidad Sectors were two daily AAC patrols from Borinquen, St. Croix, Antigua, St. Lucia, Trinidad, British Guiana, and Dutch Guiana, protecting the sea passages and shore lines surrounding these areas. The ACC sent out special patrols from Aruba and Curacao, and these patrols were augmented by Navy PBYs from Puerto Rico, St. Lucia, and Trinidad. Nonetheless, while the Antilles formed a geographic defense barrier to air and sea attacks from the east, these daytime patrols over the Lesser Antilles were limited, considering the large area to be covered, and received no night time patrols. There were no patrols between Cuba and Haiti, between Cuba and Florida, or in the central Caribbean south of Puerto Rico, Santo Domingo, and Jamaica; thus, the U-Boats easily slipped into the south Caribbean undetected through the sea passages between the islands and preyed unhindered on shipping there and to the south.

ASW Operations in the Puerto Rico Sector

The need for an air base in Puerto Rico had been endorsed as an obvious extension of the air defense of the Panama Canal, and in January 1939 the Air Corps sent Maj. George Kenney to conduct a survey for airfield sites. In mid-September 1939 work was started on a cane field at Point Borinquen, 60 miles west of San Juan, and by 1945 construction costs would exceed $51 million. On 3 November 1940 the air echelon of the 25th Bombardment Group was the first unit stationed there, arriving with 14 B-18As and two A-17s, 32 officers, and 44 enlisted men. At the same time the 13th Composite Wing also arrived from Langley Field. The 27th Recon Squadron arrived on 27 November with its nine B-18A aircraft, 18 officers, and 28 enlisted men, who flew in from Langley Field with stops at Miami, FL, and Camaguay, Cuba. On 5 December the remaining nine B-18s arrived, followed by boatloads of equipment and supplies. On 1 April 1941 squadrons of the 25BG were ordered to transfer aircraft and personnel to form the 40BG and its 29BS, 44BS, and the 45BS and the 5RS, and were stationed at Borinquen Field, PR, under the command of Col. William Souza. The new under-strength 40BG squadrons were equipped with seven B-18s and one A-17, and lived in tents for five months before moving into permanent barracks.

The Pearl Harbor attack put the Puerto Rican air units on constant alert under a blackout that was occasionally interrupted by inexplicable air raid warnings. Once the war began, the 25BG immediately began to fly long range, six hour patrols that took their toll on the Group's aging B-18s. On 25 February 1942 B-18A (37-479) crashed at Borinquen (all the crew survived), and five of the original seven B-18s were airworthy at the end of that month. On 1

The 27th Reconnaissance Squadron arrived at Borinquen, PR, in November 1940 from Langley with nine B-18As. The aircraft flew daily sea searches checking on shipping nationalities. (USAF)

May 1942 Borinquen came under the control of the Caribbean Air Force, and the four squadrons of the 40BG (29th, 44th, 45th, and 395th) were redesignated to heavy bombardment status, as they were flying B-17s and B-24s, but continued to have B-18s in their inventory. That month the 40th was transferred to the Canal Zone, with auxiliary bases in Guatemala and Baltra Island, in the Galapagos Islands. Patrols flown from Panama were designed to protect the canal from Japanese aircraft carrier attack, or from attack from German commandos transported by U-Boats. On 18 September 1942 the VI Bomber Command of the 6AF assumed control of the 25BG and 40BG. After the U-Boat threat abated Borinquen gradually became a landing field, refueling stop, and aircraft service depot for U.S. aircraft flying to Europe and Africa.

Vichy France, Martinique and a French Aircraft Carrier

After the fall of France in June 1940, the aging (launched in 1920) French aircraft carrier *Bearn*, cruiser *Emile Bertin*, auxiliary cruisers (destroyers) *Barfleur*, *Quercy*, and *Esterelle*, a submarine, and two tankers were based in Martinique harbor, and the old auxiliary training cruiser *Jeanne d'Arc* was interned at Guadalupe. In early 1940 the *Bearn* had been sent to Halifax, Nova Scotia, to collect 106 U.S. aid aircraft, and was to transport them back to France. The aircraft carried by the *Bearn* were 44 Curtiss SBC Helldivers, 15 Curtiss Hawk 75A-4s (P-36s), and six Brewster B-339s (F2-A Buffalos) that had been purchased by the Armee de l'Air. The *Bearn* sailed from Halifax back to France on 16 June, and was in the mid-Atlantic when the French signed the Armistice with Hitler. The carrier was diverted to Martinique to prevent her seizure by the British. At Martinique the aircraft were off-loaded and parked to eventually deteriorate in the humid tropical weather. The cruiser *Emile Bertin* was carrying $250 million of gold from France to the U.S. when it was diverted to Martinique, along with the other French vessels.

Martinique and its capital, Fort-au-France, was the administrative center, and economically the most important of the French colonies. It was considered a hot spot of intrigue, so much so that it was the setting for the Humphrey Bogart/Howard Hawk clone of "Casablanca," "To Have and Have Not." Initially, the Royal Navy, based in the West Indies, ostensibly "protected," but in fact blockaded the French fleet to prevent the warships and gold from falling into German hands.

The presence of these warships caused concern over the safety of the Panama Canal, as the Vichy French had brazenly collaborated with the Germans since the Nazis overran France, and the French High Commissioner of Martinique, RAdm. Georges Robert, was zealously loyal to Gen. Petain's Vichy government. The U.S. State Department immediately began secret negotiations with RAdm. Robert over the disposition of the French warships that soon reached an impasse. To make matters worse, a crewman on the German U-156 had become a casualty, and according to the laws of the seas, its commander requested hospitalization at Fort-au-France. The French complied, but the U.S. press reports gave the impression that Martinique was to become a German U-Boat base. The situation became so critical that President Roosevelt considered sending a Marine brigade and two Army regiments to invade the French islands. The U.S. felt that if the French warships decided to escape to Dakar, Africa, it would present them with a convenient excuse to seize Martinique to be used as a well-located air and naval base between Puerto Rico and Trinidad. However, an agreement was reached in July 1940, and a U.S. task force took over the blockade from the Royal Navy to prevent an Anglo-French confrontation in the Caribbean. As part of the deal the U.S. promised to maintain the administrative status quo, with the French promising to impose stringent limits on the warship's movement. The U.S. Consul was to act as the naval observer over these warships, and B-18s and LB-30s from 1BS were to fly daily reconnaissance flights during April and May 1942. The U.S. air bases in the British West Indies (Coolidge Field on Antigua, and Beane Field on St. Lucia) were the bases assigned to supervise the French warships. The two almost identical island bases bracketed Martinique and Guadalupe, with St. Lucia being only 40 miles directly south of Martinique.

Shortly after Pearl Harbor the U.S. negotiations with France were reopened, as intelligence reports indicated that the French warships were planning to leave, either to escape, or perhaps sortie to attack the Panama Canal. On 17 December further deliberations resulted in retaining the status quo, with the French maintaining control of the two islands, but the requirements for the movement of the ships was more rigorous. 1st Bomber Command clearly regarded the French warships and aircraft as a dangerous force, and developed a plan of operations entitled *Blue Plan No.1*, that was to locate and destroy the French warships before they could attack. *Blue Plan* deployed Trinidad and Puerto Rico based bombers to make a coordinated attack on Martinique and Guadalupe. Once it was decided to attack the French warships, Bomber Command was to scramble every tactical aircraft to a previously determined rendezvous point to wait for orders. The 9th Bombardment Group was

In spring 1940 the aging French aircraft carrier *Bearn* collected 106 U.S. aid aircraft at Halifax, Nova Scotia, and was to transport them back to France. But after the fall of France in June 1940, the *Bearn* escaped to Martinique harbor, where she was interned. (USN)

Coolidge Field, Antigua, BWI. (USAF via AFHRC)

to supply one of its few B-17s to fly long-range search and strike sorties, while its more numerous B-18s were to be held in reserve until the French ships entered their range of action. Six B-18s, eight Army and four Marine fighters, and four Navy PBYs out of Antigua were to attack the *Jeanne d'Arc* at Guadalupe, while B-18s from the 1st and 5th Bomb Squadrons, ten Army and four Marine fighters, and six Navy PBYs were to attack Martinique and destroy the French (American aid) aircraft on the ground. They then were to assist bombers flying out of Puerto Rico in attacking and destroying the French warships. The Martinique attack plan called for the fighters to attack the *Bearn* first with 500 pound bombs, and then strafe the seven French seaplanes and French fighters on the ground, and also to neutralize the AA positions, which were expected to be light. The Marine dive bombers were then to attack the *Bearn*, followed by the B-18s, which were to attack the *Emile Bertin* and finish off the *Bearn*, *Barfleur*, and tankers. The PBYs were to arrive last to attack any vessels still afloat. After the attack, all aircraft were to return to St. Lucia to refuel and rearm. One of the B-18s was to have a cameraman aboard to record the action. If any of the French warships escaped into the Caribbean, contingency plans were made to send all B-18s out against the escapees, except those of the 1BS, which were to continue the attacks on any ships in the harbor. If the warships moved beyond the range of the B-18s, then the B-17s and newly arrived LB-30s were to be held in reserve at Trinidad and dispatched to attack.

In the meantime, B-18s flew numerous recon missions over the harbor, and a Navy vessel patrolled outside the harbor entrance. Six B-18s had been transferred from Borinquen to Antigua, which was 180 miles from Fort-au-France. One B-18 was sent out to patrol over the harbor and report any escape preparations or an attempt, while the others were to remain on alert back at base. If the warships attempted to leave, the B-18 was to radio Antigua and stand by for orders to attack according to *Blue Plan No.1*. The deteriorating French lend-lease American aircraft were sabotaged, allegedly by American agents, to prevent their use, and were no longer a worry. To annoy the Americans, the French seaplane from the *Jeanne d'Arc* frequently took off at high speed, low over the water, and vanished for several hours and then returned. Negotiations continued into mid-May, with State Department officials and V.Adm John. Hoover meeting with V.Adm. Robert. On 15 May 1942 the negotiations were completed, and all American air units were to stand down, as the French naval and merchant vessels were to be immobilized in place. Thus, the B-18 lost a great chance to make a name for itself early in the war by sinking an aircraft carrier and other warships. If this would have occurred, perhaps the bomber would have been given an early role in the bombing of Europe.

After the resolution of the warship impasse, the vessels continued to be tracked by various B-18 units, and the blockade of the island continued. The patrolling aircraft were to stay outside a three mile limit unless it was essential to make a low sweep to observe something unusual. The pilots often were required by "unusual circumstances" to buzz at least one of the French vessels on every patrol! The blockade of the island cut food supplies, and by mid-1943 the population was faced with famine. At the time French Guiana was about to go into the Allied camp, but it was feared that V.Adm. Robert might use the impounded French warships as interference with its transfer to the Free French. Again the Allies reacted swiftly with a show of military strength, moving the new carrier *Essex* and the cruiser *Omaha* into the Windward Islands, and ASW aircraft were armed with bombs and torpedoes to attack any movement of the French fleet from Martinique. American amphibious units and parachute forces practiced an invasion plan, and with the threat of an Allied invasion, V.Adm. Robert was forced to resign on 30 June 1943, turning the island and warships over to V.Adm. Hoover, so that they were available for the Allied war effort. The *Bearn* was soon sent to New Orleans for refit as an aircraft transport, a task she continued after the war for the French Navy in Indo-China.

ASW Caribbean Operations: January 1942-August 1942

Immediately after Pearl Harbor preparations were made to expand the air security of the Caribbean and South America. On 8 December 1941, three B-18s and seven P-40s were allocated for defense of the valuable bauxite shipments from Surinam (Dutch Guiana). At the end of December the base on the Galapagos Islands was begun, and was ready by early May. By year's end, discussions for building bases at Salinas, Ecuador; Guatemala City, Guatemala; and Talara, Peru, were underway. Completion of these bases was important, as the poor weather on the South American Pacific Coast made each base an alternate landing field in case the others were closed due to weather. A rudimentary field was hacked out at Salinas, and the first U.S. aircraft (four B-17s) landed there on 16 January 1942, but it was not until September 1942 that Talara was completely operational.

In January 1942 the Caribbean Air Force, after taking into consideration the Naval and Marine aviation components available, determined that only two bomb groups were required to defend Puerto Rico and Trinidad.

On 23 January the CAF, in its *Estimate of the Situation*, stated that due to the number of patrols and the distances involved, "medium bombardment aircraft were not considered desirable in the Caribbean Theater." However, at the time the B-18 medium bomber was the most numerous model available, and would continue to fight the greater part of the air war in the Caribbean, which would become the major U-Boat hunting grounds.

Operation Neuland

After abandoning East Coast U-Boat operations, Doenitz assigned a large force from the new U-Boat bases on the French coast to the Caribbean to begin operations on the first new moon in February. The new U-Boat offensive, named *Operation Neuland*, was the redeployment of U-Boats to the Trinidad area, off Venezuela, where targets were numerous and vulnerable. Most South American oil originated either from the Venezuelan Maracaibo Fields or from Trinidad, from which it was shipped to refineries on the nearby Dutch islands of Curacao and Aruba (which was the largest oil refiner in the world). Also bauxite, the base mineral for aluminum, was mined in nearby British and Dutch Guiana, and transported to Trinidad for trans-shipment. Almost two million tons of bauxite, two-thirds of the U.S. total supply, was involved. Because of the importance of bauxite and the possible threat of an Axis invasion of the Brazilian Hump, a secret contract was negotiated on 2 November 1940 with Pan Am to build an airfield in the dense jungle 30 miles southwest of Paramaribo, the Dutch Guiana (Surinam) capital. The airfield, named Zanderij (later Anglicized to Zandery), was located on the Surinam River, and was only accessed by a narrow gauge railroad from Paramaribo. When the U.S. entered the war the Dutch granted a "duration of the war plus six months" for its use. The base was plagued by filariasis and Yellow Fever, and the nearby bauxite deposits caused erratic compass readings. However, the railway tracks, river, and the nearby reddish bauxite mines aided navigation.

The Bold Attack by Werner Hartenstein and U-156 on Aruba

The Germans ordered five U-Boats operating under the codename *Westindien* to be in position to attack the Trinidad area by mid-February 1942. The first Caribbean merchant losses occurred in the early morning of 16 February when at 0200 U-156, captained by Werner Hartenstein, boldly entered the inner harbor at San Nicholas Harbor, Aruba, and fired two torpedoes at two tankers. The SS *Oranjestad* and SS *Pedernales* were sunk, as Hartenstein used Aruba's unique "lights on" black out air raid defense system to his advantage to enter the harbor. Since the refineries always had fires going in the refining process, it was found that during a black out of the city, the refineries could be easily identified by their fires. If the lights of the city were left on, the refineries were much more difficult to distinguish, and more lights were added to increase the effect. Smoke from the sinking tankers covered the harbor, making another torpedo attack impossible, and Hartenstein reasoned that the smoke and confusion onshore would protect his U-Boat, and he decided to wait for some clearing to make another torpedo attack.

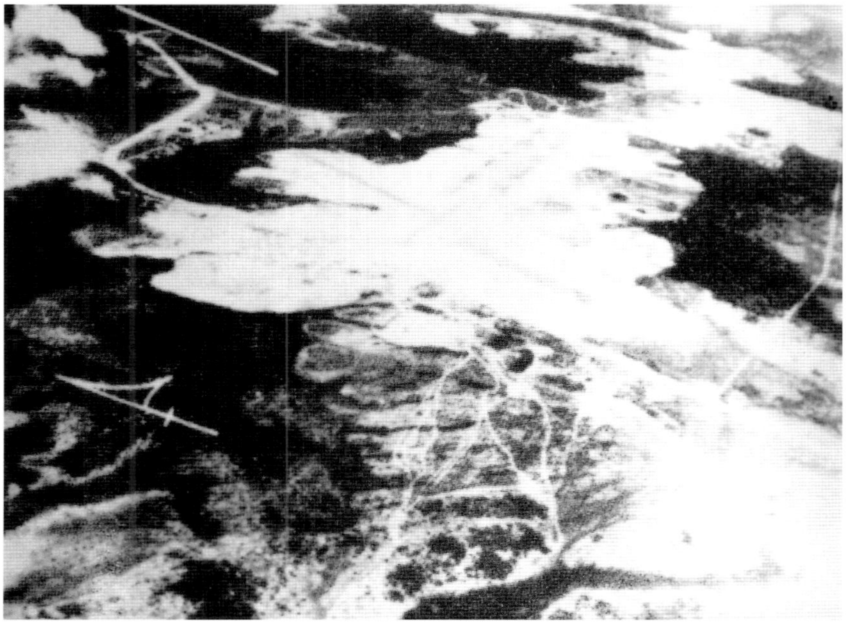

Zandery Field, Dutch Guiana (Surinam), was built to protect the vital bauxite and Venezuelan oil shipments, as well as the oil refineries at Curacao and Aruba. (USAF via AFHRC)

After an hour the U-Boat Captain fired two torpedoes, with one hitting the SS *Arkansas* docked at the Eagle Refinery, and the other missing another vessel, running ashore on the beach. Hartenstein turned his vessel and eased out toward the mouth of the harbor. Through the new smoke he saw the Eagle Refinery still lit up, and he could not resist the temptation to do damage to the large refinery with his 105mm deck gun. The gun crew hurriedly got their gun ready as shells from an American 155mm shore battery came overhead. But in the excitement the U-Boat gunners had forgotten to remove the gun's muzzle cover, and the gun exploded, killing two men. Hartenstein was not about to give up, and had his engineer cut off the damaged end of the gun, and lobbed two shells at the refinery that landed within the grounds. By this time the shore defenses were fully alerted, and the morning sun was on the horizon, so Hartenstein reluctantly decided to escape the harbor and submerge into deep water beyond the surrounding reefs, as anti-submarine air and sea patrols were closing in. The B-18 side of this episode follows:

The first Caribbean merchant losses occurred in the early morning of 16 February, when U-156, captained by Werner Hartenstein, boldly entered the inner harbor at San Nicholas Harbor, Aruba, and fired torpedoes, sinking two tankers. (Author/B.C. West)

On 15 February, after escorting a convoy of four shallow draft oil tankers for six hours from Venezuela to Oranjestad Harbor, Aruba, a B-18 piloted by 40BG CO Lt.Col. Ivan Palmer and Lt. Ira Matthews landed at Aruba Airport to refuel, catch some sleep, and have breakfast. Late that night bombardier Sgt. James Dozier was woken by distant explosions, ran down to the harbor, and saw a fully surfaced German submarine through the heavy smoke. The U-Boat had brazenly torpedoed harbor shipping, and intended to shell the refinery and town. The U-Boat crew climbed down the conning tower ladder and tried to fire the deck gun, but the gun malfunctioned; it was repaired, and fired on the refinery. Dozier ran back to the barracks and found that Matthews and his crew had heard the shots, and everyone raced to the runway to their B-18, which was only now being refueled. Matthews ordered the fuel truck to move out of the way and climbed up through the forward entrance, followed by navigator Lt. David Snow. Sgt. Dozier climbed into the bomb bay, pulled the pins from four depth charges, and entered the aircraft through the forward entrance, followed by Sgt. Bennie Slonina. Lt.Col. Palmer was absent, and Lt. Matthews took command. Matthews was only five months out of flying school, and had only about 100 hours on the B-18 and five landings. He decided to take off downwind over the harbor, located at the end of the runway, in order to fly directly over the U-Boat as soon as possible. As he was applying full power for take off Matthews saw Palmer running across the runway in front of the bomber, throttled back, and applied the brakes. Palmer climbed on board, and disagreed with Matthew's plan to take off with the strong tail winds, and taxied to the opposite end of the runway toward the harbor and U-Boat, but Palmer would have to make a turn at the end of his take off to attack the U-Boat. The B-18 climbed slowly into the wind, and was into its turn as the U-Boat was moving toward the harbor mouth, with its conning tower almost totally submerged in a crash dive into the deeper water just outside the harbor entrance. Matthews ordered the bombardier to begin strafing the U-Boat with its nose gun, but Palmer quickly countermanded this order, as he felt the machine gun's range could not reach the U-Boat. Palmer decided to bank to drop the depth charges, but did not react quickly enough, and the B-18 closed slowly on the U-Boat's fading wake. Valuable time passed as the bombardier determined his release data and opened the bomb bay doors. He instructed the pilot to hold his heading and speed, and to keep the aircraft level. He then lost the U-Boat's wake in deep water, and had to estimate the time to release the depth charges. They exploded far behind the wake, and no debris or oil slick was seen, as Palmer circled around the harbor mouth. Palmer again landed at Aruba to be debriefed, and to take on more fuel and depth charges, and then headed out on the afternoon patrol.

Operation Neuland Continues

In the frenzy of these U-Boat attacks air patrols were increased, and over the next two days five U-Boats were reported to be attacked by air. Even though these air attacks were unproductive, they were enthusiastic enough to cause the Germans to call off continuous attacks during full moon periods directly off the main south

Caribbean ports, and to concentrate their future attacks offshore. The aircraft were not equipped with ASV radar, and the crews and their aircraft, mostly 59BS A-20s and a few B-18s armed with 300 pound bombs, inflicted no damage on the three U-Boats in the area. During the next week the trio of U-Boats sank 21 ships.

In addition to Hartenstein's bold attack, in February U-161, under Albrecht Achilles, penetrated the Gulf of Paria, a large body of water virtually surrounded by the Venezuelan coast and a peninsula hooking west, and separated by a narrow passage into the Gulf. The "safe" Gulf was used as a major training area for the U.S. Navy, and was the end of the North Atlantic convoys. Achilles sunk two ships in Trinidad's Port of Spain harbor, and caused the War Department to greatly strengthen the area as an anti-submarine base, so it was carefully avoided by future U-Boat patrols.

During February and March 1942 the *Kriegsmarine* sent the second wave of six *Neuland* U-Boats to the Caribbean Sea on two to three week patrols. *Operation Neuland* had been a great success, and in less than a month 41 vessels totaling 222,651 tons had been sunk, including 18 valuable oil tankers, and threatened to cut off the oil and bauxite shipments from the Trinidad Sector. During the crisis Naval air aviation did not have enough aircraft or air stations available, but Air Corps aviation was in a better position, even though its main aircraft was the outdated B-18. In the first five months of the war the only threat the U-Boats had to contend with was air attack by Army aircraft. When the bauxite routes were almost cut off by a few U-Boats, the B-18s of the 99BS were sent from the U.S. to Zandery Field, Dutch Guiana, and began operating there only several days later.

After a brief respite in March, in April the next group of U-Boats arrived in the area after being refueled and resupplied by milch cow U-Boats off the Azores. The milch cows were a valuable asset to extending the U-Boat's operational range in the Western Atlantic and Caribbean, as they could provide fuel for 12 Type VII or five Type IX U-Boats, and thus the Type VIIs were able to operate for six weeks before returning to France. The next wave of U-Boats concentrated their attacks on three areas where merchant shipping was forced to concentrate due to geography, and the routing of oil and bauxite. These areas were the Windward Passage, where U-Boats entered the Caribbean from the north, the Aruba-Curacao oil refinery/transfer area, and the Trinidad raw oil-bauxite route. Only 14 vessels were sunk in April, as many of the U-Boats remained in the Puerto Rico Sector. But the results were better in May, as 20 U-Boats were operating in the Caribbean area, and sunk 53 ships totaling 364,000 tons. Seven U-Boats sank 18 ships in the Trinidad Sector, three U-Boats sank 15 vessels around Cuba, 20 vessels were sunk in the Puerto Rico Sector (as U-Boats destined for other Sectors used this area for transit), and five U-Boats sunk an additional 19 ships in the Gulf of Mexico. The carnage continued in June, as 64 ships were sunk in the Caribbean (along with 12 lost in the Gulf of Mexico) for 378,000 tons of precious cargo, especially oil and bauxite. The main result of the U-Boat Caribbean Campaign was to cause delays in global operational planning. Oil reserves in England used for the war effort in North Africa dwindled, as did Royal Navy oil reserves. Both the British and American aircraft industries felt the loss of bauxite for aluminum production. It became apparent that the Allies had to overcome the U-Boats, but the Germans did not have enough U-Boats with enough torpedoes available to sustain the offensive. So in July, through lack of numbers and exhaustion, the U-Boat success totals declined, as the third wave of U-Boats accounted for only 24 ships in the Caribbean, and 13 more in the Gulf of Mexico. But of this total only three were tankers, due to the institution of the convoy system in the Caribbean in July. Trinidad was the ideal location to assemble convoys to Europe and Africa, as it was at the junction of merchant shipping between North and South America, and from the Caribbean. But despite regular convoy operations through the Caribbean, it was estimated that nearly half of the merchant vessels passing through the Trinidad Sector continued to sail alone. Doenitz was unable to take full advantage of this situation due to the lack of U-Boats, as he was forced to return his attention and 113 operational U-Boats to the North Atlantic, where for reasons of logistics they could operate for longer periods, being closer to their French coast bases. However, Doenitz continued to send a small number of U-Boats to the Caribbean to tie up the large ASW forces stationed there. For their small numbers these U-Boats were relatively successful in sinking shipping, but also in tying down 200,000 troops and hundreds of aircraft and patrol ships that were sent out on continual, mostly fruitless, patrols. Millions of dollars were spent on base con-

The 99BS was sent to Zandery Field in February/March 1942 when the U-Boats threatened to stop bauxite and oil shipments in the Trinidad area. Three 99BS B-18As are operating over the railway tracks leading from Zandery to the ocean. (USAF)

struction from the Bahamas, south through the Caribbean and Central America, and into South America. This construction would serve as the post war infrastructure for Caribbean and Latin American aviation.

Air Force Units in the Cuban Area

Operations flown in the Cuban area were conformed to the terms of an agreement reached between Cuba and the U.S. on 19 June 1942, and were controlled by the HQ, 26th Antisubmarine Wing based in Miami. The 22nd, 8th, 17th, and 15th Antisubmarine Squadrons successively were assigned to operations, augmented by squadrons based in Florida and the Caribbean. These units were kept in the area to contain the few U-Boats that sank only a few ships.

Air Force Units in the Antilles

The Antilles-based AAF units, as compared to the Panama-based units, were more peripatetic, moving from area to area and from crisis to crisis, depending on the flux of the anti-submarine war. The majority of AAF units at this time were clustered in the south Caribbean and Antilles chain, while the Navy forces were concentrated in the north Caribbean. In the Trinidad Sector there was more of a joint use of facilities, as there were many Navy aircraft based on Air Force fields in Trinidad, Curacao, Aruba, Surinam, and British Guiana.

During early 1942, the Caribbean Air Force conducted daily patrols over the Puerto Rico and Trinidad Sectors using scattered and under strength air units aided by Navy PBYs based in Trinidad, St. Lucia, and Puerto Rico. These units searched the windward and leeward sides of all islands of the Caribbean Coastal Frontier, especially the passages between these islands, and also the north coast of South America from Trinidad to the Dutch-French Guiana border. There were no nighttime patrols, and surface patrol vessels were very limited in number. The CAF realized that it was not difficult for enemy U-Boats to slip through any of the passages in the Lesser Antilles, and since there were no patrols in the Central Caribbean south of Puerto Rico, Santo Domingo, and Jamaica, U-Boats could move unmolested through that area. At the beginning of March the CAF assigned B-18s to conduct continuous night patrols over the valuable and vulnerable Trinidad, Curacao, and Aruba harbors, and their oil refining and distribution facilities.

Because of the increased U-Boat threat in the first months of 1942, it became evident that an AAC force would have to be created to meet this menace, and on 16 February 1942 HQ, Caribbean Air Force formed the Antilles Air Task Force (AATF), under the command of the Commanding General of the 6th Interceptor Command. The AATF was a provisional air force based throughout the Caribbean at Trinidad, Curacao, Aruba, St. Lucia, Surinam, British Guiana, Puerto Rico, St. Croix, and Antigua. It consisted of 40 B-18 medium bombers, seven A-20 light bombers, and several fighter aircraft. These aircraft were not radar equipped, and few U-Boats were sighted or attacked. The AATF included: the 6th Interceptor Command; 9th, 25th, and 40th Bombardment Groups; and 5th, 27th, and 40th Reconnaissance Squadrons. Although this force included almost all of the tactical units based on Puerto Rico and south through the Antilles chain, it was not as impressive in fact as on paper. The bomb groups were operating at squadron strength, and the recon squadrons often had only two or three operational aircraft.

Since the AATF was under the control of the 6th Interceptor Command, its self-serving CO, Maj.Gen. Edwin House, had the 36th Pursuit Groups transferred from Puerto Rico to Waller Field, Trinidad, and the 16PS transferred from Panama to Zandery Field, Surinam. This was a wise move, as the aircraft were moved to more vulnerable and active areas. On 21 July 1942 the 6th Interceptor Command was renamed the VI Fighter Command, AATF, and in the continuing name game it was renamed the Antilles Air Task Force, and on 4 September renamed the VI Fighter Command. In December 1942 all units of the AATF were placed under the operational control of the Commander (Navy) Caribbean Sea Frontier in an attempt to coordinate its limited resources against the U-Boats in the Trinidad Sector.

To keep track of all the name changes in the Antilles Command the following chart should be of help:

Date	Command
3 Sept 1940	13th Composite Wing (Puerto Rico)
3 June 1940	Caribbean Interceptor Command
18 Sept 1940	Caribbean Air Force Interceptor Command & Caribbean Air Bomber Command
25 Oct 1940	6th Bomber Command and 6th Interceptor Command
19 Dec 1940	Caribbean Naval Coastal Command
26 Dec 1940	Army Air Command, Caribbean Coastal Frontier
5 Feb 1942	6th Interceptor Command, Caribbean Sea Frontier
16 Feb 1942	6th Interceptor Command, Antilles Air Task Force
21 July 1942	VI Fighter Command, Antilles Air Task Force
4 Sept.1942	VI Fighter Command
1 March 1943	Antilles Air Task Force
27 May 1943	Antilles Air Command

ASW Caribbean Operations: August 1942-January 1943
B-18 Deployment in the Caribbean
1st Bomber Command Becomes the AAFAC

In June 1942 the Sea Search Attack Development Unit (SADU) had been established to research ASW equipment and tactics, particularly ASV radar. Soon its talented leader, Dr. Edward Bowles, recognized that the development of these new weapons was in itself not sufficient if they could not be used effectively. Bowles saw that the immediate problem was not the lack of unity of command between the Army and Navy, but the question of organization within the Army Air Force itself, so that it could conduct an offensive campaign using the new radar that went beyond coastal patrols, and extended east into the Atlantic. Bowles proposed that an "Air Antisubmarine Force" be established under the command of a general officer of the AAF, who would control the whole land based air element of the ASW forces, including the Navy's land based inshore patrol aircraft, and also the research and training units that were to be formed. Bowles hoped his plan would implement the allocation of land based aircraft without upsetting the unity of command (e.g. the Navy rules). This new force would be under the Commanding General, AAF, and would restrict its operations to

U.S. coastal waters, but would be able to dispatch "detachments or task forces to other parts of the world." Bowles' proposal, while not accepted by the Navy, did lead to the Army Air Force Antisubmarine Command (AAFAC), which would integrate the First Bomber Command into a ASW force that could test and employ ASV radar.

By the fall of 1942, the 1BC assumed the unofficial capacity as the AAF anti-submarine force, and was spread to bases over the Caribbean; in August a few B-18s of the 40BS were detached to Trinidad and Puerto Rico. Soon, this type of detachment to outlaying bases would prove to be administratively unsatisfactory. By October 1942, with a year's experience fighting the U-Boats, plans had been laid for a larger anti-submarine force. On 15 October 1942 the Air Force activated the U.S. Army Air Forces Anti-submarine Command (AAFAC) to replace the 1st Bomber Command, but in name only. During 1942 the 1BC had remained a long-range bombardment unit, and its anti-submarine mission was considered secondary and temporary, but now, with a new name, it was constituted as an anti-submarine unit. With the name change also came a much expanded responsibility, but with the same inadequate forces. In fact, the strength of the 1BC was 216 aircraft in July 1942, and at the time of the formation of the AAFAC in October combat strength had decreased to 146 aircraft (including 14 B-18s, 12 B-17s, 35 B-25s, and 3 B-24s) with 27 mounting radar. Despite the lack of aircraft, personnel of the AAFAC, while serving in the 1BC, had gained experience and leadership abilities that would be invaluable in executing the tactical ASW doctrine once sufficient and suitable aircraft, especially the B-24, and ASV radar equipment were available. It should be noted that only 230 B-24s had been delivered to the entire AAF by June 1942. The AAFAC was to provide the personnel, aircraft, and equipment for the new Command, which by 20 November was divided into the 25th and 26th Anti-submarine Wings, with headquarters in New York and Miami, respectively. Despite remaining under the operational control of the Navy, the AAFAC afforded greater unity of command of U.S. anti-submarine forces, increased flexibility, and yielded more effective operations. In January 1943 the command had only 19 squadrons and only 20 B-24s. Most of these squadrons operated off the East and Gulf Sea Frontiers, but encountered only a few U-Boats, and made even fewer attacks. The Navy had mandated that the AAFAC commit large numbers of aircraft to patrol these areas where Doenitz had committed only a few U-Boats, with the objective being to put continued pressure on the area, and cause this over commitment of American ASW forces.

The B-18Bs and the B-18 ASRONS

As the few valuable new B-17s, LB-30s, and B-24Ds arrived they were transferred to the influential Panama command, which had experienced virtually no U-Boat activity, while the unwanted B-18s were transferred to areas of U-Boat concentration elsewhere in the Caribbean. Beginning in June 1942, the B-18s were replaced by the superior B-18As. In August 1942 the U-Boat war had reached a perilous phase, and the Air Force began an ASAP program to convert 122 B-18As to the B model with the ASV radar conversion configuration. The B-18As were ferried to San Antonio for the conversion, which removed the glass in the upper bombardier's position and replaced it with a bulbous ASV radome, as well as the installation of a rudimentary Metallic Anomaly Detector (MAD) boom located aft of the tail. By September 1942 three ASV B-18Bs were being delivered per week to the Caribbean. By December 1942 there were 45 B-18Bs and nine old B-18s in the Antilles, and only three B-18s and two B-18As in the CZ. Some 43 B-18Bs served with the 6AF and Antilles Air Command, and probably even more with the assorted ASRONs of the Anti-submarine Command.

Because of the peripatetic nature of the ASRONS (AS) there are few records of their activities. The following is a synopsis of the various ASRONs.

4th Anti-submarine Squadron

The 4AS was originally constituted as the 40BS(M) on 20 November 1940, and was activated on 15 January 1940 at Langley Field. As the U-Boat threat became apparent the unit was selected to serve as a dedicated anti-submarine squadron, and was designated as the 4th Anti-submarine Squadron (M) on 2 November 1942 (it referred to itself unofficially as the 4th ASRON). The specialized unit was self-contained, highly mobile, and nomadic. It was initially based at Guantanamo Bay Naval Air Station, Cuba, and subsequently moved throughout the Antilles in response to U-Boat sightings. The unit moved from Guantanamo on 14 September 1942 to Vernam Field, Jamaica, until 6 October, and then to Edinburgh Field, Trinidad, for a very short time. The unit was moved to Zandery Field, Surinam, where it was attached to the 99BS for only eight days, when it returned to Jamaica for a month before it returned to the States, where it reequipped with B-24s that served off the Northeast U.S. coast and Newfoundland. (Note: The B-18B at the Pima Air & Space Museum is restored in the markings of the 4th ASRON)

7th Anti-submarine Squadron

The 7AS was formed from the 78 Bombardment Squadron (M) on

4th Anti-Submarine

29 November 1942, and spent its first days stationed at Jacksonville, FL, patrolling fruitlessly off the Florida coast in their B-18s. The 7AS and its ten B-18s were attached to the 25BG at Edinburgh Field, Trinidad, to counter the active U-Boats there. However, when they arrived the U-Boats left, and the unit was assigned to convoy duty. Several aircraft were detached to the islands of St. Lucia and Barbados. Barbados was considered a lush duty station, as the AAF did not have a base there, and the crews were billeted in luxury resort hotels. The crews sent to St. Lucia were assigned to fly patrols to watch the French warships at Martinique. The unit left the Caribbean on 20 July 1943.

23rd Anti-submarine Squadron (H)

The 23AS was originally activated on 15 January 1941 as the 76BS. As the 76BS it moved from McChord Field, WA, to Jacksonville, FL, in late May 1942, equipped with one B-18, one B-23, one B-25, one P-38, and ten A-29s. At Jacksonville the unit flew close inshore patrols, and was credited with one confirmed whale destroyed by depth charges. The unit moved to Opa Locka, and then to Tampa, where it flew patrols in the Gulf of Mexico until it was redesignated, changing to its anti-submarine designation on 3 March 1943, and stationed at Batista Field, Cuba, where it was reequipped with B-25s and returned to the U.S. in mid-December 1943.

The U-Boats Continue to Harass and be Harassed

In August the Wolf Packs had returned to the North Atlantic, but Doenitz sent seven U-Boats to challenge the new Caribbean convoy system, which he felt would be vulnerable at several geographical "choke points" where they would have to concentrate. Compared to the Atlantic convoys, these convoys would be escorted by a smaller number of less experienced surface vessels, but to offset this deficiency they would have the advantage of almost continuous air support. The U-Boats arrived in the Caribbean in mid-August and began their patrols, and were immediately harassed by Allied air patrols. On the 15th a Hudson of 53 Squadron shook U-108 with its depth charges, and the next day the 53rd would have a red letter day, when another Hudson battered U-155, helmed by top Caribbean ace Adolf Piening, causing him to leave the area for repairs to his damaged boat. Later that day other 53 Squadron Hudsons would attack U-511 and U-173, causing them to hurriedly crash dive.

On 17 August 1942 a 1BS B-18 piloted by Capt. R.M. McLeod was sent to scout the Windward Passage, looking for the veteran U-108, which was on its ninth patrol. Kaptain Klaus Scholtz had sunk 25 vessels totaling almost 125,000 tons, including the American tanker *Louisiana* that morning. McLeod spotted a U-Boat lurking about 20 miles from a convoy and set up an attack. Just as the German crash dived McLeod dropped four depth charges without visible result. McLeod circled for 45 minutes after the attack, and was relieved by another B-18. However, there was no further sign of the U-Boat, which had been slightly damaged.

Meanwhile, on 19 August 1942 Piening had made his repairs on U-155, and was cruising on the surface east of Trinidad when a B-18 flying at 1,800 feet, piloted by Capt. James Barlow of the 1BS, sighted the U-Boat about five miles away. The U-Boat lookout's warning came late, and the crash dive was so hurried that the lookout was still in the conning tower as the submarine crash dived. Barlow dropped four depth charges, the last of which hit close to the vessel, causing the hull to lean into the sea, shaking the crew, and causing light bulbs to explode. Piening was lucky to escape, but the harassment was not over, as the next day U-155 was again caught on the surface by a 53 Squadron Hudson that so severely damaged the U-Boat that it was unable to dive, but again Piening escaped. He guided U-155 to safety, but had to be escorted across the Atlantic, and then make the dangerous approach into Lorient on the surface. Ironically, both U-155 and U-108 limped into Lorient on the same day, and later Piening and Scholtz could compare notes on the tenacious air patrols of the 1st Bomb Squadron B-18s and 53rd Squadron Hudsons.

Early in the ASW campaign the inexperience of the ASW pilots probably saved several U-Boats from destruction. The U-Boat commander relied on his lookouts for the visual warning of an approaching attacker, and many times after spotting the attacker and crash diving, the U-Boat's conning tower would still be awash as the depth charges fell. So the vulnerable U-Boat's survival was contingent on the accuracy and stealth of the aircraft's pilot, who would in time become more proficient with experience. On 9 July 1942 Type VIIIC U-654 of the 1st U-Boat Flotilla, captained by Lt. Ludwig Forster, left Lorient on its third long patrol, and was refueled near the Azores by milch cow U-463. After patrolling north of Panama without success, on 20 August Forster made a radio transmission, asking the BdU to transfer him to a more lucrative area, but the message was intercepted by Allied intelligence, which plotted U-654's position. The 45th Bomb Squadron, based at France Field, Panama, was one of the first ASW squadrons in the Caribbean, and had accumulated many hours flying fruitless patrols over empty seas. To break the monotony, pilots would hone their skills by approaching and diving on a hypothetical target—a floating barrel or other flotsam that would stand in for a U-Boat. The eager 45th was alerted to the possible presence of a U-654 near their base, and while patrolling 150 miles north of Colon, Panama, on 22 August 1942 the B-18 flown by Capt. P.A. Koenig caught Forster by surprise at conning tower depth at two miles out. Forster immediately began a crash dive near Isla de San Andres. Koenig began his diving attack, dropped all four of his MkXVII depth charges, and bracketed the U-Boat, causing a large oil sick and debris to float to the ocean's surface. Koenig called for support, and soon five more 45BS B-18s arrived, piloted by Captains Edward Glass and Robert Moss, and First Lieutenants Ira Matthews, C.A. Woolsey, and Marvin Goodwyn. The oil slick and debris gave the B-18 pilots a good aiming point, and the five bombers dropped 20 more depth charges on it before returning to base at nightfall. It was Koenig's attack that had sunk U-654, the second U-Boat sunk in the Caribbean, but it was not confirmed to be U-654 until after the war. Forster and all 44 crewmen were killed in the attack. The 45BS continued to fly its B-18s until November, when it was re-equipped with B-24s and transferred to Davis Field to fly Pacific side Panama patrols.

Through September 1942 the Germans had lost only six U-Boats: four in the Gulf of Mexico and two in the Caribbean, in exchange for sinking 287 Allied vessels. But this loss-to-kill ratio could only deteriorate, as there were now ten ASW air squadrons operating between Antigua and Dutch Guiana, and the arrival of more was pending. The Caribbean anti-submarine battle had become a struggle between Army aircraft and U-Boats. Initially, the Navy did not have enough surface craft or suitable aircraft, and used single engine scouting floatplanes, particularly the OS2N Kingfisher, until enough PBY Catalinas arrived. The Army Air Corps used the B-18A, which proved to be an efficient stopgap until enough radar-equipped B-18Bs and then B-24s could be spared for ASW duty. In July 1942 the Gulf of Mexico ceased to be a sphere of U-Boat activity, as the area was virtually land locked and heavily defended by land and sea based ASW units. Despite the large number of targets, the Caribbean was becoming a dangerous hunting ground for the U-Boats, as airpower made any U-Boat movement, day or night, hazardous. In the Caribbean there were no longer any "air gaps," areas that were out of aircraft range. The U-Boats could no longer move toward a convoy on the surface at night due to airborne radar, and daytime movement could be suicide, as the convoys had at least three or four aircraft escorting them.

During September, the Germans moved all Caribbean U-Boat operations toward Trinidad, despite the area's heavy air cover, and a convoy system that decreased the opportunities and increased the risks. Nevertheless, ten boats did well, as they sunk 27 ships of 127,800 tons, mainly by concentrating their attacks on independent vessels rather than convoys.

After sinking a freighter, the new U-516, skippered by Kaptain Gerhard Wiebe, arrived off Trinidad on its first patrol. On 18 September Wiebe, apparently not believing the reports of the heavy ASW air cover in the area, was waiting on the surface in daylight for a convoy to appear. Wiebe soon became a believer when a B-18 dove on his vessel, causing a hurried crash dive that was accompanied by depth charges that jolted the U-Boat. Wiebe remained submerged, and slowly moved his U-Boat away from the surface vessels that the B-18 had alerted to search for him. As he moved north to a less heavily defended area he happened upon an inter-island convoy made up of small coastal vessels. As he set up his attack he was again stalked by a B-18, and had to cancel it and move east, farther away from the air cover. After sinking a 6,000 ton American freighter on the 19th, Wiebe decided to return to Trinidad, and at

ASW patrols were nothing to smile about, as they were mostly boring and non-productive, with hours and hours of looking intently at nothing but empty ocean. (USAF)

noon surfaced to attack another inter-island convoy, but was driven into another crash dive by a St. Lucia based 5BS B-18 that placed its depth charges so close that the U-Boat was damaged, and had creep back to the east for repairs.

The October U-Boat tally showed a decrease in merchantmen sunk, not only because of the fewer number of U-Boats in the area, but also less because the Caribbean ASW defenses were more experienced and numerous. There were seven boats in the Caribbean, but two groups of three U-Boats each were on their way to reinforce the Trinidad offensive.

The ASW patrols were mostly boring and non-productive, but rarely a patrolling bomber came across the survivors of a torpedo attack. On 15 September 1942 the Norwegian freighter M/S *Sorholt*, sailing from Buenos Aires to New York, was torpedoed in the evening by U-515 off Trinidad, and 30 of the crew scrambled into three life boats. The U-Boat surfaced, and the Kaptain asked a few questions about the cargo, then left some canned food and cigarettes for the frightened crew before it submerged. After sunrise a patrolling B-18 flew over the survivors and radioed the position of the life boats to rescue vessels. When there was a reported sinking, aircraft were sent out on dedicated search and rescue missions. On 9 January 1943 Capt. Charles Ross was flying a 35BS B-18 in very poor weather when he found 42 survivors of a torpedoing, and flew over the area of the survivors at 50 feet for over six hours, signaling a Navy patrol vessel to the area for the rescue.

On 18 September, U-175 under Heinrich Bruns was on its first patrol when it arrived in the Trinidad Sector, and sunk the bauxite carrier *Norfolk* north of Georgetown. Two hours later Bruns came across a freighter moving erratically, and Bruns feared it could be a Q-ship, a heavily armed U-Boat hunting decoy; he took no chances and submerged to attack. Bruns sent off two well-aimed torpedoes that appeared to hit the ship but failed to explode, and the ship continued on its erratic way. Bruns surfaced and chased the mystery ship, and hit the ship with shells from his deck guns, again without apparent result. The reason for Bruns' lack of success was that the ship was the bauxite transport *Sanvangin*, whose cargo smothered the deck gun's accurate shell explosions and torpedo hits. The terrified merchantman captain took off in an exaggerated zigzag, heading for the coast, when a 430BS B-18 appeared and drove off the stalking U-Boat. The *Sanvangin* continued with its erratic, high speed escape, and finally its captain ran his ship aground, and the B-18 began organizing a rescue mission. Bruns continued his patrol, and on 1 October 1942 at 0340 U-175 sunk the British freighter SS *Empire Tennyson* about 70 miles off the Trinidad coast. Even though the small ship sank quickly, its radio operator managed to transmit a message that alerted 1st Bomb Squadron B-18s and VP-53 Catalinas. That morning B-18 patrols twice contacted U-175, which was forced to crash dive, sustaining only minor damage, and Bruns and his crew were happy that their tour was over.

On 1 October 1942, a large Type IXC boat was operating on the surface off the French Guiana coast, as its captain believed he was safe, since he was operating too far southeast from Trinidad-based search aircraft. Unknown to the U-Boat captain, the 99BS had been transferred to Surinam to protect the bauxite route, and he did not realize that he was now within range of the Squadron's B-18s operating out of Zandery Field, which was only 100 miles to the east. Meanwhile, pilot 1Lt. Robert Lehti took off in radar-equipped B-18B (37-597) from Zandery just before midnight, and patrolled at 1,000 feet along the French Guiana coast on a moonless night. At 0400 Lehti's radar operator contacted the enemy submarine through broken clouds at about 20 miles from the infamous Devil's Island, and 50 miles from capitol Cayenne. When the U-Boat was spotted about 12 miles away from the B-18, Lehti realized the only way he could successfully attack before the Germans dove was to make a surprise diving attack, using cloud cover and the darkness to his advantage. Lehti dropped the Bolo to 300 feet, following the U-Boat's moonlit wake until he saw the slowly moving surfaced U-Boat at about a quarter mile, with its surprised crew still on deck with the hatches open. Lehti was in good position to drop two depth charges from 50 feet before executing a sharp turning pullout. Two of the depth charges exploded very near the U-Boat, which then dove to about 140 feet, sustaining heavy damage. Two thirds of the crew were killed by the initial flooding, and when the surging sea water reached the batteries in the forward torpedo compartment it caused chlorine gas to leak, incapacitating the 16 crewmen there. There were not enough of the Drager escape equipment available, and only two men escaped from the stricken vessel by swimming out through the flooded torpedo tubes using the escape gear. Upon circling back over the area the B-18 crew saw an oil slick and floating debris, and a lone survivor from the crew of 51. One of the B-18 crew dropped a partially inflated life raft near the survivor who climbed aboard. On the next pass a jug of water was dropped, and the B-18 returned to Zandery. Ten days later the destroyer USS *Ellis* picked up the German survivor, who had endured by killing and eating sea birds that tried to attack him. The veteran submariner was questioned in Trinidad, and divulged important information. The survivor revealed that the B-18 had sunk U-512 of the 10th U-Boat Flotilla commanded by Capt. Wolfgang Schultze. The U-Boat had left Kiel on 8 August 1942 on its first patrol, and had been replenished by milch cow U-462 near the Azores in early September. On the way to the Caribbean the U-Boat had been spotted in the Sargasso Sea and strafed and bombed, sustaining some minor damage. Before its destruction, U-512 had sunk two ships and damaged another. Lehti and his crew were awarded the Distinguished Flying Cross, which was rarely awarded in this theater.

On 9 October, U-332 of the 3rd U-Boat Flotilla out of La Pallice, commanded by Johannes Liebe, was instructed by BdU to move to Trinidad. Liebe had sunk six vessels during two previous patrols off the U.S. coast, and was eager to add to his score. At 1130 he was cruising on the surface about 300 miles southeast of Galeota Point when the lookouts spotted an attacking B-18 of the 99BS. The conning tower was just submerged as three depth charges were dropped and exploded at a depth of 25 feet next to the diving U-Boat. The blast jolted the boat and damaged sensitive equipment, and after an hour of taking cover Liebe felt it was safe to surface to investigate the damage. Liebe found that the attack periscope was bent, and would be useless for any submerged attack. After an hour the lookouts again spotted a B-18, and the repair crews and deck watch scrambled through the conning tower hatch. Liebe crash dived quicker this time, and the charges exploded a little further away from the boat, which was only rattled. Liebe remained submerged, and waited until nightfall to move away from the dangerous area. As soon as he surfaced the U-Boat was attacked by a 53 Squadron Hudson that dropped four more depth charges, fortunately too late to do significant damage. Despite the damage to his boat Liebe did not return home, continuing his patrol, and sunk a 5,000 ton freighter on 19 October. U-332 suffered several more frightening air attacks before leaving the Caribbean for good on the 22nd.

On 13 October U-514, under Hans-Jurgen Auffermann, was ready to return home from a patrol that netted him five ships sunk and one damaged. The U-Boat was given a disheartening send off when it was first sighted by a VP-53 PBY that radioed its position to a 99BS B-18, which made a rattling but unsuccessful depth charge attack.

2Lt. Benson Munro's attempted landing at Atkinson Field in January 1943. (USAF)

Casualties

Few air casualties were due to combat. On 14 October 1942 a 35BS B-18 piloted by Capt. Kenneth Carlsen, flying a patrol from Edinburgh Field, Trinidad, lost both engines (there are unsubstantiated reports that the engines were shot out by a U-Boat). Carlsen managed to hold the stricken bomber steady as the crew bailed out, but Carlsen was killed as he attempted to ditch. After an extensive week's search only the radar operator was rescued from his life raft. Edinburgh Field was renamed Carlsen Field in his honor.

On 23 January 1943 2Lt. Benson Munro was attempting to land at Atkinson Field when his engines failed, and he was forced to land in the Delmerara River, which flowed at the end of the runway. Munro made a successful water landing, and the B-18's renowned amphibious qualities allowed the aircraft to be salvaged.

Summary of Late 1942 Caribbean U-Boat Operations

While the majority of U-Boats continued operating in the North Atlantic, Doenitz sent a small number of them west to harass and tie down Allied Caribbean ASW forces. In September ten U-Boats sunk 27 ships totaling 127,000 tons, as the convoy system had significantly reduced the opportunity for effective U-Boat operations; consequently, the U-Boats preyed on the solo merchantmen. In October 13 ships totaling 52,013 tons were sunk by an average of four operational U-Boats. In November there were only eight U-Boats in the area, as many were in the Mediterranean approaches in response to *Operation Torch*, the Allied invasion of North Africa. Just as Doenitz' previous widespread U-Boat offensive, encompassing the North Atlantic and North American coast through the Caribbean, had diluted the Allied ASW defenses over a wide area, this Mediterranean Allied offensive now spread his U-Boats thin. In November Caribbean U-Boats sunk 24 ships totaling 146,820 tons, and in December 1942 they sank only six ships. But 1942 had been the premier year for U-Boat operations, as 1,160 ships totaling 6.7 million tons had been sunk worldwide, more than in 1939, 1940, and 1941 combined. Of the 1,160 losses, the Caribbean accounted for 337, totaling 1,870,000 tons, or about one-third the total.

ASW Caribbean Operations: January 1943-May 1943

In January 1943 the overworked and aging B-18s had already reached their peak numbers, but gradually attrition reduced these figures. The B-18 soldiered on with 10BS and 23ASRON B-18s based at Carlsen Field (Edinburgh Field), Trinidad, from early January 1943 until August 1943. Gradually, these aging aircraft were augmented with LB-30s and B-24s. At Vernam Field, Jamaica, B-18s covered the southern approaches to the Windward Passage, while 5BS B-18s continued to operate out of St. Lucia, with a detachment stationed at Antigua to protect these islands. During the critical periods in January and March 1943, when U-Boat activity again heightened, the 35BS(M) again was in the forefront of ASW action, and was reinforced with aircraft from other units. In late January it received three B-18Bs from the 90BS, and during the second week of March received two B-18Bs from the 10BS, and three B-18Bs from the 8th and 9th ASRONs. Also, VB-130 of the USN was attached to, or operated with, the unit as "tactical emergencies" dictated. In addition to the AAC air units, the Trinidad Sector based five USN Catalina and Marina flying boat squadrons at NAS Chaguaramas, along with four land based ASW squadrons at Edinburgh Field, and the airship ZP-51. The Royal Navy operated over 300 training aircraft at Piarco Field, many of which were used for ASW duties. Also, USN carrier based aircraft trained from an airfield in the area. Trinidad had 16 runways that could operate 700 aircraft. From Puerto Rico to French Guiana there were 21 ASW squadrons in operation, compared to only three during the productive days in 1942. The Gulf of Paria was a training area for USN warships, and there was a dedicated ASW PC and destroyer unit based there, along with Royal Navy patrol craft and an ocean escort group. As many as 30 convoys consisting of 1,000 merchant vessels per month were plying the Caribbean, escorted by 300 or more warships, and all safe from U-Boat attack.

In an otherwise quiet January a pack of 12 U-Boats. code named *Delphin*, and led by ace Johann Mohr of U-124, would savage two convoys that were attempting to run oil directly from Trinidad to the Mediterranean, instead of the Trinidad-Miami-New York-Liverpool-Mediterranean route. Convoy TM-1 would lose seven of nine tankers, and TB-1 was decimated, and was forced to flee for shallow water off Dutch Guiana. The renowned German ace Hartenstein returned to the coast of Trinidad in U-156, accompanied by U-510, but neither boat was able to get near the heavily escorted convoys, and they were forced away from the area by incessant aerial harassment. On 3 January 1943 1Lt. William Smith, flying a B-18B (37-519) of the 12BS, acquired a radar contact with a surfaced U-Boat in the early dawn at 0640. Smith maneuvered the bomber to drop three depth charges, and then made a second run to drop another depth charge on the U-Boat, which was still surfaced. The crew was unable to make a visual confirmation of the attack results, and dropped a flare that revealed nothing, then dropped a second flare that fell into the water before it ignited. Except for the TB-1/TM-1 convoy disaster, only a schooner and freighter were sunk in January, and by the end of the month there were no U-Boats present, not only in the Caribbean, but also off North or South America; a far cry from the halcyon days of May-June-July 1942. No ships were sunk in February, even though three U-Boats were in the area. U-68, under Albert Lauzemis, discovered that their *Metox* radar warning equipment was not functioning, and withdrew to the Bahamas to wait for spare parts. What Lauzemis did not know was that the Allied aircraft were now equipped with the new centimetric radar, and his *Metox* was useless against it, not malfunctioning.

In November 1942 the United States had been forced to take action against the Vichy French in the Caribbean, and had blockaded Martinique. By March 1943 the situation there became so strained that Vichy Adm. Robert had been forced to resign. After his resignation Robert announced that French Guiana would join the Allied camp, but at the time the leadership of the Free French was being contested by Generals de Gaulle and Giraud. This dispute placed the Allies in a tenuous position, as both the de Gaulle and Giraud factions of the Free French claimed French Guiana. Using the convenient pretext that he was the senior officer, Giraud's representatives decided to take control of the colony. The Americans were concerned about Guiana's La Gallion airfield, and on 18

18s, were the first to use La Gallion Field as an advanced anti-submarine base, beginning on 31 March 1943 with a convoy escort mission.

March 1943 was to see an increase in U-Boat activity, but not in the Caribbean. Since many of the Allied escort vessels remained in the Mediterranean, Doenitz decided to mass 40 U-Boats in the North Atlantic for a decisive battle to destroy the North Atlantic convoy route. In the first ten days of March 41 merchantmen were sunk, with 53 more destroyed in the next ten days, and 108 for the month for the loss of 15 U-Boats. These losses concerned the Allies, who up to this point felt that they were finally winning the Battle of the Atlantic. The Germans were placing five new U-Boats into combat each week, and even the effectiveness of the escorted convoy system was being questioned, as two-thirds of the shipping lost in March was sailing in convoys. The British were firm believers in the convoy system, and Adm. Ernest King continued his belief in the hunter-killer ASW approach, but there had not been enough escort vessels to serve both the convoy and hunter-killer concepts. King acquiesced to the British and their tried and tested convoy system, and lent his available escort vessels to continue to shepherd the convoys. In March the Allied convoy system was reorganized, and the British and Canadians were given responsibility for the North Atlantic convoys, and the U.S. Navy for the mid-Atlantic, American coastal, and all Caribbean and South American convoys. Now King was free to pursue his hunter-killer proclivities, but the decision would initially backfire on King, as the Caribbean convoys would suffer disproportionate losses to the small number of U-Boats Doenitz had allotted to the theater. The reason was that the Caribbean convoys continued to be escorted by Patrol Craft (PC) that had no sonar or radar, and not by destroyers that had been assigned to escort "Jeep" escort carriers in the new Task Groups stationed in the mid-Atlantic and Azores. Soon the British introduced a similar system using small carriers in Support Groups, and incorporated them into their well organized convoy system. Finally the Allies had an effective, integrated ASW system in place that made life ever more difficult for the U-Boats. After leaving their U-Boat pens on the French coast, the U-Boats had to cross the Bay of Biscay submerged or in groups, mutually protecting each other with anti-aircraft fire to escape the canopy of patrolling Allied aircraft. Once safely across the Bay of Biscay, the U-Boat either headed for the Americas, entering their refueling areas off the Azores, or headed to the North Atlantic, where they faced well organized convoys protected by radar equipped ASW aircraft and British Support Groups. The replenished American bound U-Boats then continued on to their mid-Atlantic crossing routes, where the American Task Groups waited. If these U-Boats survived their ocean crossings and then the battles at their destinations, they had to face the same dangers on their return journeys to France.

By early 1943 the U-Boat continued as the principal threat in the Caribbean, and the not long forgotten, ill-perceived threat from German paratroopers or Axis aircraft carriers, or aircraft attacking the Panama Canal. But it was not until March that Gen. George Marshall called a conference in Miami to make an important command ASW decision. Since before the war the USN and AAF had overlapping ASW operations in the Caribbean, and since it was a naval theater of operations, Marshall transferred control of Army ASW assets, including aircraft and crews, to the Navy. This prudent decision would alter ASW warfare in the Caribbean.

Hartenstein Returns to the Caribbean and Finds Things Have Changed

Meanwhile, on 2 March Werner Hartenstein's U-156 returned to the Caribbean for the first time since his audacious attack in Aruba's harbor in June 1942. Hartenstein was desperately trying to catch Convoy TB-4, which had left Trinidad earlier in the day, and to gain more speed the German surfaced his boat just at sunset, only 17 miles off Boca Grande, Trinidad. Almost immediately U-156 was forced to crash dive when it was attacked by a B-18B of 9th Antisubmarine Squadron that was returning from a convoy escort mission. The Bolo made a strong radar contact at 17 miles and closed through haze and approaching darkness that restricted visibility to one mile. The pilot lost altitude, and at three quarters mile sighted the U-Boat visually, and flew over the surfaced U-Boat at 400 feet. The pilot made a quick turn and attacked at 90 degrees to the U-Boat's coarse, dropping two Mark 13 depth charges and two Mark 17 depth bombs at 50 feet into the dive wake and shaking the submarine, which crept away.

Hartenstein was determined to catch TB-4 and resurfaced, but the B-18 had radioed the position of the attack. A radar equipped B-18B of the 80BS out of Edinburgh Field was scrambled 30 minutes later, and made a radar contact off the north coast. The pilot closed at 200 feet altitude, and at one mile turned on his Leigh light and drew the fire of Hartenstein's deck gun, with the tracers causing the pilot to turn off the Leigh Light and move out of range. The Bolo turned to set up an attack run, but Hartenstein crash dived and escaped. The U-Boat captain found that the tactical situation had changed over the past ten months since he left, as the Americans had established an air patrol system that ran in relays, and the Allied pilots had gained experience. The Americans set up a square search and, soon after Hartenstein resurfaced, a patrolling B-18 made a radar contact with his U-Boat at 11 miles, 60 degrees to port. The B-18 homed on the target at 200 feet, but had to make a second pass to line up his attack. Finally at 50 feet, the B-18 dropped two Mark 13 depth charges and two Mark 17 depth bombs spaced at 20 feet that detonated at 25 feet under the boat. If the boat had been in a crash dive the charges would have probably sunk her. As the B-18 was banking to make another attack, the distressed U-Boat went into a crash dive and again escaped.

Hartenstein surfaced later that night, and escaped detection to recharge his batteries and verify the extent of the damage. Before daylight he submerged and headed away from Trinidad and to safety. Unknown to Hartenstein was that the second B-18 attack damaged his U-Boat's fuel tank, and he was trailing a long oil slick on the surface. The ASW command on Trinidad organized a search for the U-Boat, and the next morning at 1045, Airship Z-17 of ZP-51 was patrolling off the northeast coast and discovered the oil slick trailing to the east. The airship captain dropped down and followed the slick to its source, but was unable to get a MAD contact. He began a square search pattern of the area, and soon made a MAD contact and set up an attack. He dropped three depth charges that did no harm, but alerted U-156 that it had been discovered. The airship radioed the U-Boat's position, and a PBY of VP-53 arrived, followed by several patrol vessels that were unable to make contact. Hartenstein was trying to take the submerged U-156 slowly eastward out of harm's way when he accidentally crossed the path of convoy TE-1, which had left Boca Grande, Trinidad, that morning protected by four escorts. The escort, SS *Nelson*, dropped nine depth charges that exploded around U-156, which again escaped, but suffered further damage. The cunning Hartenstein had escaped again, but was forced to surface in early afternoon to recharge his batteries and assess the new damage. Once on the surface, his conning tower lookout soon spotted a patrolling aircraft in the distance, and Hartenstein took U-156 down in a crash dive. The radar equipped B-18B out of Edinburgh had a radar fix, but lost it as its pilot set up his attack. The determined B-18 pilot conducted a further dozen radar runs without result. Hartenstein remained submerged until nightfall and then surfaced, hoping to quickly clear the area without being detected. During the morning of 5 March, after 30 hours of radio silence, Hartenstein felt he was in a safe area, and needed to transmit an important situation report on the dangerous state of operating a U-Boat in the Trinidad area. He reported that the Allies were using a new radar apparatus that could not be picked up by *Metox*, and that aircraft could make night interceptions without using searchlights. His report was intercepted at both Edinburgh and Seawell Airfields, allowing its source to be fixed, and relays of search aircraft were sent along the U-Boat's main course. Hartenstein eluded the search for another three days, and on the morning of the 8th he felt that he was finally safe, nearly 400 miles away from Trinidad, and surfaced, running east. A PBY of VP-53 piloted by Lt. John Dryden was patrolling through clouds when his radar operator detected U-156 20 miles out. Dryden's stealthy approach caught the U-Boat with four depth charges dropped pointblank, lifting it out of the water, and cutting it in two. Eleven of the crew survived the sinking, and Dryden dropped them a life raft, but six of the injured crew soon perished. Later that afternoon the destroyer, SS *Barney*, was sent to the area, but no trace of the survivors was found.

Black May 1943: The End of the U-Boat Threat

In the North Atlantic, the battle for the westbound Convoy ONS 5 was to be the turning point in the Battle of the Atlantic. To decide the battle, the U-Boats needed to overcome the Allied ASW defenses and sink a substantial number of merchantmen, while the Allied convoy escorts and aircraft needed to protect the merchant ships and sink U-Boats. The titanic battle raged from the time ONS 5 sailed from the north coast of Scotland into the Atlantic on 22 April until 6 May, when the U-Boats broke off their attack. The convoy consisted of 43 merchant ships protected by the Royal Navy Escort Groups B-7, comprised of two destroyers, a frigate, four corvettes, and two trawlers, and on 2 May were joined by five destroyers of the 3rd Escort Group. At various times 40 U-Boats made attacks on the convoy, and the BdU had employed 28 U-Boats alone in the culmination of the struggle between 1200 hours on 4 May and 1200 hours on 5 May, when nine ONS 5 vessels were lost. At the end of the battle four U-Boats were lost, and although the convoy lost 13 merchantmen, 20 separate U-Boat attacks were repelled, with the U-Boats forced to crash dive. Adm. Max Horton, CinCWA (Western Approaches), wrote to the Lord Commissioners of the Admiralty: "It may well be that the heavy casualties inflicted on the enemy have gravely affected his morale and will prove to be a turning point in the Battle of the Atlantic."

Horton's assessment proved to be true, as the tide in the North Atlantic had turned, and from this point during the remaining two years of the European war, the U-Boats would never dominate the seas. Westbound convoy ONS 6 followed ONS 5 and was heavily escorted by surface Escort Group B6, while air cover was provided first by aircraft from Iceland, and was reinforced later by aircraft from the escort carrier HMS *Archer*. The heavy escort and judicious routing resulted in the loss of only one ship. ON 181 left Britain a day after ONS 6, and made no U-Boat contacts by evading patrol lines with prudent routing around the Wolf Packs. The next large convoy was Eastbound HX 237, consisting of 42 ships, escorted by two escort groups consisting of four destroyers, four corvettes, and most importantly, the escort carrier HMS *Biter*. HX 237 left St. Johns, Newfoundland, on 7 May, and Doenitz ordered 10 U-Boats to meet it on a patrol line named *Rhein*, and 17 more on the patrol line *Elbe*. The U-Boats on these patrol lines were separated by 20 miles and under radio silence. Meanwhile, slow convoy SC 129 left Halifax, Nova Scotia, on 2 May, and was escorted by two destroyers and five corvettes under the command of famed U-Boat killer Cdr. Donald MacIntyre. This convoy followed HX 237, but on a more southerly course, and was not discovered by the Germans until 8 May. The crucial battle began on 9 May, and continued until 14 May. The Germans dispatched a number of U-Boats from the HX 237 *Rhein-Elbe* group patrol line to intercept SC 129, but the convoy evaded the first patrol line. The BdU ordered another patrol line, which was to be reinforced by U-Boat Group *Drossel*, to move into position to intercept both convoys. The convoy and U-Boats battled over several days, and when the struggle ended on 14 May, HX 237 lost only three vessels, while SC 129 lost only two vessels. The escort carrier *Biter* destroyed two U-Boats, while a patrol aircraft and surface vessel sunk two more, respectively, for the seemingly small loss of only four U-Boats, but Doenitz could not sustain a one-for-one U-Boat to merchantman loss ratio. The next convoy, SC 130, did not lose one vessel, but the attackers lost four U-Boats. East bound HX 239 lost none of its 44 vessels, as it eluded the first patrol line of 21 U-Boats, and was then protected the remainder of the way to Britain from many U-Boat contacts by a heavy escort, particularly from aircraft off the escort carrier HMS *Archer*, which sank U-572. Westbound convoy ON 184 arrived safely without losing a ship, also protected by an escort carrier (the USS *Bogue*), which sank U-569. The day of the escort carrier had come, and the U-Boats were doomed.

After losing four U-Boats in April, Doenitz lost 18 U-Boats in May in exchange for only 29 Allied merchant ships in 12 convoy battles, and now for each U-Boat lost the Allies only lost 10,000 tons of shipping. On 19 May Doenitz lost his son Peter to a hedge hog attack by a Royal Navy corvette while he was serving onboard U-954 on its first patrol. The tide of war had turned for Germany, first in North Africa, and then at Stalingrad, and now for the once successful U-Boat campaign. By mid-1943 the American shipbuilding colossus had replaced the entire tonnage lost in the previous four years of the war, and the Allied convoys were able to arrive in England intact for the build up of the huge masses of equipment and supplies for the invasion of France in 1944. In July Doenitz was forced to call off the Battle of the Atlantic, and to re-examine the reasons for the reversal of fortune of his U-Boat operations.

Doenitz concluded that the main reason for the setback was Allied radar. British developed aerial radar (AVS) and undersea sonar (Asdic) had been improved by the Americans, and with experience proved to be the Allies' leading anti-submarine weapons. On the sea, better ASW techniques and detection equipment were developed, and when combined with better depth charges and the hedge hog rocket propelled charges made the survival of the U-Boat problematic. The BdU reported that during the recent North Atlantic battles Allied aircraft "played an important part in causing such high losses," as it estimated 20 U-Boats definitely lost to Allied aircraft, six possibly destroyed by aircraft and eight sunk by surface vessels. The BdU drew up a memorandum entitled "Effects of Radar used by Enemy Forces in U-Boat Warfare and Necessary Countermeasures," which believed that the Allies had a "type of radar and/or infrared apparatus" that gave aircraft the capability to detect a surfaced U-Boat from a long distance as it approached the convoy, and also to detect a U-Boat patrol line, and thus give the Allies the ability to reroute the convoys around them.

Unknown to Doenitz was that his supposedly unbreakable Enigma U-Boat codes had been deciphered, and allowed the routing of convoys around U-Boat patrol lines, as well as exposed the location of the Wolf Packs. The breaking of the German codes gave the Allies vital information on the location of the milch cow fleet, which was subsequently destroyed in a priority concerted air and sea attack on these vital supply vessels. However, after breaking the German Enigma codes the British Admiralty treated their own convoy codes rather cavalierly, which allowed the Germans to decode them, giving the BdU vital information on convoy routes, and causing avoidable merchant losses to the alerted U-Boats. In this ongoing cat and mouse intelligence game, fortunately the Allies were faster and better, while the Germans never discovered that their codes had been broken.

The "fight back" U-Boats had their large deck gun replaced by a semi-automatic 37mm cannon, and an anti-aircraft platform was added by extending the aft end of the conning tower, while twin 20mm cannons replaced the single 20mm gun. (USN)

The treacherous mid-Atlantic air gap had been closed by hunter-killer teams of escort carriers and their radar-equipped aircraft, which operated in conjunction with improved ASW surface vessels and their sonar to cover thousands of square miles of open ocean that had previously been out of the limit of long range aircraft. The patrolling aircraft could spot surfaced U-Boats from miles away and call in the ASW vessels for the kill. The mere appearance of the now more plentiful escort vessels around a convoy was deterrent enough to discourage the U-Boats from approaching.

Ultimately, the *Kriegsmarine's* real problem was U-Boat technology, which needed updating to overcome Allied weapons, intelligence, and tactics. The Germans had the revolutionary Walther U-Boat on the drawing boards, but failed to develop this U-Boat, which could stay underwater indefinitely at high speeds and depths, and would have made them difficult targets for Allied sonar and radar. But instead of prioritizing the Walther U-Boats, the BdU countered the Allied convoys and ASW defenses with new types of torpedoes, equipping the U-Boats with heavier AA defenses, and developing specialized anti-aircraft U-Boats. The new "fight back" U-Boats had their large deck gun replaced by a semi-automatic 37mm cannon, and an anti-aircraft platform was added by extending the aft end of the conning tower, and twin 20mm cannons replaced the single 20mm gun. Doenitz changed U-Boat tactics, and ordered preferably two or three of the AA enhanced U-Boats to remain surfaced when attacking aircraft approached. The problem was that once the U-Boat(s) decided to remain on the surface they could not dive, and became vulnerable until they completely submerged. It took attacking pilots a while to learn the hard way to stay out of AA range and call for reinforcements for a coordinated attack to overwhelm the U-Boat defenses, which did not carry enough AA ammunition to defend this type of attack. Although the U-Boats would be successful, shooting down a few aircraft, by this point in the war the Allies could easily afford to lose them.

ASW Caribbean Operations: May 1943-July 1943
Increase in American Airpower in the Trinidad Area

The decreased U-Boat activity in the north Caribbean was an opportunity for the U.S. ASW forces to rest, regroup, reinforce, train, and build new auxiliary bases that would allow ASW aircraft to be deployed to the more active Trinidad area. This period was accompanied by confusing deployments, redesignations, and sometimes the disappearance of units.

35th Bomb Squadron

The 35BS from Borinquen, PR, replaced the veteran and efficient 99BS at Zandery Field, Dutch Guiana, and Atkinson Field, British Guiana. The 35BS had a varied and complicated organizational history. The 35BS (Heavy) had initially flown B-18s out of Langley Field before it was transferred to Borinquen, PR, in late October 1940. After a year in Puerto Rico the unit was sent to Coolidge

Field, Antigua, BWI, with its four B-18s and three B-18As. The four old B-18s were replaced with six B-18As in February 1942. In May 1942 the unit was redesignated as the 35BS (Medium), which more correctly described its aircraft and role. In late October the unit's four well-worn B-18As were replaced with four B-18Bs equipped with the new ASV radar. The rudimentary radar sets proved to be very difficult to operate and maintain, and thus initially were of little use. At the end of October 1942, the personnel and equipment of the 430BS was transferred to the 35BS(M). The 430BS(H) had been the result of the reorganization of the troubled 44RS on 22 April 1942. The 44RS had served in the CZ from April 1941 (as the 44th Observation Squadron) at Albrook, and then shortly at Howard Field, where it flew its five B-18s, a B-18A, and a B-17B on long range patrols. In early November 1941 the unit was transferred to Atkinson Field, and attached to the Trinidad Base Command under the 9BG(H). In British Guiana the unit had its B-17 taken from it, and operated with three B-18s and two B-18As, but by mid-February an accident and maintenance problems reduced its operating strength to one B-18A, and the unit had only one trained combat crew! On 22 April 1942 the 44RS was disbanded. and formed into the 430BS(H). At the time it received the 430BS, the 35BS also received a portion of the equipment and personnel of the 99BS(M), based at Zandery Field, Surinam. The 99th was divided into A and B flights, which were to transfer to the 35BS, and C and D flights, which were to remain as the 430BS. The 430BS was later detached from Atkinson and sent to Brazil to meet the increased U-Boat activity there. Meanwhile, by December 1942 the 35BS had five B-18Bs and one B-18 at Zandery, one B-18B at Atkinson, and five B-18Bs and a B-18 at Coolidge Field, Antigua, to join its original complement of Bolos that had been reduced to a B-18, and two B-18Bs. In January 1943 the 35th Headquarters was returned from Coolidge to Borinquen, PR, while the B and D flights of the 417BS(M) were attached to the 35th at Zandery.

The 1BS, the pioneer Caribbean unit, left Edinburgh Field, Trinidad, and was replaced by the three squadrons (10BS, 80BS, and 417BS) of the 25BG, which were equipped with 28 B-18Bs. A British air base, Edinburgh Field had grown to two runways, and was the designated bomber base, while Waller Field handled the fighter contingent. By October 1942 the 6AF had reached its maximum numbers, with 115 bombers: 65 B-18s, 20 B-24s, 17 B-17s (six had crashed or were scrapped due to wear), and 13 LB-30s. Occasionally, to augment AAF patrols, the 6AF Command was lent 25 Navy PBYs when they were available.

9th Antisubmarine Squadron

The 9th Antisubmarine Squadron was the first ASW squadron to operate from Trinidad, when 42 officers and 72 enlisted men of the air echelon under the command of Maj. Glendon Overing arrived with ten B-18Bs at Edinburgh Field on 2 December 1942. The Squadron joined the 10BS, a flight from the 417BS, and a few Navy planes to conduct ASW patrols. However, once the 9AS arrived the BdU began to withdraw their U-Boats from the area, and after sinking 13 in November and five vessels in December, none were sunk in January. However, before its return to the U.S. at the end of March 1943, the unit reported seven sightings and two attacks, but the unit flew many convoy protective patrols and accelerated training. During the time the unit gained experience in using its radar, and once back in the States it rejoined its ground echelon for transitional training in the B-24 ASW aircraft.

7th Antisubmarine Squadron

Ten ASV equipped B-18Bs of the air echelon of the 7th Antisubmarine Squadron, and part of its ground echelon, were sent to Trinidad to replace the 9th Antisubmarine Squadron when it left. They continued convoy patrols both to the west, and the oil and bauxite shipping lanes along the South American coast. Sporadically, convoys would sail to the north and northeast of Trinidad, and to furnish inclusive coverage several aircraft were sent to St. Lucia and Barbados. Normally, depending on the weather, four to ten daily patrols were flown, three quarters using radar. From the end of March until the middle of July, when it was relieved by the 8th Antisubmarine Squadron, the Squadron made only one contact and no attacks. The Squadron did detach two B-18s to St. Lucia to patrol the French warships at Martinique. By July 1943 the Squadron received a few B-24D replacement patrol bombers, and unsuccessfully participated in the three attacks made that month by the Squadron, but the Liberators did contribute in two successful killer-hunts. On 31 August the Squadron was replaced by the 33rd Antisubmarine Squadron, which experimentally flew 75mm cannon equipped B-25s.

Carribean U-Boat Campaign Renewed

After the early 1943 U-Boat respite in the Caribbean, Doenitz decided to renew the onslaught against the attractive Caribbean convoys by transferring 30 U-Boats there piecemeal from the now hazardous North Atlantic. The offensive was to be maintained by a number of milch cow fuel tanker U-Boats. The voyage across the Bay of Biscay and into the Atlantic proved to be hazardous, as by the third week in July eight had been sunk on their way west, and four had to return to France due to damage. This left 11 U-Boats gathering in the Caribbean, with seven more soon to join them.

While crossing Biscay, two or three U-Boats grouped together had remained on the surface, and fought attacking aircraft using their enhanced AA defenses. A warning was issued on 14 July to aircraft in the Caribbean theater that these heavily armed U-Boats were to be expected to remain on the surface and fight it out. The warning was not taken too seriously, as thus far U-Boats had not been known to stage a respectable defense, and it was worded in a way that made it sound as though the pilots were expected to attack the U-Boats, and not take the prudent option and call for air support from the nearby bases. Four days later the slow, unwieldy Navy airship K-74 was shot down by U-134 far to the north in the Florida Straits. ASW bombers, particularly the Catalina and Mariner flying boats, were not fast or maneuverable, and in their straight on diving depth charge attacks were particularly vulnerable to the enhanced AA U-Boat armament directed by bold U-Boat captains.

The next day a B-24 of the 35BS flying 220 miles out of Zandery Field would find that the U-Boat had teeth. At 1010 the B-24 spotted a U-Boat on the surface, and banked into a diving attack before the enemy could dive. But instead of diving, the gunners raced to their AA guns and put up a curtain of flak, causing the bomber to break off its attack. For the next half hour the B-24 pilot tried to force the U-Boat to dive, but was driven off by AA fire. The frustrated pilot decided to disregard the AA fire and attacked, dropping four depth charges near the U-Boat, but the Liberator was heavily hit, with one engine on fire, all its propellers holed, and its tail controls damaged. The pilot had no choice but to return to base, beaten by the tenacious and apparently undamaged U-Boat, which was U-662, captained by Heinz-Eberhard Muller. Three hours later, a B-18B of the 35BS piloted by 1Lt. Paul Crandall and 2Lt. J.E. Kinney located a surfaced U-Boat that was trailing an oil slick. Crandall made a quick diving attack, but had to retreat in the face of intense AA fire that put several holes in his aircraft. Crandall circled, looking for an opening to drop his depth charges while the birdcage turret gunner fired at the U-Boat. Finally, the pilot maneuvered into a favorable position to attack the U-Boat and dove into the heavy AA fire, dropping five depth charges, none of which exploded close enough to cause any damage. The B-18 was now heavily damaged, and limped back to base, while the U-Boat (probably Muller) escaped. That day two other U-Boats (U-415 and U-572) put up a strong AA defense, damaging two Mariners (of VP-204 and VP-205). On the 20th, Muller's U-662 was attacked at 2300 by a PBY of VP-94 that was able to drop its depth charges, but suffered serious AA damage and several wounded crewmen. U-662, though extensively damaged, was operational, but Muller's problem was he had used most of its AA ammunition after three duels with attacking aircraft, which it had won. Muller was ordered to pick up more AA ammunition during a rendezvous with U-516, which was scheduled to return to France. The next day on his way to the meeting, a PBY of VP-94 surprised Muller, who did not dive, but ordered his gunners to withhold fire to conserve precious ammunition, and only to fire at the last second to destroy the attacker, or at least deflect his aim. The PBY held its course and dropped its depth charges accurately, sinking the U-Boat in a matter of seconds. Muller and two other men on the deck boarded a raft, and Muller and a lookout endured an epic 17 day, 500 mile voyage before being rescued by the SS *Siren* after being sighted by a patrolling B-24 looking for U-Boats.

U-466, under Gerhard Thaeter, left La Pallice on 29 June, and arrived off French Guiana on 19 July to begin his patrol. During the next four days his U-Boat was prevented from approaching merchantmen sailing off the coast by Navy PBYs of VP-94, and Army B-18Bs and B-24s of the 35BS based out of Gallion and Zandery Fields. As U-466 was cruising on the surface at about 120 miles off the coast it was approached by a VP-94 PBY, which was barreling in for an attack. Thaeter refused to dive, and his gunners filled the sky with AA fire. Despite damage to his aircraft, the pilot pressed his attack and dropped four depth charges, but Theater quickly changed his course enough to escape the full brunt of the blasts. The damaged PBY could not linger, as it was over 600 miles from its base, and left the area. Three hours later, while Theater's engineers were repairing the damage, a B-18B arrived piloted by 1Lt. Paul Crandall, who had attacked Muller's U-662 on the 20th and suffered AA damage. Crandall and co-pilot C.E. Glasener had been alerted by navigator Capt. Joseph Cohen that radar had picked up a surfacing U-Boat five miles away. The U-Boat saw the B-18 immediately, but this time it went into a crash dive. Crandall released four depth charges that detonated under the U-Boat, damaging the hull and equipment. Turning back Crandall saw an oil slick, but after two hours of circling no further contact was made. After waiting until after dark Theater surfaced, and began to repair the damage, which would take well into the next day. At 1430 the next day another patrol aircraft was spotted, and Theater's gunners were forced to fight it out with an attacking 35BS B-24 from Zandery. The gunners opened intense fire, causing the Liberator to drop the depth charges too early, and when the bomber was almost on top of them the 20mm shells tore open its nose. As the B-24 staggered away the U-Boat gunners scrambled inside as the U-Boat started its crash dive. Despite having one engine feathered, numerous holes in the fuselage, a cabin fire, and several wounded on board, the distraught Liberator pilot turned to attack the now submerged U-Boat, and in revenge dropped a knowingly ineffective depth charge before he turned back to base. U-466 was also badly damaged, with two seriously wounded gunners, and after only four days in the Caribbean was forced to return to base. Nearly a year later, on 5 July, U-466 was holed up in Toulon harbor, in Southern France, when she was damaged during a B-24 bombing raid. As the Allies invaded Southern France in August the boat was scuttled on 19 August to prevent her from falling into Allied hands.

During the early evening of 26 July veteran U-653, under Hans-Albrecht Kandler, was on its seventh patrol 90 miles north of Paramaribo when a B-18B flown by Capt. Erskine Berry and 2Lt. R.P. Hall of the 35BS made a radar contact at 2300 at nine miles. The Bolo closed and dropped a flare before it was forced to break off its attack by AA from the U-Boat's deck gun. The B-18 returned and dropped a second flare, but contact was lost, as the U-Boat crash dived. The B-18 continued its patrol, and two and a half hours later got radar contact. Another flare was dropped, and the bomber climbed to get a better view of the search area. As Erskine reached 2,000 feet the wake of the diving U-Boat was spotted, and because he was so high and out of time the optimistic pilot dropped the five depth charges from that altitude! No damage was done to the U-Boat, but Kandler requested permission to leave the unproductive and very dangerous area.

By the end of July the situation for the *Kriegsmarine* hardly looked promising, as 32 U-Boats had been committed to the offensive, but five had already been sunk in the Bay of Biscay on their way to the Caribbean, another five were sunk once they arrived, and five turned back due to damage. To make matters worse, the valuable milch cows were being hunted down and sunk—four in five weeks—and once in the Caribbean the U-Boats were under constant harassment from Army and Navy airpower. During a two week period in July VP-32 had sunk three U-Boats. But the U-Boat offensive was not finished, as there were 14 U-Boats still operating in the area that were going to have to complete their patrols and return to France, when two more would be sunk. The last days of July were going to be chaotic for the beleaguered U-Boats.

U-406, under Horst Dieterichs, left the 7th Flotilla base at St. Nazaire, France, on 26 June on its seventh patrol, and after being in the Caribbean for several fruitless weeks, Dieterichs decided to move to better hunting 15 miles off the Dutch Guiana coast. During the early evening of 29 July, U-406 was approached by a B-18B of 35BS piloted by Capt. Richard Mansfield and 2Lt. W.A. McGrew, but the Bolo passed over the U-Boat twice failing to identify it. On his third pass Mansfield dropped flares that lit up the U-Boat, but Dieterichs' gunners were waiting for the flares, and the AA guns opened fire on the exposed bomber. The undaunted pilot dove through the fierce barrage and dropped five depth charges that exploded out of lethal range. The bomber was peppered by the AA fire, and was damaged in both wings, causing Mansfield to scurry back to base as the U-Boat dove to safety. The following night at 1100, as Dieterichs was surfaced, he was pounced upon by a Chaguaramas based Mariner of VP-204 piloted by Lt. L.D. Crockett. The Mariner dropped flares that illuminated the submarine. As the Mariner closed on its attack, a 20mm shell entered the cockpit and wounded the co-pilot in the stomach, and the attack had to be broken off. The dying co-pilot was tended to as Crockett dropped two more flares on the resolute German, who would not dive, but stayed to fight. Crockett circled, waiting for an opening, and finally made a bow attack, dropping four ineffective depth charges before he returned to base with his dead co-pilot. A 35BS B-18B arrived to continue the hunt for the submerged U-406, whose captain was not about to push his luck by resurfacing.

The Epic Struggle of Ralph Kapitsky and U-615

On 12 June 1943 type VIIC U-615, under Ralph Kapitsky, left La Pallice, France, for the Caribbean on what was to become its epic fourth patrol. Kapitsky had joined the *Kriegsmarine* in 1935, but soon transferred to the *Luftwaffe*, and by the time of the invasion of Poland he served as a co-pilot of a He-111 bomber and was shot down. He recovered from his wounds, and then flew over a hundred missions during the Battle of Britain in a Ju-88. In December 1940 he transferred to the 7th U-Boat Flotilla, and served two patrols with U-93 before taking command of U-615 of the 3rd U-Boat Flotilla. After sinking two vessels on his second patrol, U-615 was sent out in March 1943 to track Convoy HX229 on his next patrol. U-615 was badly damaged by seven depth charge attacks, but

Captain Ralph Kapitsky (left) commanded U-615 in an epic struggle with American ASW forces when his boat endured 14 attacks from nine aircraft from six different units and a patrol vessel—shooting down one aircraft and damaging two others. (Author/B.C. West)

Kapitsky continued his patrol and followed the convoy, sinking the ammunition ship *Edward B. Dudley*, which violently exploded, and a piece of debris wounded Kapitsky. After its March patrol U-615 was repaired, and modified to the anti-aircraft gun requirements of the new "fight back" air attack order. In the company of U-600 and U-257, the three U-Boats endured concerted air attacks in the Bay of Biscay, but all three suffered damage, and dead and wounded. On one of the attacks U-615's gun crews shot down a Whitley bomber, but one of the conning tower crew died of wounds suffered in a strafing attack. The three U-Boat captains conferred, and decided to make the remainder of the Biscay crossing submerged, before the Coastal Command bombers could concentrate their attacks. U-615 was scheduled to be refueled off the Azores by milch cow U-119, which had been sunk, and tanker U-487 was the only one available. To meet the increased fuel demand the tanker had already been supplied by three operational U-Boats that had their patrols cancelled and returned to France. Since this tanker was surrounded by a number of U-Boats requiring fuel, U-615 was ordered to take its fuel from operational U-535, which would then return to France after the fueling procedur, which slowly transferred 20 tons of fuel through a fire hose. While their refueling process was in progress the two U-Boat captains could hear the depth charge attacks that destroyed U-487. On 5 July, during its return U-535 was subsequently sunk with its crew in the Bay of Biscay by No.53 Squadron of the Coastal Command.

Kapitsky arrived in the Caribbean through the never before used Guadeloupe Passage, which was shallow, lengthy, and under the route of the U.S. air squadrons from Antigua, supervising the Vichy French at Guadeloupe. Kapitsky had orders to attack oil tankers sailing from the Venezuelan Lake Maracaibo oil fields to the refineries at Aruba and Curacao. Kapitsky decided to infiltrate the heavily guarded Gulf of Paria, but after a very brief stay he wisely decided to leave, and moved west toward Curacao to attack the oil tankers that formed the convoys that sailed to New York or Halifax. As he lay in wait a large number of tankers passed by, but Kapitsky was unable to attack due to the substantial surface escort and air patrols. U-615 was forced to stay submerged most of the time, surfacing only for brief periods during the night to recharge the batteries and ventilate the boat. On 27 July 1943, U-615 sunk the small Dutch tanker SS *Rosalia* with two torpedoes ten miles off Curacao's Willemstadt harbor, which set off an alert and search.

During the night of 28/29 July a 12BS B-18B out of Aruba piloted by 1Lt. T.L. Merrill was directed 60 miles northwest of Curacao to the suspected location of the U-Boat, and made a radar contact at 0040. Merrill dropped a flare, the U-Boat was illuminated on the surface, and the pilot dropped the Bolo to 100 feet at 140 mph and released a string of four depth charges through AA fire from the U-Boat's deck gun. The bomber crew observed large spouts of water from the explosions as the bomber took evasive action. No further visual or radar contact was made during a further search, and Kapitsky escaped without damage.

Kapitsky decided to head toward safety, but the next day he accidentally encountered convoy GAT-77, and the periscope wake was spotted by a patrol craft (PC 1196) that raced in and dropped depth charges. The U-Boat suffered slight damage that was repaired after the U-Boat escaped and resurfaced. Kapitsky moved into the open sea, and sent off a long message to Adm. Doenitz at the BdU, which was intercepted by the Allied code breakers, and a massive air and sea search was launched for the talkative U-Boat. The attack and the message prompted the Allies to intensify the search between Trinidad and Curacao for the active U-Boat.

On 5 August, at 1700, the U.S. Navy destroyer USS *Biddle* was acting independently when it made an asdic contact and turned to attack. Kapitsky ordered a decoy to be launched that allowed his boat to escape the inaccurate depth charge attack. Kapitsky crept away as the destroyer called for air support, and continued the search without any success, except that the contact gave Intelligence the means to narrow the boundaries of the search.

That night U-612 was running east on the surface in a light rain about 40 miles north and west of Blanquilla Island, when it was contacted by a USN Mariner flying boat of VP-205 piloted by Lt. J.M. Eskine. Despite having his Mariner shot up by U-572 on 19 July, Eskine was not gun shy. He climbed to 3,000 feet to drop flares to illuminate the submarine, which was running with decks awash, but at only six knots, so as not to leave a long wake. Eskine banked the flying boat in a 180 degree turn and closed at 1,600 feet, dropping two HE bombs that exploded without success. Kapitsky did not dive, hoping that the flares would burn out, and the attacker would then lose him in the darkness. After a while Kapitsky believed he had made the right decision, but then the Mariner reappeared directly overhead and released four depth charges. The charmed U-Boat again escaped destruction when three of the four depth charges did not release from their shackles. The determined Eskine turned into the attack again, came over a conning tower that was about to disappear in a crash dive, and dropped another depth charge that shook the U-Boat. Kapitsky guided his submerged U-Boat slowly eastward, waiting to surface and escape at high speed.

Having located the U-Boat, Eskine radioed Chaguaramas NAS for help, and meanwhile tried a number of tricks to get the U-Boat to surface. Another Mariner (P-6) of VP-205 arrived about an hour later, and the two aircraft flew grids over the area to contact the U-Boat as soon as it surfaced. At 0200 Eskine's radar operator made a contact, and the pilot immediately turned into the attack in the darkness, dropping depth charges on a dark shadow below. He then climbed to release flares, and in their light saw an island schooner pitching in the wake of the explosions, but was apparently unharmed. The wily Kapitsky had seen the schooner through his periscope and surfaced and moved toward it to use its radar shadow as a decoy to escape. By this time the ASW command was thoroughly frustrated by Kapitsky, and was determined to sink U-612; they allotted relays of aircraft to search the area. The first to arrive was a VP-130 Harpoon from Edinburgh Field that joined two VP-205 Mariners that had continued their search. Two B-18Bs of the 7ASRON arrived from Edinburgh Field, and the five aircraft prevented the U-Boat from surfacing that evening.

At sunrise, 6 August, Kapitsky was still in serious trouble; the five aircraft had left to return to base, but only one Mariner (P-4) from VP-205 arrived, piloted by Lt. A.R. Mantuski and a crew of ten. There were five convoys operating off Trinidad, and ASW was forced to assign air cover to them, and many aircraft were operating to the east, searching for U-Boats retreating from the Caribbean. The one for five substitution of his tormentors gave Kapitsky the respite he needed. The Mariner was unable to make a radar contact, and was forced to begin a new search of the area. Kapitsky watched the Mariner through his periscope, and when it disappeared he surfaced and hurried off to the northeast, trying to charge his batteries and ventilate the boat. The German tried to remain on the surface as long as possible, but at 1330 began to submerge just before the Mariner was overhead, and no one aboard the U-Boat saw it. Mantuski dropped four depth charges that bracketed the unsuspecting U-Boat, which went into a steep uncontrolled dive. Kapitsky worked feverishly to halt the dive at 800 feet and bring it back to the surface. U-615 was severely wounded, with both diesels and the port electric motor out of commission, and gallons of sea water were flowing on board, but it was not quite ready to perish, as it was still afloat and proceeding very slowly, stern down. With no electric motors the U-Boat could not run submerged, and Kapitsky had no alternative but to stay on the surface. Once surfaced the gunners raced to their positions, which had been modified in France to add an extra 20mm cannon on the conning tower, and a quad mount of MG 34 machine guns on the deck. The U-Boat gun crews shot the Mariner down, but not before its radio operator managed to transmit a message that alerted VP-204 at Chaguaramas.

It became obvious that the U-Boat was mortally wounded, but Kapitsky did not scuttle the boat, and ordered his mechanics to start repairs, and to try to keep the ship afloat. Kapitsky ordered that all the ammunition be brought on deck to defend the boat. VP-204 could send only one Mariner (P-11 of VP-205), as its others were either damaged in recent attacks, undergoing maintenance, or out on patrol. Chaguaramas recalled other patrolling Mariners for refueling, and other bases on Trinidad were alerted to transfer their aircraft to the area. After an hour the Mariner flown by Lt. L.D. Crockett arrived at the reported position of the attack at 1523 and began his search. Crockett was the same pilot who lost his co-pilot in the attack on Dieterichs' U-406 six days earlier. Fourteen minutes after the Mariner's arrival a radar contact was made at 12 miles to the west, and at seven miles visual contact was made. Crockett did not rush to the attack, but circled at 3,000 feet, three miles out. The U-Boat opened fire, and Crockett had to retreat and continue his surveillance, radioing his base of the situation. Crockett dropped down to 1,500 feet and closed on the U-Boat, walking .50 caliber machine gun fire from his nose guns toward the U-Boat. Kapitsky withheld his AA fire, and at 300 yards opened fire, hitting the Mariner as it released two Mk-17 HE bombs that exploded off the U-Boat's port quarter. The AA fire hit the Mariner's starboard wing root, setting its fuel lines on fire, but Crockett ignored the possibility of an explosion, and incredibly turned to make another diving attack into the face of a curtain of AA fire. Trailing smoke and fire, the Mariner released four depth charges that stunned the U-Boat crew and cracked the already damaged pressure hull, which immediately began to ship water. The U-Boat was doomed, as it slowed to two knots with its rudder jammed, and the U-Boat could only move in slow circles. Soon aircraft from Trinidad, only an hour away, would swarm the area, and a truly *Gotterdammerung* struggle was about to be staged.

The courageous Kapitsky was the epitome of an elite U-Boat captain, and still did not take the easy way out of the predicament by scuttling the boat and surrendering, but encouraged his crew to fight on. The U-Boa's stern began to ship water, and the bow was out of the water, but the machinists got the pumps working, and the ship was trimmed. The AA guns were readied, and ammunition stacked nearby to take down as many of the attackers as possible before the end came. Meanwhile, on Mariner P-11 a machinist mate managed to extinguish the fire, and Crockett flew it out of range and called for help, and would wait on the scene to direct the reinforcements. Sector Head Quarters at Port of Spain had followed the protracted battle for days, and sent a Harpoon of VB-130 from Edinburgh Field; at 1630 it arrived, and Crockett directed a coordinated attack with the Lockheed. The Harpoon was to make the primary depth charge attack from astern, while Crockett swung his Mariner around to the starboard, hoping to divert the U-Boat's attention by strafing and drawing its fire. However, the much faster Harpoon, firing its nose guns on the way in, closed before the Mariner, and the U-Boat gunners ignored the fire from the Mariner. But the Harpoon was so fast that it flew ahead of the AA gunfire, and remained unscathed as it dropped four depth charges that straddled the U-Boat. If the U-Boat had been diving it would have been destroyed, but since it was on the surface the blast pushed it underwater, sweeping the deck crew into the water. Crockett's flying boat continued to strafe the U-Boat as the swamped crew clamored back on deck. Kapitsky was hit in the thigh by a .50 caliber bullet and was bleeding profusely, while his Senior Petty Officer was also seriously wounded. Both wounded men refused to be taken below, and directed the restoration of the U-Boat's trim. Although the boat could only cruise slowly in a circle, it was ready to take on the circling enemy aircraft. The U-Boat had endured four hours of continuous attacks, including four depth charge attacks, one bombing attack, and heavy strafing. Crockett had one depth charge remaining, and was about to close to finally finish off his nemesis when he was driven off by a wall of AA fire to wait for more reinforcements.

Soon the next reinforcement arrived on scene, which was another VP-204 Mariner (P-8) flown by Lt(jg). J.W. Dresbach, who had attacked U-415 on 24 July. Crockett again set up another coordinated attack, with Dresbach making the depth charge attack, while the Harpoon would strafe from the starboard, and Crockett from the port side, slightly ahead of the newly arrived Mariner. At 1815 the attack began, and the U-Boat gunners withheld fire until the attackers were within range. When they opened fire they ignored the two strafing attacks and concentrated their fire on Dresbach, who was mortally wounded in the chest, but his dying act was to release the depth charges. The Mariner co-pilot, Lt(jg). Oran Christian, regained control of the aircraft from the dead pilot and eased the damaged aircraft away, while the four depth charges exploded just aft of the U-Boat, lifting the stern out of the water and further damaging its steering mechanism. The Mariner's nose was smashed and twisted, and the inside of the Mariner was covered with blood from the dead pilot and four wounded crewmen. Ignoring Crockett's orders to break off, the vengeful Christian turned the damaged Mariner back toward the U-Boat and crossed over the target, which was bristling with AA fire. At 1,500 feet Christian dropped two HE bombs that exploded harmlessly 300 feet off the U-Boat's port. Crockett ordered the badly damaged P-8 back to base, and Christian complied and limped off, unsure of arriving safely back at base.

During the attack another Mariner (P-2) of VP-205 arrived flown by Lt.Cdr. Hull, but was ordered by Crockett to stand off until P-8 finished its attack. Crockett dearly wanted to finish off the U-Boat, and again set up a similar attack, with his Mariner and the Harpoon strafing the U-Boat's flanks, and the new Mariner making the depth charge attack just as the sun set. The gunners again ignored the strafing attacks, and concentrated on P-2, whose depth charge release malfunctioned, and the bombs exploded 600 feet short. In frustration Hull continued his attack using his .50 turret guns while flying into the accurate U-Boat AA fire. The Mariner was heavily damaged, and several crewmen were wounded, but the angry Hull did not listen to Crockett imploring him to wait and circle. The rash Hull put the Mariner into a climb to 1,500 feet, and since the Mariner did not have a bombsight, he maneuvered over the target and dropped the bombs by eye, missing by 500 feet, but suffering further significant AA damage, with several more crewmen wounded. At 1640 the Harpoon only had enough fuel to return to base, and reluctantly left for Edinburgh, followed by Hull's wounded P-2.

Now there was standoff between the wounded Kapitsky, lying in his conning tower, and watching Crockett in his circling Mariner. Soon the Mariner would have to return to its base, but the heavily damaged U-Boat was barely afloat after having withstood 12 depth charge and bombing attacks from six different aircraft. Kapitsky and his Senior Petty Officer were severely wounded, and the crew was exhausted from warding off the attacks, and keeping their boat from sinking. Kapitsky realized that he could no longer effectively give commands, and relinquished control to his First Watch Officer. The U-Boat crew hoped that the approaching darkness would give some reprieve, and offer a chance to escape for repairs, and perhaps rescue by another U-Boat.

The Captain of Airship K-68 from ZP-51, Lt(jg). Wallace Wydean, had heard calls for help just as he finished a patrol and was low on fuel. Without hesitation he turned his lumbering airship westward at 75 knots, and arrived on the battle scene soon after the Harpoon and P-2 left. Crockett refused to let Wydean make what would be a suicidal attack and ordered the airship to stand off.

At 1900 it began to get dark, and a weather front was approaching with low clouds and rainy weather; it appeared as though the U-Boat would be given another opportunity to escape. A 10BS B-18B named *Robust Man*, piloted by 1Lt. Milton Wiederhold and 2Lt. P.E. DeWeerdt, raced to the area of the U-Boat, and was contacted by Crockett, who again set up an attack plan. The B-18 was unable to find the U-Boat in the deteriorating weather, and visibility was diminishing to zero. Crockett radioed a course to Wiederhold, and was able to join him to begin a coordinated attack. But somehow U-615 had disappeared, and all the two hunters could do was to continue the search by dropping flares. After about an hour the weather had deteriorated to such an extent that the low clouds and rain prevented the B-18 from getting high enough to drop useful flares. Finally, after having been in battle for over six hours, Crockett was running low on fuel,, and was forced to return to base. Lt.Cdr.

Joster, flying P-15 of VP-205, had recently arrived on the scene, and was given command of the battle by Crockett. After leaving the battle Crockett still could not relax, as it would require all his skill to get his badly damaged aircraft home with only one compass working to guide him through bad weather.

The poor weather made it difficult for the patrolling aircraft to see the ocean below, and they were unable to find and track the U-Boat. However, at this point Wydean's airship K-68 did what airships did best; fly low and slow below the undercast to search for the U-Boat. At 2115 Wydean radioed the B-18 that he had found the U-Boat dead in the water. Despite the final barrage of AA fire, Wiederhold put his B-18 into a diving attack and dropped depth charges that caused more damage to the hull. The weather closed and concealed the U-Boat, and Wiederhold's B-18 was running short on fuel, and was forced to return to Waller Field. Despite running low on fuel himself, Wydean continued K-68's patrol that made the discovery and final attack on the U-Boat possible. By doing so K-68 did not have enough fuel to fly the 180 miles back to Trinidad, and had to try to make an emergency landing on uninhabited Venezuelan Blanquilla Island. Wydean just had enough fuel to reach the island, but once he arrived there were no mooring facilities, and the airship was destroyed in its landing attempt, the last victim of U-615. In the previous week the U-Boat had endured 14 attacks from nine aircraft from six different units and a patrol vessel, shooting down one aircraft and damaging two others.

On U-615 the starboard engine finally quit, leaving the boat dead in the water, and soon the pumps stopped. Kapitsky and two other wounded men were placed in a life raft on the deck in case of emergency. A large wave washed the raft overboard, and the raft disappeared. The Watch Officer in charge ordered a search, which was difficult with the faltering electric motors. Nonetheless, the raft was recovered after an hour with Kapitsky still alive. He would die shortly, and after a brief ceremony his remains were committed to the deep by his dedicated crew. His last act was to order that the grievously damaged U-Boat be scuttled, and the Watch Officer ordered the 43 crewmen onto life rafts as U-615 went down in the dark predawn at 0450.

The USS *Walker*, on its shake down cruise, had been dispatched hours before from Trinidad, and reached the courageous survivors of the U-Boat shortly after it sunk. As the Germans prepared to scramble up the *Walker's* hull nets an explosion was heard, and the *Walker* discontinued the rescue and raced off to join a circling Mariner in the distance that directed the destroyer to make two depth charge attacks. After a while the *Walker* returned to complete the rescue operation.

Kapitsky's determined struggle convinced the American ASW command that there had to be more than one U-Boat involved, and ordered a rigorous search of the area that continued for four more days, tying up scores of warships and numerous aircraft, including a dozen Mariners and many B-18s. This concentrated search provided seven retreating U-Boats in other areas a respite to escape on the surface and head back to France. On 20 October 1943 Brig.Gen. Lyons awarded the crew of the *Robust Man* Distinguished Flying Crosses, and shared credit for the epic sinking of U-615. (There are varying accounts and chronologies of this battle that I have tried to clarify, mainly using as sources; Gaylord Kelshall, *U-Boat War in the Caribbean*, and Theodore Savas, *Silent Hunters*, along with various USN and AAF squadron histories and German U-Boat sources.)

The Army/Navy Contest for Administrative Control Finally Ends

The Army and Navy continued to battle for control until Army Chief of Staff Gen. George C. Marshall intervened in April 1943, and suggested that the Joint Chiefs of Staff take control of anti-submarine operations. However, Navy Commander-in-Chief Adm. Ernest King steadfastly rejected the proposal; resulting in the establishment of the Tenth Fleet on 20 May. The Tenth Fleet was a centralized anti-submarine organization under Navy Command that left the AAF even more firmly under Navy operational control, a circumstance that was wholly intolerable to the Air Force. On 9 July 1943 the Army and Navy met, and it was agreed that the AAF would withdraw from anti-submarine operations altogether. By 6 October 1943 the AAF would turn over its 77 ASV equipped anti-submarine B-24s to the Navy, and the Navy would transfer an equal number of unequipped B-24s to the AAF, so it could equip heavy bomber squadrons in the redesignated 1st Bomber Command of the First Air Force.

July 1943: U-Boat Caribbean Swan Song

In July 1943 the BdU made one last determined offensive in the Trinidad area, with the U-Boats sinking four ships per day. As a consequence there were 13 attacks on U-Boats: four by B-18s of the 35BS at Zandery Field, two by B-24s of the 8AS and Waller and Zandery, and the remainder by Navy aircraft. In ten of the thirteen attacks the Germans chose to fight back on the surface.

July 1943 was the swan song for the German U-Boat force, as 37 had been destroyed world wide, including 18 of the 32 connected with the Caribbean offensive, plus the four milch cow tankers that were supporting them in the mid-Atlantic. The Caribbean payoff for the Germans was only three tankers, two freighters, and two schooners sunk, and damage to two more tankers. During six weeks in the Bay of Biscay, the RAF had sunk 19 U-Boats, many of which tried to fight it out on the surface, causing the loss of 57 aircraft. Biscay now had to be either crossed submerged, which took far longer but was much safer, or crossed by three to five U-Boats grouped on the surface, and whose AA defenses could ward off a single attacking aircraft, but not multiple air attackers.

B-18 Caribbean Swan Song

By November 1942, B-18Bs had flown nearly 50% more ASW patrol hours than any other AAF aircraft; made more U-boat sightings, and carried out the bulk of bombing and depth charge attacks on U-Boats. With the U-Boat threat subsided, and the insidious tropical mosquitoes being the only enemies remaining, the War Department decided to use the 6AF and the Caribbean area as a training ground for melding combat trained crews into tactical units using available aircraft. In mid-August 1943 the old guard passed the baton to the new. The able V.Adm. John Hoover, who had commanded the Caribbean ASW campaign through its most difficult time, was transferred from the Caribbean, and was replaced by V.Adm. A.B. Cooke. At the time the Bolos were being replaced by radar-equipped B-25s and B-24s, but by mid-1943, Puerto Rico and the Antilles sectors still had 54 B-18 and B-18As available (designated as RB-18s). Because the B-18s could not economically be converted to radar-equipped B-18Bs, some were relegated to target towing duties by having C-5 windlass gear installed. The first B-18 to be converted was 36-324, and was stationed at Borinquen Field, PR. The 23rd Tow Target Squadron (TTS) was formed on 1 November 1943 from the incorporation of the 108th Reconnaissance Squadron (Special) and 1st Reconnaissance Squadron (Special), which had previously been target tug units. It was based at Howard Field, Panama, and was equipped with cast off aircraft RB-18s (#36-294, 36-275, and later 36-282), O-47s, and A-20s. The B-18s averaged 4,000 hours additional flying time, became dilapidated, and were abandoned in 1944.

In September 1943 most of the B-18s in the Caribbean were stationed in the Trinidad Sector: six B-18Bs of the 10BS at Edinburgh Field, Trinidad, and 12 B-18Bs of the 35BS at Edinburgh Field, Trinidad, and Zandery Field, Surinam. Of the total 34 B-18s and 61 B-18As assigned to the Caribbean, by December 1943 there were only 15 B-18s known to remain (one with the 10BS, one with the 4TRS, one with the 417BS, one with 4OS as a target tug, two with the 12BS, one with the 35BS, three with the 80BS, two with the 99BS, and three with the 23TTS). The 35BS continued to actively use the B-18B well into 1943. The B-18Bs and their replacement B-25s and B-24s flew hundreds of monotonous hours over empty ocean, and by the end of the war in August 1945 there were three B-18s and five B-18Bs in active service in the Caribbean.

B-18/B-18A/B-18Bs in the Caribbean

B-18s (34)

36-269 36-275 36-279 36-280 36-282 36-283 36-285
36-291 36-294
36-295 36-297 36-299 36-300 36-301 36-302 36-307
36-308 36-309
36-312 36-315 33-318 36-319 36-320 36-321 36-323
36-324 37-9
37-10 37-13 37-24 37-27 37-31* 37-32* 37-33*

B-18As (62) B-18B/C conversions in *italics* (43)+

37-458 *37-462* *37-463* *37-464* *37-465* 37-467 *37-473*
37-474 *37-478*
37-479 37-481 *37-482* *37-483* *37-485* *37-486* *37-487*
37-489 *37-495*
37-497 *37-499* *37-504* *37-507* *37-508* *37-512* *37-513*
37-514 37-516
37-519 *37-530* 37-532 *37-547* 37-548 *37-551* *37-560*
37-563 37-565
37-567 37-568 *37-589* *37-594* 37-601 37-602 37-603
37-604 *37-605*
37-607 37-608 *37-617* *37-621* 37-623 *37-624* *38-587*
38-593 38-602
38-603 *38-604* *38-605* *38-607* *38-608* *39-12* *39-13*
39-24

* First three B-18s to arrive 37-31/32 at France Field, CZ, and 37-33 at Albrook Field, CZ, in January 1939.

+ On 1 April 194C 122 B-18As were to be converted to B-18Bs.

The anti-submarine war was an extremely exasperating experience for B-18 pilots and crews, as they spent weeks, months, and even years over miles of open ocean and off tropical coastlines without ever spotting a U-Boat, much less attacking one. Nevertheless, the B-18 crew had to be constantly ready to dive into an attack on a

In September 1943 many B-18s in the Caribbean were stationed in the Trinidad Sector, with 12 B-18s of the 35BS at Edinburgh Field, Trinidad, and Zandery Field, Surinam. Pictured is a 35BS in a typical pose for the crew to send home. (Pima)

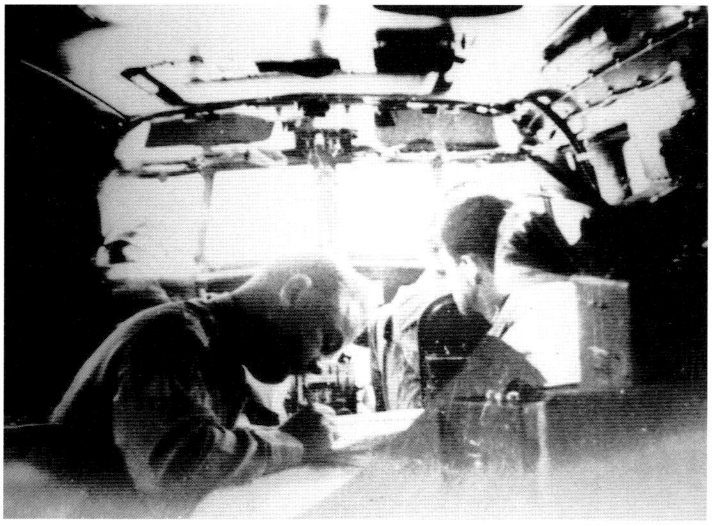

A last look at a B-18 crew that spent hours of fruitless searching, not fully realizing the important role they had played in keeping the U-Boats submerged day and night, thus reducing their effectiveness. (Pima)

U-Boat that was bristling with AA guns that could shoot down several B-18s. Between Pearl Harbor and VJ-Day 28 B-18s were lost, mostly to flying accidents and mechanical failure, a few to enemy action, and at least four disappeared without a trace. However, the plight of the B-18s and their crews paled in comparison to the 863 operational U-Boats, of which 754 were lost, and to the 39,000 German submariners, of whom 27,500 went down with their ships, and 5,000 were taken prisoner.

Summary of the Caribbean U-Boat Campaign

In conducting the U-Boat campaign in 1942, the BdU had concentrated its ultimately limited U-Boat force too thinly in an immense area that extended from Scandinavia, the British Isles, western Europe, south to the west coast of Africa, and across the Atlantic to North, Central, and South America, with excursions into the South Atlantic, and even the Indian Ocean. The Caribbean area alone was very large, and extended from the Gulf of Mexico 3,200 miles south to the north Brazilian border, and was longer than the North Atlantic convoy route that extended from the U.S. east coast to Great Britain. Doenitz' initial U-Boat strategy using a limited number of boats was successful in the North Atlantic and Europe, as the weak Allied ASW defenses were also spread thin. Later, as the Allies instituted the convoy system, the BdU gathered its still relatively small U-Boat forces and developed the successful Wolf Pack concept in the North Atlantic. To be successful in the Americas, the U-Boats needed to be concentrated with as much zeal as they were in the Wolf Packs against a single North Atlantic convoy. A single critical Caribbean area—the Gulf of Mexico, the area between Aruba and the Windward Passage, the oil-rich Trinidad area, or the bauxite route—should have been singled out by Doenitz to bear the brunt of a Caribbean Wolf Pack system to defeat the defenders, and bring shipping in that region to a halt. Once Allied defenses were concentrated against the attackers in that area, Doenitz could then have moved the Caribbean Wolf Pack to another. Instead, he operated his Caribbean U-Boats individually, making piecemeal attacks.

The defeat of the German U-Boats in the Caribbean was finally due to the Allies' realization that the Panama Canal was safe, and that the U-Boats were the real threat to the Caribbean. Then the powers-to-be greatly expanded Allied airpower: Navy PBYs and Mariners, and Air Force Bolos and Liberators that worked with surface vessels, which were all directed by an increasingly efficient ASW command center as the Army and Navy slowly resolved their differences. Allied air resources were concentrated in critical areas, such as the Gulf of Mexico coasts, Florida Straights, Windward Passage, and Trinidad. Doenitz sent his U-Boats piecemeal to these same heavily defended areas, with the result being that Caribbean ASW airpower overwhelmed the U-Boats, which were shadowed constantly. This air surveillance was not theater-wide, but the Germans did not take advantage by patrolling the 500 mile long Aruba to Windward Passage convoy route, which could only be covered by long range aircraft, and the combat (not weather) conditions there would have been similar to those in the North Atlantic, and favorable to the U-Boat. The Aruba and Curacao areas, through which convoys were routed, were for a long time weakly defended by aircraft, and could have been overwhelmed by a concentration of U-Boats. The southern portion of the bauxite route was defended by air bases that only stationed a squadron or part of one, and the U-Boats that did operate there were sent in small numbers, whereas their concentration could have been more successful. Allied Caribbean convoys were notoriously poorly escorted by surface patrol vessels that were in short supply, and were PC vessels, not the more effective destroyers. After ships in several weakly escorted convoys sailing too far off shore were picked off by single U-Boats, subsequent convoys stayed close to shallow shorelines under air cover, continuous supervision by shore based radar, and independent hunter-killer groups, and thus sustained fewer losses. These convoys in due course did have to traverse areas of deep water that were out of range of shore based radar and heavy air cover, but the BdU nonetheless chose to fight in the most strongly defended areas. Later the convoys were escorted by more surface vessels as they became available, and canceled out these areas as relatively safe areas for attack. The loss of the milch cows for refueling in the mid-Atlantic particularly reduced the Type VII's patrol time, and also that of the Type IX. The bottom line of the Caribbean U-Boat offensive was that, while it was extremely productive, it incurred a staggering 61% loss rate, and when combined with the six of eight boats lost in the later Brazilian operations came to an unsustainable 64%. By mid-1943 Doenitz had no choice but to relegate the theater to secondary status, sending a few U-Boats there to continue to tie down the large ASW forces based in the area.0

The final issue of the *Antisubmarine Monthly Intelligence Report* (August 1943) stated "Öperhaps the lasting contributions of the Antisubmarine Command and its battle against the U-Boats are various tactical and technical improvements, either developed by the organization or stimulated by it and completed by special research agencies." But the fact was that most of the special projects had not been completed, or were not operational by that time. The U-Boats had been defeated because they had been overwhelmed by air and surface vessels, and not because they had been hunted to destruction by "special devices which might have done the job more speedily."

Balance Sheet on the U-Boat Campaign in the Americas

During the eight and a half months from 18 December 1941 to 31 August 1942, the BdU sent out 184 U-Boat patrols to the Americas: 104 by Type VII U-Boats, and 80 by Type IX boats (for an average of 20 sailings per month). These patrols sunk 609 vessels (including 143 oil tankers) totaling 3.1 million tons, including those sunk on the way to and from their patrol areas (for a monthly average of 68 vessels sunk of about 350,000 tons). The American theater was the major U-Boat campaign of the war, if one takes into consideration that in its relatively short period of time of eight and a half months, the 609 vessels/3.1 million tons sunk constituted over one quarter of the total of Allied shipping sunk by U-Boats during the entire five and half years of war. The paradox in these figures was that the Type VII U-Boats, which Doenitz and the BdU favored, and produced in much greater numbers, were not the most successful U-Boat Type in the Americas. In their 80 patrols the Type IX U-Boats sank 384 boats (63%) totaling 2.0 million tons (65%), while 104 Type VII U-Boats accounted for 225 vessels (37%) of 1.1 million tons (35%).

Although the 3.1 million tons of shipping sunk in the Americas was a significant U-Boat success, it accounted for only about 10% of the available Allied merchant shipping total of 30 million tons, and by mid-1942 the losses were being more than replenished by new construction. In conducting the American U-Boat campaign Doenitz did not realize his strategic goal, as the Allies continued to transport men and materiel in new ships from America to battlefields not only in Europe, but also the Pacific. Despite the Army-Navy infighting, experience, improved ASW techniques and equipment, and the implementation of a successful convoy system forced Doenitz to withdraw his flagging U-Boats from the Americas, but not before forcing the Allies to consign massive resources to protecting shipping sailing the Eastern United States, the Gulf, the Caribbean, and Latin America. After the U-Boat withdrawal, the U-Boat threat compelled the Allies to continue to devote men, vessels, and aircraft to guard against the U–Boat possibility.

10

Canadian and Foreign B-18s

B-18s in Canadian Service

The Weir Mission was sent by the British government to Canada and the U.S. between 25 April and 25 May 1938 to investigate potential aircraft procurement to supplement the urgent RAF rearmament proposals. Funded with $25 million, the Mission was headed by British industrialist James Weir, and included Air Commodore Arthur Harris, who would later head Bomber Command as the notorious "Bomber" Harris. The Mission was to consider twin-engine general reconnaissance aircraft, long-range bomber aircraft, advanced training aircraft, and fighter aircraft for the Fleet Air Arm. The long range bombers investigated were the Lockheed Model 14, Boeing B-17, Consolidated PBY-1, and the Douglas B-18. The B-17, while preferred, was not available, and the PBY-1 was judged as unsuitable, as it was not a long range bomber per se.

When the Mission visited the Lockheed plant in Burbank, the company did not have a Model 14 Super Electra available, as all were shipped to Japan as part of an order! Lockheed's Chief Engineer at the time was Hall Hibbard, who assigned the talented C.L. "Kelly" Johnson to compress three months work into five days, and produce a full scale wooden mock up of the Model 14 as a medium reconnaissance bomber. The Mission initially turned down the design proposal but, impressed with Lockheed, sent it to England for evaluation and redesign. The Mission next visited Douglas to examine its B-18, which was in production, and received a good evaluation from RAF test pilot S/L Horrex. Douglas had gained eminence for its DC series of airliners and transports, and the company's reputation for workmanship had impressed the Mission. When the Mission visited the Douglas Santa Monica factory it was equipping the B-18 with an engine upgrade the USAAC had ordered, and the Mission would request this improvement, along with the installation of British turrets. The proven Douglas design and workmanship dictated a high price tag, but to offset this outlay the aircraft would potentially meet both the bomber and general reconnaissance requirements. However, the Mission did not order the B-18, but Kelly Johnson had flown to England and incorporated the numerous changes demanded by the RAF on the Lockheed Model 14. On 23 June 1938 the British ordered 175 smaller and less expensive ($90,000 vs. $120,000) Lockheed Hudsons (named after explorer Henry Hudson), and 50 more on 17 December 1938 (along with 400 North American Harvard Advanced Trainers). The Hudson was to serve in a maritime role with the Coastal Command.

In September 1938, Canada recognized that when the inevitable European war finally broke out, Britain would be unable to aid members of the Commonwealth with arms. In mid-September the Canadian minister in Washington, DC, made inquiries about American aircraft that were immediately available for purchase. Chief of Air Staff AVM George Croil realized that it would take time for the Canadian Parliament to increase the nation's defense budget and then appropriate these funds, so he procured a $5 million purchase warrant from the Governor-General. He sent a small group of procurement agents led by Air Commodore E.W. Stedman to the Americas to meet with U.S. War Department officials and industry representatives about acquiring aircraft. The Canadian requirements were derived from the conclusions of the Weir Mission, and Stedman was to procure aircraft that would be available in three months. This time constraint eliminated classified aircraft not yet released for export, and meant only U.S. aircraft in production could be purchased. In long and intensive meetings the Canadians were advised of U.S. aircraft in production, and their specifications were studied for suitability. Orders for several aircraft types were prepared, including one for 18 Douglas B-18 bombers, which were the only U.S. bombers in production at the time. The contracts were not executed, as when the Canadians returned home, British Prime Minister Neville Chamberlain came back from Munich, and proclaimed that there would be "Peace in our time"—there would be no European war.

Almost a year later, in August 1939 the situation had deteriorated, and a European war was now unavoidable; the Canadian procurement agents urgently returned to America with the intention to buy aircraft. During this visit, only on the recommendation of senior AAC officers and the assurance that the bomber was readily available, the Canadians bought 20 B-18 bombers sight unseen. After all, the B-18 was America's frontline production bomber. The Lockheed Hudson was also suitable for Canadian purposes, but could not be delivered until December at the earliest, as British contracts for the aircraft had to be fulfilled first. The American B-

17 and B-24 heavy bombers were in the developmental stage, with no production timetables established, and no domestic, much less foreign, commitments having been made.

On 31 August 1939, the Defence Purchasing Board of Canada contracted Douglas for 20 aircraft at $117,330 each for a total order amounting to $2,346,600. The aircraft to be ordered through Douglas were to be the standard AAC models with the same engines, communications equipment, armament, and bombs and bombing equipment. Normally the Vickers .303 machine gun was used by the RCAF, but the gun could not be obtained until July 1940, and the American .30 caliber machine gun was to be installed until the Vickers guns became available. Except for a short time in September 1939, there never was a shortage of .30 caliber ammunition, and the Canadian B-18s retained their .30's throughout the B-18's service. The Canadians intended their B-18s to be used in maritime reconnaissance, and non-standard equipment was added: two bomb bay auxiliary fuel tanks, an auxiliary oil tank, wing floatation compartments, and wing and propeller de-icing apparatus.

The Defence Purchasing Board of Canada ordered 20 standard AAC B-18 models through Douglas. Vickers machine guns were RCAF standard, but the American .30 caliber machine gun was to be installed until the Vickers guns became available. Except for a short time in September 1939, there never was a shortage of .30 caliber ammunition, and the Canadian B-18s retained their .30s throughout the B-18's service. (RCAF via Stevens)

Before the first aircraft were delivered, Squadron Leaders Gordon and Carscallen and six ground crewmen spent a week in the Douglas Santa Monica factory to familiarize themselves with the bomber. The first five B-18s were to be delivered in December 1939, the next ten in March 1940, and the final five in May. The 20 aircraft were given the RCAF serial numbers 738 through 757.

In fall 1939, the 10 (BR) Bomber Reconnaissance Squadron was selected to operate the Digby. The 10(BR) had been originally formed as the 3 (Bomber) Squadron on 1 June 1937, when it was assigned to fly Westland Wapitis out of Halifax, Nova Scotia. 3(B) moved to Calgary, Alberta, on 1 September 1937, and four days later it was divided into the 10 and 11 Bomber Squadrons, but soon the 10(B) was re-redesignated as the 10 Bomber Reconnaissance Squadron (10BR). The squadron went on to be known unofficially as the "North Atlantic Squadron."

B-18 delivery plans changed, and on 18 November 1939 four pilots led by Squadron Leaders Annis, Carscallen, Small, and Gordon, and three crewmen left for Coutts, Alberta, to take delivery of the first two Digbys. At the time neutrality laws prohibited aircraft from being flown directly into Canada from the U.S. The border at Coutts was a barbed wire fence that traversed a small rise slanting downward to the north into Canada. The Canadians, dressed in civilian clothes, waited as the two unmarked B-18s were flown to Sweet Grass, MT. The aircraft landed out of sight, just beyond the top of the rise, and soon taxied to the border. The American pilots from the Douglas Company got out of the aircraft, quickly shook hands with the Canadians, and walked back over the hill. Once the Americans left, Annis photographed the transfer with a movie camera provided by RCAF HQ to prove that U.S. citizens had nothing to do with the transfer. The Canadians cut the barbed wire and threw a rope across the border, and somehow it inexplicably became tied to the undercarriage of the first aircraft, and then became attached to a team of horses that started it moving. The empty aircraft was photographed ethically rolling downhill into Canada, but the Canadians weren't anticipating the aircraft would get out of control on its downhill journey. The panicked Canadians had to race after the wayward aircraft and climb into the cockpit to apply the brakes. Once the two aircraft were across the border the pilots had to spend the night reading the flight manuals, as none had ever flown such a sophisticated aircraft. The next morning the pilots faced taking off into a high crosswind on a "field" that had been recently mowed by a farmer, probably supplying his cattle with a meal of hay from their pasture. After a daunting take off the Digbys proceeded to Winnipeg, Manitoba, where a 10(BR) detachment had been set up for transition training. After the Coutts border transfer fiasco, the remaining 18 Digbys were brought across the border at Emerson, Manitoba, where the topography was more suited for the transfer.

At Winnipeg, the four original pilots familiarized themselves with the aircraft through check flights, and then taught incoming pilots. The 10(BR) B-18 detachment stayed at Winnipeg for a month while the new pilots and crews became familiar with their modern aircraft, which were a great improvement over the Wapiti. In all the transition was smooth, except that the Wright engines presented some cold weather problems with the starting motor overloading, then overheating, and burning out. Canadian mechanics developed

The RCAF received 20 B-18s that were given the RCAF serial numbers 738 through 757, and on 3 January 1940 the name "Digby" was officially adopted after a Lincolnshire, England, airfield built at the end of World War I. (Pima)

a degreasing method to remedy the starting motor problem. It was at Winnipeg that on 3 January 1940 the name "Digby" was officially adopted for the Canadian B-18s. Digby was the name of a Lincolnshire, England, airfield built at the end of World War I that became associated with the RCAF, particularly the No. 112 City of Winnipeg Squadron, in early World War II.

In January the cold Manitoba weather was making training difficult, and the detachment moved to St. Hubert, Quebec, where there was more hangar space and better accommodations. Training continued there until 2 April 1940, when nine 10(BR) Digbys joined No. 11 Squadron Hudsons and No. 5 Squadron Stranraers at the partially completed base at Dartmouth, Nova Scotia, across the harbor from Halifax. After the Wapitis were discarded the remaining Digbys were stored: five at Trenton and six at Malton. During the next two months the whole unit began transitioning to the new bomber that was based at Dartmouth, but flown to Halifax for training while the base facilities there were being completed. The four-man Digby crew consisted of the first pilot, second pilot/navigator, radio operator, and air gunner, who was also the bomb aimer. During training it was determined that the air gunner was unable to man all the guns during an attack, and two more gunners were added to the crew.

With the invasion and impending fall of France, the Canadian government feared a German attack on the airfield at Gander, and its seizure for use for further enemy attacks on Canada. On 27 May the Joint Service Committee Halifax recommended that a 10(BR) detachment be sent to Gander. The five bombers (744, 749, 752, 753, and 754) and their crews, designated "A" Flight, received orders to transfer to Gander, Newfoundland. They left for Gander the following day in poor weather, and were forced to return to base, but tried again the next day and arrived safely. At the time Gander consisted of a combination control tower and administration building, a small hangar, and a few houses, but there were four enormous intersecting asphalt runways, one measuring 1,200 feet, and three 600 foot wide. By mid-1943 Gander would become the largest RCAF operational airbase, with 15,000 personnel of the RCAF, USAAF, RAF Ferry Command, and ancillary personnel.

After the departure of "A" Flight, the Squadron was left with three aircraft, as #743 crashed at Dartmouth on 20 April. Eleven aircraft were in storage, and the five at Trenton were ordered to be immediately ferried to Dartmouth. When the Trenton Digbys arrived 10(BR), mechanics found them to be in a very poor state, as they had been stored in the open and not maintained. The six stored at Malton were flown to Ottawa, and did not arrive at Dartmouth until late September, except for #751, which was allotted to 12 (Communications) Squadron at Rockcliffe. Continued training lapsed during June 1940, as the 10(BR) was busy organizing their move to Dartmouth, which was completed on 17 June. During the month 10(BR) personnel also assisted in the loading of American built aircraft and equipment on ships bound for France in a vain effort to prevent that country's defeat.

The Digbys were assigned to the 10(BR) and trained at Halifax before half the unit was sent to the new base at Gander, and would then rotate back to Halifax to be replaced by the other half of 10(BR). (RCAF via Stevens)

During June B-18 pilots would have their first encounter with the French aircraft carrier *Berne*, when 10(BR) personnel were enlisted in loading the carrier with American lend-lease aircraft. Some of the airmen sailed with the ship, which was diverted to Martinique after the fall of France. The RCAF crew on board were taken off at Martinique by the cruiser HMS *Fiji*. Later the American B-18 pilots would fly patrols over Martinique to insure that the carrier and other French warships remained in port.

During the summer the two components of 10(BR)—one at Dartmouth, and the other at Gander—were active, the former flying training and familiarization flights, while the latter flew two or three long daily patrols and shorter reconnaissance missions, depending on the weather. The operational headquarters were in Gander under W/C H.C. Gordon. The "A" Flight Digbys returned to Dartmouth for maintenance, and were replaced by others from Dartmouth. The Digbys also flew as transports between the two bases, carrying personnel and supplies, along with the occasional VIP. As the BCATP (British Commonwealth Air Training Program) was not fully established, the Squadron had to train its own gunners.

An interlude in 10(BR) training and operations during this period occurred when the Digbys took a supporting role in the British Ministry of Information propaganda film the *49th Parallel* (called the *Invaders* in the U.S.). The hit film by noted director Michael Powell had an impressive cast, including Leslie Howard, Eric Portman, Raymond Massey, and Laurence Olivier, and went on to be nominated for an Oscar for Best Picture, and won the Oscar for Best Screenplay. The plot concerned the flight of six survivors of a German U-Boat, led by their dedicated Nazi captain, after it was sunk by RCAF Digbys in the Gulf of St. Lawrence. The film was intended to inflame neutral American sympathies, but was not released in the U.S. until March 1942, four months after Pearl Harbor. The Dartmouth Digby's role was for interior aircraft shots, while the Gander Digbys flew several "missions" to the Bay of Islands, on the western coast of Newfoundland, where they made diving attacks on a replica U-Boat, sinking it.

The Newfoundland based 10(BR) spent most of its time flying patrols and escorting shipping. Until the Douglas PBY "Cansos" became operational, the Digbys were Canada's longest range aircraft, and with their bomb bay fuel tanks could stay aloft more than 12 hours, giving them a maximum range of over 2,000 miles. On 6 November there was a report that the German Battlecruiser *Admiral Scheer* had been located 600 miles off the Newfoundland coast. Three Digbys left Gander at 0730 to intercept the raider, but found nothing after flying through intermittent rainy weather. All three returned safely, with two landing in darkness.

While based in Newfoundland,Digby #741 was equipped with an F-24 aerial camera, and was assigned to photograph the island for the first time from the air. The task proved to be a trying one, as the weather did not cooperate with many clear days. #741 flew several missions that began in good weather, but ended in harrowing searches for the base, and then treacherous landings in deteriorating weather.

As described earlier, in the early evening of 18 November #749 returned to Gander from a long patrol to find the field socked in by bad weather. The pilot, F/O J.G. Richardson, headed toward Montreal, hoping to find a clear landing field along the way. Unable to find a field, and almost out of fuel, the pilot ordered the crew to bail out. Three of six crewmen survived and were rescued after an intensive search aided by American B-18s.

At Dartmouth, on 12 December #746 crashed on take off, but the crew escaped safely from the aircraft before its two 600 pound bombs and pyrotechnics exploded. By the end of 1940 15 of the original 20 Digbys were operational, with ten usually based at Gander, and five at Dartmouth, where two were rotated out of service for periodic maintenance. Although some patrols were flown from Dartmouth, Gander was the operational headquarters.

The onset of winter in Newfoundland came with very high winds, sleet, and wet snow off the nearby ocean, while the mixing of the warm Gulf Stream Current and the cold artic Labrador Current formed an almost continual fog on the Grand Banks, which was at the outer limits of Canadian patrols. There was only one hangar at Gander, but the Hudsons being ferried to England had priority, so the Digby mechanics had to service their aircraft outdoors. The maintenance situation became untenable, and finally the 10(BR) insisted on at least some hangar time so the unit could remain operational. Even if the Digby engines could be started after being frozen by the combination of high winds and wet snow, the runways were covered with snow, and take offs and landings were always interesting. The weather was so bad in January 1941 the Gander Flight was grounded for 22 days, and only able to fly nine patrols and complete four. The weather in February and March was also generally bad, but more patrols per month were flown in the brief periods of good weather. On 22 February a convoy reported that it had lost five vessels to the German raiders *Scharnhorst* and *Gneisenau*, which were now operating within range of 10(BR)'s Digbys. During that day 10(BR) was searching for a missing Hudson, in which Sir Frederick Banting, the Canadian co-discoverer of insulin, lost his life. At the time of the communication concerning the German raiders was received, the bombers were low on fuel and had to return to Gander to refuel in order to conduct a reconnaissance, but on their return darkness fell, and the Digbys lost their chance for an interception. After sinking or capturing 16 ships from two convoys, on 15-16 March the two *Kriegsmarine* battlecruisers returned to within 350 miles southeast of St. Johns, and were continuing their harassment of two convoys. Two Digbys were enroute to escort the outer convoy when they heard a radio call that one of the convoys was under attack. For an unrecorded reason, neither aircraft inquired about the position of the attack, and both flew on to escort the wrong convoy. The result was that several more ships were sunk, with the Digbys missing the choice German targets by only a few miles. The German warships escaped unscathed to Brest, France, and the Digbys lost an opportunity for a degree of distinction due to serious operational negligence.

Finally, on 11 April, the Dartmouth Digby air and ground crews joined the Gander contingent at their new RCAF Station, named Newfoundland Airport Station, under new CO W/C H.M. Carscallen. That same month the USAAF 41st Reconnaissance Squadron arrived at Gander, and began to fly patrols with its B-17s. Finally all 17 Digbys were together at Gander, and the newly arrived Dartmouth unit gained valuable experience, and often a third of the 10(BR) Digbys could be patrolling off the coast into the spring and summer of 1941. The patrols were uneventful, as the German U-Boats had not yet extended their operations across the Atlantic. On 18 May the German pocket battleship *Bismarck* and the heavy cruiser *Prince Eugen* sortied from the Baltic, and six days later sank the Royal Navy battleship *Hood* in the Denmark Strait, between Iceland and Greenland. The Admiralty believed that the two German raiders would head toward the convoy routes to the west and, on 26 and 27 May, 10(BR) Digbys were ordered to extend their patrols far to the east, anticipating the Germans' move to the west. In fact, the *Bismarck* moved back toward Europe, and was sunk off the southeast coast of Ireland. On the 28th, the aircraft of the Eastern Air Command continued their search for the *Prince Eugen*, which had cruised into the western Atlantic, but out of range of air search.

The increased patrolling began to take its toll. On 21 July #752, piloted by F/L R.A. Ashman, was forced to ditch off the coast of Cape Race. As he descended to find a place to ditch, Ashman spotted a schooner and put down close by. The aircrew waited on the floating Digby until they were rescued by the schooner's dory. The floating Digby was shelled by the Canadian Navy, but after the bomber was reluctant to sink the RCAF decided on salvaging it, and towed it back to land, but it never flew again. Five days later #742 was returning from a long patrol in bad weather and crashed three miles from the field, killing all six crewmen. Accidents continued into the early fall thanks to one pilot, F/O Robert Butts. On 11 September #750 was assigned to practice take offs and landings, and Butts took off without checking to find if the rudder and elevator chocks had been removed. The chocks were designed to keep these control surfaces immobile in the winds when the aircraft was parked, but functioned likewise during take off. The Digby did not get airborne, and ran off the end of the runway into a drainage ditch. Although the crew escaped injury, the bomber's landing gear was thrust into the main wing spar, and the Digby was sent back to Dartmouth for repair; it did not return to 10(BR) for a year. Then on 3 October, Butts was assigned to fly #748 on an early morning patrol. The negligent pilot failed to have the heavy frost that had accumulated overnight removed from the wings, and as the aircraft took off and reached 75 feet it slowly descended over the end of the runway into the soft cushion of a thick stand of low fir trees. Again, as the aircraft settled its landing gear was driven into the main spar, and this aircraft was also lost to the squadron for almost a year. Because of the shortage of aircrew Butts was again pressed into service, and on 11 October he was assigned to fly #754. The take off and patrol went fine for ten hours, but as he approached Gander Butts did not check on the change in wind direction, and attempted to land downwind on a wet runway. The Digby careened down the runway, ending up in a familiar drainage ditch, and again ramming the landing gear into the main spar, putting the bomber out of commission for eight months. So within a month one pilot single-handedly reduced 10(BR) strength 20% to 12 aircraft.

In fall 1941, German intelligence learned that Allied convoys passed through the Strait of Belle Isle, and sent four U-Boats to this choke point. A U-Boat had been reportedly sighted on 30 September, but disappeared before the Digby could attack. However, it was not until 24 October 1940 that the U-Boats began their offensive in Canadian waters, and at sunrise the next morning seven Digbys were sent out on patrol: two to escort Convoy ON 26, and the rest on search patrols. Aircraft #740 was piloted by S/L Clare Annis who, because of the shortage of pilots, was called to duty from his post as Eastern Air Command Armaments Officer. Annis took off at 0730 with his co-pilot acting as the navigator, and three gunners searching the sea's surface. The search was conducted at 900 to 1,000 feet, and consisted of a pattern, with the initial north-east outward leg extending 40 miles from the coast, and continuing on another leg heading 270 miles north, followed by an 18 mile westward hop, and concluding with an almost due south leg back to Gander. The winds during the entire patrol were somewhat stronger than the usual North Atlantic gale, averaging about 45 knots, which made for a rough flight, making navigation difficult. Late in the patrol, at about 1450 hours, Annis looked out of his salt spray covered windshield through a thick 1,000 foot cloud base and spotted a surfaced U-Boat base pitching violently in 60 knot winds. The navigator assumed his normal second pilot's position and put the engines into the manual rich position, while the pilot increased the boost and opened the bomb bays. The bomber went into a dive that was slowed to about 60mph by the strong head winds, and arrived just as the U-Boat submerged. At 300 feet Annis tried to keep an eye on the spot where the U-Boat submerged and dropped the Digby's two 600 pound bombs on the calculated U-Boat course, straddling the area fore and aft. As the crew looked back there were no explosions, as the bomb aimer had disarmed the bombs without the pilot's knowledge. The bomb aimer's action probably saved the Digby's crew, as these bombs had not been designed for anti-submarine use, and if dropped below 500 feet their detonation could have destroyed an aircraft flying above the explosion.

In early November the weather had deteriorated, and the U-Boats used the situation to their advantage, sinking four ships without fear of air attack. By late in the month three of the 12 10(BR) Digbys were out of service for maintenance, and there were only seven crews available for the nine other aircraft. Fortunately, for the RCAF on 22 November Hitler forced Doenitz to withdraw the North Atlantic U-Boats into the Mediterranean to help stop the British offensive in North Africa. With no U-Boats in the area the remainder of the year was uneventful, with 10(BR) flying patrols, convoy escorts, and search and rescue missions, usually for aircraft lost in the chronic fog, rain, snow, and high winds that dogged the Canadian eastern coast. On 29 December 1941 #744 disappeared without a trace with the loss of its six man crew. Then, on 2 January 1942, #738 crashed in flames into Freshwater Bay, but the crew was able to swim ashore; now the original Digby fleet was reduced in half to 10 aircraft.

At the end of the year various improvements were made on the Digbys. The most important as far as the crew was concerned was cabin heating, which was not a problem for the American B-18s operating in more southerly tropical waters. When the Canadian Digbys were sitting on the ground the water in the heating system froze, and it did not function. To remedy the situation, a small amount of alcohol was added to the water to keep it from freezing. Once the

The Canadian Digbys carried two U.S. supplied 600 pound anti-submarine bombs (pictured) that were later replaced with the more efficient Mark VIII 250 pound amatol-filled depth charge. (RCAF via Stevens)

engines were warmed up before take off the alcohol/water mixture was drained, pure water substituted, and the crews were soon warm. To bolster long-range navigation the astro compass, astrograph, drift recorder, and some radio apparatus was replaced with more contemporary equipment. However, the new drift recorder was too large for the installation in the nose, and the bombsight continued to be used for drift navigation. The original bombsight in the Digbys was the Estoppey Sight, which was scheduled to be replaced with the MkIX type. Like the drift recorder, the MkIX and its bracket were too large, and would crowd the bomb aimer, and the thick metal nose window frames would block the line of sight of the device. By this time Digby bomb aimers had considered themselves proficient in Estoppey use, and the conversion to the new, more complex sight would require more training, so the bomb aimers were allowed to keep their old bombsights. The more efficient Mark VIII 250 pound amatol-filled depth charge was delivered to replace the 600 pound supposed anti-submarine bombs supplied by the U.S. However, these weren't the powerful Torpex-filled depth charges used later, but were nevertheless considered to be much better than the bombs. Also, fittings were ordered from Britain to use the 450 pound Mark VII naval depth charge for aerial use. Several of the Digbys were equipped with the first ASV radar sets, and after a while the operators became somewhat proficient in its use. The ASV radar was not only valuable for locating U-Boats, but also for rendezvousing with the convoys the Digbys were scheduled to escort.

After Pearl Harbor the U-Boats moved west into North American waters, and U-Boat sightings off Newfoundland increased. On 19 January 1942 a Digby piloted by F/L J.M. Young was patrolling in heavy snow squalls due east of Gander when it discovered a U-Boat (U-86) with its conning tower and deck awash. Both antagonists dived at the same time, and Young attacked at right angles, dropping a salvo of three 250 pound depth charges set to detonate at 50 feet on the still visible U-Boat. He made a second attack run, dropping three more depth charges set at 100 feet in the vicinity of the U-Boat's wake. The attack was well executed, but Young orbited the area looking for oil or debris until heavy snow forced him back to base. After the war U-86 logs reported damage to welded seams, but the sub did not sink.

Several of the Digbys were equipped with the first ASV radar sets, which were not only valuable for locating U-Boats, but also for rendezvousing with the convoys the bombers were to escort. Shown is the early British ASV antenna array. (RCAF via Stevens)

On the 22nd F/L E.M. Williams' Digby was returning from a convoy escort when he spotted a U-Boat (U-84) in the same area as Young's attack three days earlier. The U-Boat was moving quickly on the surface, about three miles ahead on the port bow. Williams was able to close to within a mile before the U-Boat began to dive. At 1,100 feet, with the U-Boat's stern still visible, he salvoed the trio of depth charges set to detonate at 50 feet. Unfortunately two depth charges hung up in the bomb racks, but the third hit near the center of the U-Boat, causing it to shudder violently. In the excitement of the attack the gunner/bomb aimer had forgotten that the 12 volt distributor had not been installed before the mission, and each depth charge had to be dropped manually. Williams made two more passes, dropping three depth charges, and then returned to drop his last two depth charges. He dropped floating flares to mark the U-Boat's last position, and continued to patrol until he was obliged to leave due to poor weather and low visibility.

The next several months were uneventful, as the U-Boats moved further south along the more lucrative U.S. coast, but the Eastern Air Command continued to patrol through the dreadful weather. The Digby roster dropped to nine aircraft, as #753 crashed landed near Gander on 26 March, with the crew escaping safely. The Digby personnel roster also dipped, as crews were diverted to fly newly arrived PBYs stationed at Yarmouth, Nova Scotia. During this period several more Digbys were equipped with the centimetric ASV radar, and with training and experience the crews began to fly night patrols. The 250 pound depth charge was replaced by the 450 pound standard naval depth charge adapted for air drop. This period was uneventful, with a Digby being hit by AA fire from a friendly merchantman on 1 April with a crew member suffering a slight wound to his hand.

In May the U-Boats returned to eastern Canadian waters, and 10(BR) patrolled not only its Newfoundland sector, but also the Gulf of St. Lawrence and off Cape Race. Despite the increase in patrols only one U-Boat was seen, on 3 May by P/O E.J. Padden, but it crash dived before an attack could be made. Long patrols were the order of the day, as P/O Padden set an endurance record of 13 hours and 10 minutes on 7 June, and then four days later he extended the record to just under 15 hours. Padden was lost on 14 July when #739 disappeared over the Atlantic without a trace, leaving eight Digbys. In June 10(BR) flew patrols and convoy escort from Goose Bay, Labrador, and Gander. Convoy escort was not without its dangers, as in June a merchantman fired on F/L J.S. Sanderson's Digby, hitting an engine and the fuselage with AA shells, but Sanderson skillfully brought his bomber safely back to base.

On 3 September two convoys were entering the Strait of Belle Isle, and were trailed by U-517, captained by Paul Hartwig, and by U-165, under Eberhard Hoffman, who had already sunk several ships. At dawn, a Hudson from Sydney and a B-17 and Digby from Gander were scheduled to fly convoy escort. F/O J.H. Sanderson of 10(BR) was escorting a convoy through the Strait of Belle Isle when, just after noon, he spotted U-517 surfaced on a converging course about nine miles south of the convoy. The U-Boat had been spotted on the surface several times previously that morning and forced to dive. Sanderson accelerated, and dove from 900 feet, arriving just seconds after the U-Boat dove, and dropped his depth charges from 150 feet on the dive wake. Sanderson's aircraft received some damage from the premature detonation of the depth charge, but continued to patrol the area until he was relieved by another Digby and a Royal Canadian Navy corvette.

At the end of October there were 13 Type VII U-Boats on station off the Grand Banks, along with three large Type IX boats. On 30 October two 145 Squadron Hudsons were sent on an anti-submarine sweep ahead of convoy SC107, which was eastbound. Upon reaching the limit of his Hudson's endurance F/O E.L. Robinson made a textbook attack on a surfaced U-Boat (U-658) and sank it. Several minutes later Digby #747, piloted by F/O D.F. Raynes, took off at 1225 from Gander on an outer anti-submarine patrol over ON 140. After its escort was completed #747 headed back to Newfoundland at 3,200 feet, and at 2002 hours spotted a U-Boat on the surface, 115 miles due east of St. Johns, sailing in the fog with its decks under water. Raymes dove to attack along the U-Boat's track directly astern, and dropped four 450 pound depth charges just seconds after the conning tower submerged. When the aircrew looked back they saw a dark body rise to the surface and disappear, followed by oil and bubbles. Raymes landed at Gander at 2355 and claimed a U-Boat "probably destroyed." It was the seventh confirmed U-Boat attack for the Eastern Air Command, and its third kill. After the war the U-Boat sunk by #747 was determined to be the Type IXC U-520 of the Second U-Boat Flotilla on its first patrol, which went down with its captain, Volkmar Schwartzkopf, and 52 crewmen.

In November 10(BR) moved back to Dartmouth, but before it did there was one exciting mission. On 3 November F/O Sanderson spotted a U-Boat (either U-106 or U-183) as it was crash diving, and successfully escaped the Digby's depth charges. After the move to Dartmouth, #751, under F/O M.L. Foster, was conducting an anti-submarine sweep of a convoy and spotted a surfaced U-Boat through low haze. The U-Boat dove before Foster could drop his depth charges, and it escaped, but at least it did not bother the convoy while the Digby circled during the next several hours. This was the last attack on a U-Boat by a 10(BR) Digby in World War II. During the last five months of the Digby's service with the 10(BR),

The 20 Digbys were reduced by attrition to three war weary examples by November 1943, as shown by beat up PBL in the photo. On 8 April 1943 the last 10(BR) Digby flew its last operational mission, an ice patrol. (Pima)

the maturing Digbys were encountering maintenance problems that were exacerbated when many of the experienced mechanics were left behind at Gander. But after a while a number of Digby mechanics returned to extend the life expectancy of their charges. As more PBY Cansos arrived to take over Digby medium range patrols, the Digbys were used increasingly for transport chores, carrying freight and personnel to Gander and Goose Bay. In December 1942 the Digby anti-submarine days were definitely numbered, and the next month EAC conducted a study on converting the Digby back to its Douglas transport roots, which was approved. Over the winter and spring of 1942-43 the 10(BR) flew second line patrols and sweeps with its Cansos, while the Digbys flew numerous transport flights, along with search and rescue missions and an occasional patrol. On 8 April 1943 the last 10(BR) Digby flew its last operational mission, which was an ice patrol looking for ice bergs and floes that could be navigational hazards.

In the meantime, 10(BR) received 15 valued U.S. Douglas B-24s, and became a frontline unit once again. The long range four engine Liberators finally closed the so called "North Atlantic Air Gap," and the U-Boats had nowhere to operate without detection. The prevailing winds over the North Atlantic were westerly, and so strong that the Digbys and the Cansos, with their great endurance, were reduced to low cruising speeds. The Digby cruised at 140mph, and the usual westerly wind was 45mph, limiting their range to 800 miles off the Canadian coast, thus leaving a large gap in the mid-Atlantic where aircraft from Canada and Britain could not patrol. The B-24s, with their higher cruising speed, were able to close this gap significantly. Throughout the year 10(BR) pilots were transferred for transitional training to the B-24s. In May and June the RCAF enlisted two veteran Trans Canada Air Lines pilots to instruct Digby pilots. Before beginning their instruction the two pilots spent over 700 hours learning not so much the workings of the Liberators, but the intricacies of flying in the perilous conditions of the North Atlantic, with which the Digby pilots were already familiar.

When the 10(BR) received its new lease on operational life so did the Digbys, as 161(BR) was formed on 1 April 1943. The unit was to be equipped with 18 Cansos, but they were not immediately available, and eight of the 11 operational RCAF Digbys (three damaged earlier were repaired and returned) were sent to 161(BR), while two (#745 and #748) of the three remaining aircraft were sent to 167 Communications Squadron, the 167(CS). The third Digby (#747) was assigned to be equipped with radar, and on 26 July was dispatched to the Canso base at Botwood, so it could patrol well beyond its normal Gander range to search for the U-Boat Wolf Packs. The 161(BR)'s new pilots transitioned to the Digbys by flying with 10(BR) on its move back to Dartmouth, where its Liberators were ready to be flown to Gander. On 29 April the 161(BR) occupied the 10(BR)'s hangars and personnel quarters at Dartmouth. Before the unit became operational on 16 May, the pilots flew training and familiarization flights. Once the unit was operational it was assigned to patrol harbor entrances, and during the summer of 1943 flew patrols, convoy escorts, and search and rescue missions. The 161(BR), like the 10(BR), suffered from having insufficient operational crews, initially having seven crews for eight aircraft until the end of June, when three more crews were added. Once there were enough crews to man the Digbys, the old bombers were not only having maintenance problems to reduce their operational numbers, but were also sent out, one by one, to be painted in the new white format, and then several were sent out for installation of radar (one for three months!). In August #750 was the first 161(BR) Digby to become terminal, and was sent to be scrapped. In September the squadron reported three attacks on U-Boats that were not substantiated by German records. By the end of September 161(BR) had only three Digbys (#740, 756, and 757) in service, while the others were down for maintenance or painting, but the Squadron was reinforced at the end of November by three PBY Cansos. In November #740 was sent to the 4th Repair Depot at Scoudouc, New Brunswick, and remained there until it was scrapped on 16 May 1944. December would be the last month of Digby combat operations. On the 17th, #756 limped back from a patrol with engine trouble, and on the same day #757 had problems when on take off for a convoy escort one engine cut out at 15 feet and restarted, but not before the propeller tips and bomb bay belly contacted the ground. Although the pilot was able to continue the take off and return safely, there were serious doubts arising about the Digby's safety, and all were grounded. On 19 December #756, followed by #757, was sent to the 4th Repair Depot at Scoudouc, New Brunswick, but after several days both were back with the Squadron, and they were taken off of operational status. The remaining four 161(BR) Digbys returned to the unit after painting and maintenance to join #756 and #757, and they continued to fly through January 1944, but only on practice and training flights, not operations.

In February, the six 161(BR) Digbys were sent to the 4th Repair Depot at Scoudouc, New Brunswick, for storage, and would never fly again. #756 was scrapped in May 1944, and was followed by #757 on 16 June 1944. #741, 751, 754, and 755 remained in storage at the 4th Repair Depot until they were eventually sold to War Assets Ltd. in 1946. In March 1944 the radar equipped Digby #747 was also stored at the 4th Repair Depot, and was scrapped on 16 May 1944. Two Digbys of the 167(CS), #745 and #748, were spared in the February write off. #745 continued to fly until June 1944, when it was sent to Mont Joli, Quebec, for storage, while the last flying Digby (#748) was sent to storage at Mont Joli on 3 July 1944. Both aircraft were sold to War Assets Ltd. in 1946.

Assessment of Digby Operations

The Eastern Air Command assigned too many aircraft on long range anti-submarine patrols, rather than searches near threatened convoys, where U-Boats congregated. The objective of convoy patrols was not only to attack U-Boats, but to keep them submerged and unable to attack the convoy. The basis of long range searches relied on reports of U-Boats sightings far offshore, and then the subsequent needle in a haystack searches in the typically poor weather. The searches were conducted on the "square search" pattern routine, a patrol of 30 to 40 mile legs in a square, box pattern that were the cause of operational fatigue. The search altitude recommended by Coastal Command was at 4,000 to 5,000 feet, which was usually just below the prevailing cloud ceiling. But operationally 10(BR) did not fly there, instead flying at much lower altitudes to make a

U-Boat sighting more feasible. The best chance of destroying a U-Boat was to spot it less than ten miles ahead, and initiating an attack before the U-Boat could dive. White camouflage paint was not widely used in the Eastern Air Command, and lookouts on the U-Boats were often able to spot the approaching Digbys and initiate a crash dive. When radar was installed on but a few aircraft it was depended upon too much, and the U-Boats, using detection devices and lookouts, were able to dive before the Digbys approached. There was little use of photography during an attack to evaluate its accuracy and outcome.

Besides the deficient tactics, the lack of equipment was the other key problem that interfered with operations. Digby navigators had no astrodomes to shoot stars, and relied on outdated sextants, and on compasses whose accuracy was less than five degrees. The anti-submarine ordnance was initially American 600 pound bombs, and then early depth charges adapted from naval types; later the better 250 pound type arrived, which were released using a jury-rigged system that reduced their effectiveness. Communications between vessels at sea and aircraft were sketchy, and needed improvement. Even with the Digby's heating system the crew was cold, and needed better flying clothing, as the winter of 1942-43 was the worst of the war years. In January and most of February the appalling weather had more influence on warfare than the antagonists, with the elements hindering flying, battering the convoys, and leaving the U-Boats impotent. In this appalling weather, for both aircraft and U-Boats a systematic search was hopeless, and when a target happened to be located the weather hindered an attack.

B-18s in Brazilian Service

In April 1939, five months before Hitler invaded Poland, the Joint Planning Committee of the U.S. Army-Navy Board was concerned that the French West African colonies could serve as a base for the possible Axis invasion of South America. The potential invasion could be launched across the Atlantic Narrows—the closest point from the West African Coast to the northeastern tip of Brazil, called the "Hump." After the fall of France in June 1940, and with the Vichy presence in the West African colonies, especially Dakar, this concern became more of an American obsession. At the time several South American nations had shown sympathy towards the German and Italian regimes. In mid-1940, the U.S. Chiefs of Staff and the Roosevelt administration began to send Lend-Lease military armament and equipment to Brazil and other Latin American nations, and were conducting negotiations to base American ground, sea, and air forces in their countries. With the negative historical U.S. military intervention in Latin America, Brazil and other nations were reluctant to allow American troops to be based in their countries.

The Germans began a limited U-Boat offensive in the south Atlantic in the fall of 1940 in the Atlantic Narrows, when U-65, under Capt. Hans-Gerrit von Stockhausen, sank six merchant ships off Africa. Brazil then conceded to have its Navy "to interpose no objections to advance discreet operations" of U.S. naval surface and air forces in the South Atlantic. In late November Brazil also approved the construction of a series of seven U.S. Army airfields between Belem and Bahia in exchange for economic aid. In April 1941 the Joint Chiefs of Staff approved the U.S. Navy's request to also use these bases.

After U-Boat activities diminished, and became almost non-existent during the first quarter of 1941, action picked up somewhat in April when six ships were sunk off Africa, and the first merchantman was sunk off South America by U-105. During May 28 ships were destroyed below 10 degrees north, and in mid-June V.Adm. Jonas Ingram's Task Force Three (later designated as the South Atlantic Force, and then reorganized as the U.S. Fourth Fleet) began surface operations between the Hump and Trinidad. Brazil granted the use of the major ports of Recife and Bahia to the USN, and later allowed Natal to be used as a major operating base for USN air patrol units.

Soon Doenitz was forced to curtail operations in the South Atlantic to fulfill the vital need for U-Boats in the Mediterranean to counter the British offensive in North Africa. By January 1942 the U.S. Fourth Fleet was operating in an area known as the Brazilian Area, extending from the border of French Guiana south to Rio de Janeiro, and halfway across the Narrows toward Africa. The appearance of American warships in this area provoked the Germans and Italians to send submarines against Brazilian ships.

Brazil was the largest and most populous of the South American countries, and possessed the most developed air arms on the continent. In 1941 Brazil had two separate air forces: the Army's Aviacao Militar (Military Aviation), and the Navy's Aviacao Naval (Naval Aviation). In the late 1930s the Brazilian Army had purchased a number of U.S. aircraft, including Stearman A75L3s, Vultee V-11GB2s, and Lockheed 12As, and a few German and Italian types; while the Brazilian Navy had purchased the North American NA-46. In March 1941 the U.S. War Department allocated $100 million in military equipment to Brazil, and the USAAC planned to deliver 230 military aircraft to Brazil. On 20 January 1941 the Brazilian government created the Ministerial da Aeronuatica (Aeronautics Ministry), which oversaw both military and civil aviation. The Army and Navy Air Forces were combined to create the Forcas Aereas Nacionais (National Air Forces), and then redesignated as the Forca Aerea Brasileira (FAB) on 22 May 1942. The Navy contributed 99 aircraft of 15 different types, while the Army sent 331 aircraft of 20 different types. Thus, the FAB possessed about 430 obsolete aircraft made up of 35 types, with only five types in common. In the midst of the May 1941 U-Boat crisis, Gen. Marshall announced at a Standing Liaison Committee meeting that the AAC was trying to release 20 modern A-20 light bombers to Brazil from the British allotment. Gen. Arnold released 12 immediately, even though the FAB had no pilots qualified to fly them. But the British decided that they would not release these aircraft, and the Brazilians responded adversely by threatening the cancellation of base construction. At the end of 1941 substitute aircraft were immediately furnished: transport aircraft for national air mail service operated by the FAB, and the assignment to the U.S. Air Mission of a few B-18s to be used for transition training of Brazilian pilots to prepare them to operate more modern aircraft, such as the A-20. The quality of the FAB pilots was very good, as many of them were conscripted from the National Air Mail Service, and only required

Two former U.S. operational B-18Bs parked at Santos Dumont airport in February 1942, waiting to be used by a training unit based at Fortaleza, in northeast Brazil. (Nacero)

training in modern military types that were unavailable. With the help of the USAAF, Brazil planned a mass pilot training program to begin in February 1942, and since no modern aircraft would be available to the Brazilians until the fall of 1942, a large initial delivery of 60 training aircraft—primary, basic, and advanced—was promised, along with a Lockheed C-66 transport for Brazilian President Getulio Vargas. The FAB fighter force was assembled slowly, with ten Curtiss P-36s, followed by 46 P-40s, and then P-47Ds, which served in Italy with the Brazilian Expeditionary Force from October 1944 until the end of the war. Between 1942 and V-E Day, Brazil received 1,288 aircraft from the U.S. through the Lend-Lease Act of 1941. Brazilian airports and facilities were improved with American finances, as it was the Allied plan to ferry aircraft to Britain over the Brazil-Africa route.

The first Brazilian unit equipped with "modern" American aircraft was the Agrupamento de Avioes de Adaptacao (Operational Conversion Aircraft Group), a training unit created with the help of the USAAF. They were based at Fortaleza, in northeast Brazil, on 4 February 1942, and the unit was equipped with ten Curtiss P-36 fighters, six North American B-25B bombers, and two Douglas B-18Bs. These B-18s had previously been in USAAF service, and had U.S. serial numbers 36-300 and 37-32, which were redesignated to the FAB numbers 5026 and 5027.

Brazil declared war on the Axis on 22 August 1942, principally because of the German U-Boat activity off the country's coast, during which time five Brazilian ships and a number of other ships were sunk. The first FAB anti-submarine aircraft were the two B-18s, but they were soon overshadowed by the B-25s, which would make the first U-Boat attack. On 22 May 1942 a B-25 flying with a mixed USAAF and Brazilian crew depth charged a U-Boat without result. At the end of 1942 two dozen Lockheed Hudsons and a number of Douglas PBY Catalinas arrived, and the B-18s were demoted to transport duties, possibly with 1GMI, based at Sao Paulo. Later two more veteran B-18s were delivered (B-18 #36-286 and B-18B #37-596), to be used by the Aviation Technical School as ground instruction airframes. Since #37-596 was assigned FAB #5073, it probably was used in a flying capacity—probably transport—and was not relegated to ground school.

B-18s in Australian Service

In January and February 1942, after the fall of the Philippines, a small number of USAAF bombers and transports retreated to Australia, where they formed the first American unit based in Australia. The 21st Troop Carrier Squadron was activated on 28 January 1942 at Amberley RAAF Station, near Brisbane. The CO of the 21TCS, F/Lt. Edgar Hampton, was ordered to send all U.S. transport aircraft in Australia, and any flyable bombers that were unfit for combat, to the Air Transport Command, which was part of the Allied Directorate of Air Transport (ADAT) that had been created to coordinate all Allied air transport in the Southwest Pacific Area (SWPA). The ADAT was transferred to Brisbane's Archer Field on 2 February 1942, and the 21TCS was placed under the command of Capt. Paul "Pappy" Gunn. The ADAT assumed command over Australian civil airlines, and Dutch civil and transport aircraft evacuated from the Netherlands East Indies. During the first six months of 1942, the ADAT flew an amazing five million miles. The aircraft assigned to the unit were three B-18s, one C-39, one B-17, five new C-53s, and three new LB-30s. All ADAT aircraft were allotted Australian civil registrations as radio call signs. The VH-C block call was used first, and these markings were painted in white over the camouflage on their tails and rear fuselage. The three B-18s that continued their service as transports and hacks were: VH-CCB (36-343), VH-CWA (s/n unknown), and VH-CWB (37-16). B-18 (36-343), nicknamed *Damfino*, was assigned to the Australian RAAF 4th ADG Depot at Garbutt in early 1943. After its glass bombardier's station in the nose was covered by a solid metal nose it served as the personal aircraft of Col. Bertrandias, who had to make frequent flights to the 5AF Service Command at Eagle Farm Airbase, Brisbane. The aircraft must have been more than satisfactory, as indicated by its nickname *Damfino*, but the weary transport was finally scrapped in October 1943.

A B-18 (arrow) parked at Townsville AFB, Australia, in 1943 (see National Insignia), probably as part of the Allied Directorate of Air Transport (ADAT). (USAF)

11

B-23 Dragon - B-18 Upgrade

Introduction

By the late 1930s it was obvious that the B-18, whose original design left little room for development, was on the verge of becoming superseded by the B-17 and B-24 heavy four engine bombers, which were receiving the Army's focus as its bombers of the future. With the emergence of the 300mph fighter the B-18 was too slow with a top speed of 225mph, which was only 5mph faster than the Douglas C-39 Air Corps cargo plane (a DC-2 with a large door to load cargo).

B-23 Attack Bomber Proposal

In the meantime, in December 1937 the AAC had issued Air Corps Specification #98-102 for an attack bomber to be ready by March 1939. But by September 1938 the world political and military aviation situation had rapidly changed, and new specifications were issued under the designation C-103 for the construction of a single experimental attack bomber. On 12 January 1939 Roosevelt asked Congress for an immediate defense appropriation of $525 million, of which $300 million was allocated to the purchase of at least 3,000 aircraft. With the President's request, new specifications designated as C-103A as part of the Emergency Procurement Program were issued on 15 December 1938 that would lead to a production contract, instead of a single experimental bomber. There was no longer time to build the experimental models, and Circular 39-460 was issued on 17 January to establish a bid process for awarding production contracts.

Douglas had purchased Northrop in April 1937, and after a strike in September 1937 Donald Douglas simply shut down the Northrop El Segundo plant, and moved its engineers and managers across town to the Douglas Santa Monica plant. After negotiations the Northrop plant was reopened, and became the El Segundo Division of Douglas Aircraft Co. As part of the takeover Douglas inherited the Northrop Model 7A twin engine bomber design, which had been suspended, as Northrop was busy with the production of Navy BT-1s and A-17As for the Army. The AAC 98-102 specifications corresponded to those of the Northrop 7A, which for some unknown reason was redesignated by Douglas as the Model 7B. In late fall 1938 Douglas began to rework the design, especially as the

The last 38 B-18As ordered were to be built and delivered as B-23s, and would have the same B-18A serial numbers, s/n 39-27 through 39-64. (Douglas)

Although the B-23 was not revolutionary, the AAC realized that it was more advanced than the B-18A, and could be put into production ASAP in lieu of developing another adequate medium bomber design. (Douglas)

French government had shown an interest in modernizing its air force in the face of an ever more threatening Nazi regime. The French Purchasing Commission arrived at El Segundo on 23 January 1939, to observe a demonstration flight of the Model 7B that crashed, killing veteran test pilot John Cable and a French engineering officer. Despite the crash, the French had seen the potential of the 7B, and decided to purchase the bomber in quantity as the DB-7. Douglas now had no worries about its Northrop/El Segundo Division. In mid-1939 the DB-7 and the turbo-supercharged DB-7, designated the A-20, were purchased by the AAC. The A-20 design led to the development of the Douglas B-26 Invader attack bomber that, while contracted on 31 October 1941 and first flown on 10 July 1942, subsequently had many teething problems that delayed its combat debut into the spring of 1945.

Both Douglas and the AAC were eager to do something to save the B-18 program, which had a number of aircraft left on its production contract. Before the revision of the C-103A Douglas offered the Army their proposed B-18 upgrade (the B-23) as a feasible consideration for an attack bomber. Douglas initially tried to salvage the B-18A by adding a pair of 1,600hp Twin Cyclone Wright R-2600-1 radials that had been newly fitted to Boeing's Model 314 flying boats, and designated the proposed aircraft as the XB-22. While it was still on the drawing board, Douglas engineers realized that despite the increase in power, the XB-22 would not meet other specifications, and the venture was abandoned in the design stage.

The next option the Douglas engineers pursued was to revamp the B-18, which was based on the DC-2 design, by adapting the technology of the more contemporary DC-3 commercial transport. Douglas submitted their design to the Air Corps in late 1938, and because the proposal was not overwhelming in its concept the AAC agreed to a change order instead of a new contract. This Change Order (#2661) specified that the last 38 B-18As ordered under contract AC9977 be built and delivered as B-23s, and would retain the same B-18A serial numbers (s/n 39-27 through 39-64) and Constructor's Numbers (c/n 2713 through 2750). Although the B-23 was not revolutionary, the AAC realized that it was more advanced than the B-18A, and could be put into production ASAP in lieu of developing another adequate medium bomber design. To expedite production, the standard prototype and service test steps were waived, and the bombers were to be delivered as production B-23s. This waiver provision was exercised again a year later when 200 Martin B-26s were ordered directly off the drawing board and put into production. This decision was to prove detrimental, as the Marauders could have benefited from a XB-26 version, as they were too revolutionary and flawed, and the fledgling pilots found them too "hot," which resulted in numerous training accidents.

The B-23 received its name Dragon by default. North American Aviation had developed the NA-21 in 1936 as a late entry into the March 1937 bomber competition, but the bomber encountered problems with tail buffeting and high engine temperatures in testing. These problems were solved, but the aircraft had a price tag of $122,600, as compared to the B-18's $64,000, which was already under production contract. The NA-21 was returned to North American for further testing. In 1939 the AAC purchased the NA-21 from NAA for $550,000 to further develop it as the redesignated NA-39, with the intent to purchase five additional production B-21 bombers. Soon the AAC decided not to continue the project, and the name Dragon intended for the B-21 was assigned to the B-23.

In operational service the B-23 was to have a crew of six: pilot, bombardier/nose gunner, navigator, radio operator, camera operator, and tail gunner. (Note: during the 1930s the bombardier was referred to as the "bomber.") The inclusion of the cameraman projected the use of the B-23 as a long-range photo-reconnaissance aircraft in addition to its bombing role. In what seemed like a departure from the norm the co-pilot was eliminated, and only a single seat and one set of controls was located at the pilot's position. The pilot sat at the normal left station, and the co-pilot's position was replaced by an opening into the nose/bombardier's station.

The B-23 encountered no production problems, as the way had been paved by the use of B-18A lines and stock DC-3 parts. Just eight months after the change order contract approval, the first flight took place at Clover Field, Santa Monica, on 27 July 1939. Comparatively, the B-26 was authorized in September 1939, and did not fly until 25 November 1940, and then spent many months surmounting its many teething problems before being assigned to squadron service. Promptness was imperative in the B-23 program, and the

Douglas workers from the Santa Monica plant watch the first B-23 flight at Clover Field on 27 July 1939. (Douglas)

transparent nose section of the first bomber 39-27 was not ready at the time of the scheduled flight testing. The Douglas engineers quickly designed and fabricated a basic nose cone of hammered sheet aluminum that approximated the specified nose cone. However, reports are erroneous that 39-27 was delivered with an unglazed nose, because contemporary tactical thought was that the more powerful Wright R-2600 engines would give the streamlined bomber a top speed high enough to make frontal fighter attacks less probable, and the need for a forward turret unnecessary. Later production models were equipped with a glazed nose lodging the bombardier's station and a .30 caliber machine gun.

The first B-23 (39-27) came off the production line in July 1939, and was a significant improvement over the B-18A. Although the dimensions were similar (weight: 19,089 pounds empty/26,500 pounds loaded/32,400 pounds maximum (combat); length: 58 feet, 4.75 inches; height: 18 feet, 5.5 inches; wingspan: 92 feet; and wing area: 993 square feet), the portly fuselage of the B-18A was streamlined, and was even more slender than the DC-3, and the more substantial wings of the DC-3 were incorporated. The nose of the B-23 was very short compared to other bombers of the time, but its most conspicuous feature was the large tail fin and rudder, that was regarded as being sizeable by existing bomber standards. The empennage was integral to the fuselage to accommodate a tail gun position. The 1,600hp Wright R-2600 air cooled radial engines that were planned to power the XB-22 were to be used initially. This engine was 1,600hp rated for take off and 1,275hp rated at 12,000 feet, yielded a maximum speed of 282mph at 12,000 feet, and a cruising speed of 210mph, which was 66mph faster than the B-18, and only 5mph slower than the YB-17. However, the B-23 was much slower compared to its proposed medium bomber rivals: the B-25 at 315mph and B-26 at 310mph. Its rate of climb was 10,000 feet in 6.75 minutes, and it had a service ceiling of 31,600 feet (vs. YB-17: 35,000 feet). It had a normal range of 1,400 miles with 4,000 pounds of bombs, and a maximum range of 2,750 miles (one way, no bombs).

Description

Fuselage

The fuselage was of a semi-monocoque construction divided into five main compartments (fore to aft): nose compartment, pilot's compartment, bomb bay, main compartment, and tail compartment.

Nose Compartment

The nose compartment was also called the bombardier's, or front gunner's compartment. It was a removable window paneled unit with seven windows, including three on the left hand side, three on the right hand side, and one upper center. The passageway on the starboard side also had a window. Entrance to the compartment was from the pilot's compartment through a doorway located on the right hand side of the bulkhead at Station 120.

Pilot's Compartment

The pilot's compartment was located between fuselage Stations 120 and 270.5, and provided seats for the pilot, navigator, and radio operator. There was a molded window panel installed in the upper surface, a V-type windshield of stationary panels, a left and right hand sliding window, and a left and right hand stationary window aft of the sliding windows. The entrance door was located immedi-

Ten B-23s waiting for a final checkout at the Santa Monica plant. The B-23 encountered no production problems, as the way had been paved by the use of B-18 lines and stock DC-3 parts. (Douglas)

In what seemed like a departure from the norm the co-pilot was eliminated, and only a single seat and one set of controls was located at the pilot's position. The pilot sat at the normal left station, and the co-pilot's position was replaced by an opening into the nose/bombardier's station. (USAF)

ately aft of the window on the left hand side of the fuselage. An observation hatch was installed in the upper surface of the compartment aft of the pilot's overhead window. Two windows were provided in the left hand side of the fuselage adjacent to the navigator's and radio operator's seats. Another window was located on the right wall of the fuselage over the radio equipment table.

Bomb Bay Compartment

This compartment was located between the pilot's compartment and the main compartment, and extended from Station 270.5 and 380.25. Hydraulically operated bomb doors were provided, and formed the lower part of the fuselage when closed. Passage through the bomb bay could not be made when a full bomb load or partial bomb load, or a bomb bay fuel tank were carried on both racks. When no bombs, a partial bomb load, or a bomb bay fuel tank were carried only on the left hand rack, then a catwalk could be installed for passage.

Main Compartment

This compartment was located between Stations 380.25 and 513.5, and was sometimes referred to as the cameraman's, or center gunner's compartment. Two .30 caliber machine guns were installed in this compartment, and are described elsewhere. Provisions were made to mount camera equipment on the left hand side of the compartment immediately aft of Station 380.2 bulkhead. Two windows were installed for the cameraman: one located at the left hand side of the fuselage, and the other located beneath the hinged floor panel at the right of the camera installation. The floor gunner was provided with two window hatches: one located in the lower part of the compartment entrance door, and the other in the right hand side of the fuselage. The waist gunner had three sliding glass window hatches, with the upper sliding forward, and the left and right sliding aft.

Tail Compartment

The tail compartment was located aft of Stations 513.5, and terminated into the tail cone, where a .50 caliber machine gun was mounted (to be described later). The tail cone was detachable from the main fuselage at Station 662.375.

Wing

The B-23 wing was full cantilever, of monocoque stressed skin construction. It consisted of a center section with engine nacelle attached, and a left and right tapered outer panel. The wing was attached to the fuselage at the center section by eight main vertical fittings. Each outer panel was attached to the center section by fittings at each spar and skin attaching angles. A rib was placed between the center and outer panel to transfer the stresses evenly.

The wing center section was made up of a front section, left and right hand trailing sections, and left and right engines nacelles. The nacelles extended the full chord of the wing on the upper surface, and approximately to the wing flap hinge on the lower surface. The length of each nacelle was lengthened and extended downward to allow the landing gear, whose lower half of the retracted wheels were exposed on the DC-3 and B-18, to now be completely enclosed under large clamshell doors. The center section also incorporated the wing flap, extending outward from the wing fillets to the outer wing joints. The flaps were attached to the wing trailing sections by piano-type hinges.

The outer wing panel was made up of a detachable wing tip, a main section, and an inboard trailing section. The trailing section extended outward from the wing joint to the aileron, and was attached to the main section at the rear spar. Each outer wing panel

This photo of the B-23 shows its wing to be similar to the B-18's, but the length of each nacelle was lengthened and extended downward, while the fuselage tail accommodates the revolutionary turret. (USAF)

incorporated a wing flap and aileron. The wing flap extended from the wing joint outward to the aileron, and was attached to the trailing section by a piano-type hinge. The hydraulically actuated, all-metal wing flaps were of the split trailing edge type. The aileron was hinged to the outer panel at six places, and moved on sealed lubricated bearings. The ailerons were a fabric-covered metal frame, with the right aileron equipped with a controllable trim tab. Each outer panel incorporated three watertight flotation compartments that were formed between the front and center spars, and between the center and rear spars, respectively.

Empennage

Because of the tail turret, the vertical stabilizer of the B-23 was taller than the B-18. When the bombers were parked the B-23 tail extended to 18.4 feet, while the B-18 extended to 15.4 feet. The all-metal horizontal and vertical stabilizers were of a multi-cellular construction, and attached in fixed alignment to the fuselage. The horizontal stabilizer consisted of two halves bolted together at the fuselage centerline. The rudders and elevators were a fabric-covered metal frame, and were statically and aerodynamically balanced. The rudder was hinged to the vertical stabilizer at three places, and was equipped with a controllable trim tab hinged at three places. The elevators were hinged to the horizontal stabilizer at three places, and each elevator was equipped with a controllable trim tab hinged at three places.

The clam shell doors and increased diameter of the engine nacelle allowed the landing gear and tires to retract fully into the nacelle. (Pima)

Landing Gear

As discussed previously, each nacelle was lengthened and extended downward to allow the landing gear, whose lower half of the retracted wheels were exposed on the B-18, to now be completely enclosed under large clamshell doors. Landing gear consisted of two independent, hydraulically operated units that retracted into the nacelles. Doors, operated by linkage attached to the landing gear rear brace strut, completely enclosed the landing gear when retracted. The doors were attached to the nacelles by piano-type hinges. A safety latch locked the gear in the down position. A green signal light indicated that the gear was down and latched in place. A horn also would sound if the engines were throttled when the gear was not in the safe landing position. The gear was equipped with 45x20.00-18 wheels and eight-ply smooth contour tires, with two 14 inch x 4 inch duo-servo, hydraulically actuated brakes per wheel. A parking brake was provided for extended parking. Two air-oil shock absorber struts were solidly clamped at their piston ends to the axle of each wheel, and were interconnected at their upper cylinder ends by rigid trusses. A hydraulically operated strut was attached near the center of a rotating truss so that, when the strut was retracted, it rotated the upper truss forward and upward into the nacelle in an arc.

The tail wheel was full swiveling, hydraulically retracted, and fully enclosed by doors. Tail wheel extension, retraction, and latching were simultaneous with the operation of the main landing gear. A control was provided for locking the tail wheel in trailing position for take off and landing. A 22x9.00-6 tail wheel and eight-ply, smooth contour tire were used.

Surface Controls

A conventional control column and wheel was provided for controlling the movement of the elevators and ailerons. The control column was constructed of aluminum alloy tubing and castings, with the exception of the torque tube. The rudder movement was controlled by conventional rudder pedal operation, and the pedals were adjustable fore and aft. The wheel brakes were operated from the rudder pedals by toe pressure on the pedal foot rests. The rudder tab control was operated by a rod and cable drum assembly located in the leading edge of the rudder, and was controlled by a crank on the pilot's control pedestal. The elevator movement was controlled by conventional fore and aft movement of the control column. The elevator tab control was operated by a rod and cable assembly located on the main torque tube of each elevator. Tab movement was controlled by a wheel installed on the pilot's control pedestal. Aileron movement was controlled by rotation of the wheel on the control column. Only the right hand aileron was equipped with a trim tab, which was operated by a crank on the control pedestal. The wing flaps were operated by a hydraulic strut installed in each wing center panel flap enclosure. Push-pull rods, mounted in rollers, were attached to each end of the strut piston shaft. When the push-pull rods moved inboard the flaps lowered, and when they moved outboard, the flaps raised.

Engine

The R-2600 *Cyclone* 14 was Wright's first successful two row, air-cooled radial engine, having two rows of seven cylinders each, and producing 1,600hp and 2,400rpms. The cylinder design used two pushrod actuated valves per cylinder in a hemispherical combustion chamber, and was based on Wright's R-1820 engine, which was used in the B-18s. The early cylinder barrels had machined integral cooling fins, while later dash numbers had Wright's patented "W" design, which was comprised of an aluminum muff swaged onto the barrel. The R-2600 was introduced in 1937 and used by Boeing on the Pan American *Clipper*, and was a proven engine by the time the war started. The Douglas XB-22 was a proposed B-18 airframe mounting 1,600hp Wright R-2600-1 engines, but none of these aircraft were built. When the B-23 was planned, it was to mount the Wright R-2600-1, but this engine was replaced by the Wright R-2600-3, which also was being installed in the Douglas A-20A (as well as the Navy BD-1 derivative and the A-20E). The major difference between the two dash numbers were that the -1 was equipped with a single stage, single speed Wright supercharger, while the -3 had a two stage Wright supercharger and a different propeller shaft reduction gear ratio. After superchargers were installed in the B-23s and A-20s they were deleted from all subsequent A-20 models, as these medium attack bombers were to fly their missions at low to medium altitudes, where a supercharger would have no advantage. Only 164 of the -3s were built, and they were replaced in numbers by the higher compression ratio R-2600-11 in February 1940.

Propellers

The 13 feet, 6 inch diameter Hamilton-Standard Hydromatic Quick Feathering propellers were supplied by generators that gave constant speed and feathering control. (See B-18 propellers)

Oil System

Oil was contained in two tanks, one mounted in each nacelle. Each tank carried 33 gallons, and was provided with a level cock at the minimum oil quantity level of 20.5 gallons. Oil dilution apparatus for winter starting was provided. Oil pressure and temperature gauges were located on the lower right hand corner of the pilot's instrument panel. Oil temperature controls were provided to regulate the hydraulically operated oil cooler flaps. Control handle movement was synchronized with the movement of the oil cooler flaps, which could be set in any position between full open and full closed.

Fuel System

Fuel was carried in four wing center section tanks that were not self-sealing. The two forward tanks contained 229 gallons each, and the two rear tanks 215 gallon each. A jettisonable 420 gallon bomb bay tank could be carried to give the bomber a total fuel capacity of 1,290 gallons. In normal operation, each engine had its individual fuel system. The left wing tanks were used to supply fuel to the left hand engine, and the right hand tanks supplied fuel to the right hand engine, but any tank could be used to supply fuel to either or both engines. A fuel pump was mounted on the right hand side of each engine. Fuel pressure gauges and warning lights were located on the lower right hand corner of the pilot's instrument panel, and would light when the pressure dropped below 12 pounds (normal was 14 pounds). There was a fuel quantity gauge and selector switch (one for each of the five tanks) located on the instrument panel. Two manually operated wobble pumps were installed.

Hydraulic system

Two engine-driven oil pumps, one mounted on each engine, supplied pressure to the hydraulic system, which normally operated at a pressure of between 775 and 825 pounds. The 12 gallon system was of the pressure tank type, and incorporated a hand pump for auxiliary use to supply pressure to any of the units deriving their source of power from the engine-driven pumps. The following were hydraulically operated: main landing gear and tail gear, brakes, wing flaps, bomb doors, cowl flaps, and oil cooler flaps.

The B-18's Wright R-2600-1 engines were replaced by the Wright R-2600-3 that spun the Hamilton-Standard Hydromatic Quick Feathering propellers. (USAF)

Electrical System

The single wire 12 volt grounded system was used, except where instrument deflection could occur. Two type D-6, 68 ampere batteries were installed in the left nacelle. Battery charging was by either or both generators, and was controlled by the generator main line switches. The generator control panel was installed on the bulkhead aft of the radio operator's seat. The pilot's electrical panel was located at the left of the instrument panel, and the bombardier's electrical panel was located on the right side of the nose compartment.

Crew Comfort and Safety

The pilot's seat was adjustable vertically. The navigator's and radio operator's seats were provided with 90 degree rotation. A jump seat was provided, and folded upward against the bulkhead on which it was mounted. A removable seat was provided for the cameraman, and when not in use it was stowed just aft of the camera support tube on the left hand side, where it was retained by an elastic cord.

The nose and pilot's compartments were soundproofed. There was no armor at the various crew stations. A glare curtain was provided for night flying, and was attached to the fuselage above the pilot's seat. A personal baggage rack was provided, and was installed on a bomb rack.

The oxygen for the pilot, bombardier, radio operator, navigator, and cameraman was supplied from four individual E-1 cylinders and Type A-6 regulators, while the tail and waist gunners were supplied from separate E-1 cylinders and A-6 regulators.

There were two thermos bottles supplied for drinking water. One was installed on the left hand side of the pilot's compartment, with a cup dispenser installed on the left hand side of the bulkhead, and the other was mounted on the forward bulkhead of the main compartment, with the cup dispenser just below it. A relief tube was provided in the pilot's compartment and in the tail compartment, while a toilet was provided on the right hand side of the tail compartment.

Alarm bells were located in the bombardier's compartment, the main compartment, and the tail compartment. The life raft container was installed on the left nacelle for stowing a Type A-2 life raft. A Type A-12 CO_2 fire extinguisher system was installed with perforated outlets in each engine section and nacelle. Two Type A-2 hand-held extinguishers were provided: one located in the pilot's compartment, and the other in the main compartment.

The heating system was the steam heated air type. The right hand engine exhaust collector incorporated a boiler where water was converted into steam by the heat of exhaust gases. The steam was passed into a radiator, where it heated air brought in from the outside. The heated air was then circulated through ducts to outlets in various compartments in the airplane. A desired interior temperature could be obtained by regulating the mixture of heated air and air at outside temperature. Outside air could be circulated through the system in variable amounts. Two outlets were provided for defrosting the pilot's windshield.

For water ditching, three watertight compartments were located within each wing outer panel, and one below the bombardier's floor. A vent and purge line extended from each wing compartment to the interior of the bomb bay; the two lines from the fuselage compartment followed along the bulkhead that formed the rear boundary of the watertight compartment. The purge lines were provided with fittings for a bilge pump and corks for plugging the upper outlets.

Communications System consisted of:
Command Set: SCR-AJ-183
Liaison Set: SCR-187-A
Radio Compass: SCR-242-B
Marker Beacon Receptor: RC-20-A
Interphone: RC-15A

Photographic Equipment

Provisions were made for Type T-3A, or Type K-3B, or Type K-7C camera installation, with a Type A-2 Vertical View Finder installed immediately aft of the camera. The camera mounting apparatus was located in the main compartment corner formed by the forward bulkhead and the left hand fuselage wall. There was a removable door in the fuselage below the camera.

Bombing Equipment

The bomb bay was located aft of the pilot's compartment, and the hydraulic doors formed part of the fuselage when closed. Bomb racks on each side of the bomb bay incorporated six stations for carrying variations of bomb loads. The B-23 was designed to carry a useful load of one 2,000 pound bomb, two 1,100 pound bombs, four 600 pound bombs, six 300 pound bombs, or twelve 100 pound bombs. The bomb hoisting mechanism was located on the forward bulkhead of the main compartment, and operated a crane mechanism in the bomb bay.

The nose compartment was also called the bombardier's, or front gunner's compartment. It was a removable cone/window paneled with the seven windows. The small round window was the opening for the .30 caliber machine gun. The Norden bombsight is covered by canvas bag. (USAF)

The bombardier's station contained the equipment necessary for bombing operations. Bombs could be released by means of an electrical contact automatically closed within the bombsight, a momentary contact switch (firing key) operated in conjunction with a two-way switch on the bombardier's electrical panel, a Type A-1 Interval Control Unit, which was also controlled by the two-way switch, and a bomb door salvo lever. An emergency bomb salvo release was provided for the pilot. An electrical and mechanical safety locking device was incorporated with the bomb release system to prevent the release of bombs before the bomb bay doors were opened.

Armament

The defensive armament of the B-23 was considered to be unsatisfactory, as the bomber's defense was based on the Air Corp's inflexible dedication to outdated military thinking that was no longer valid once the European war began in September 1939.

After the first production model B-23, the succeeding models were equipped with a glazed nose lodging the bombardier's station and a .30 caliber ball and socket mounted M-2 machine gun with six 100 round ammunition boxes. The gun had a 55 degree conical angle of fire.

Besides the single .30 caliber machine gun in the nose, there

B-23 Armament Gallery

Nose Machine Gun. (USAF)

Aft Fuselage Wall Mounted Machine Gun. (USAF)

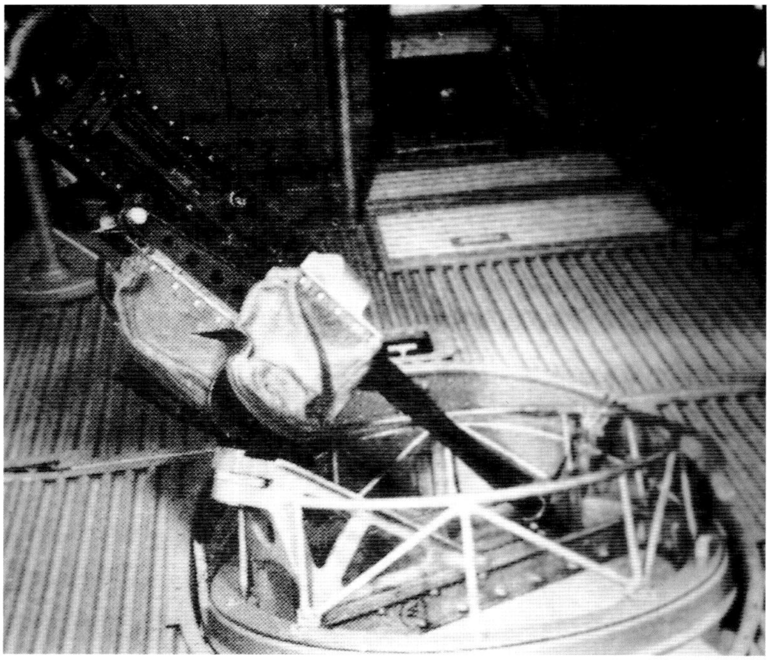

Floor Mounted Machine Gun. (USAF)

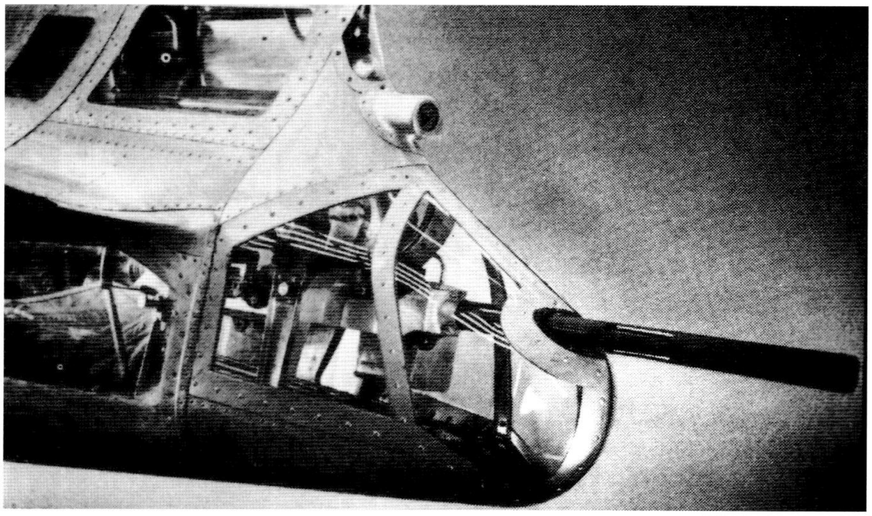

Tail Turret (closed). (USAF)

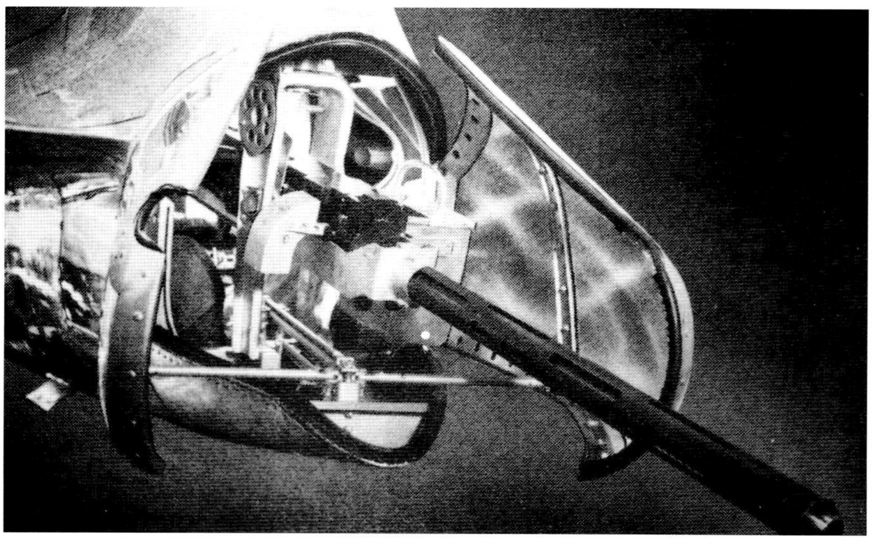

Tail Turret (opened). (USAF)

were two more .30 caliber machine guns in the mid-fuselage, each supplied with 600 rounds of ammunition in six 100 round boxes located at the aft bulkhead. One gun was located on a swivel mount fixed to the aft fuselage bulkhead, and could be fired through two sliding side windows, or through a sliding roof window in an arrangement that expected one gunner to cover three different areas. When aimed through the upper window, the gun had a 110 degree angle of fire left to right, and a fore and aft angle of fire of 140 degrees. Through the side windows it had a 115 degree up and down angle of fire and 140 degrees fore and aft.

Defense against an attack from below was provided by a .30 caliber machine gun firing through the ventral floor hatch with an 80 degree conical angle of fire. The gun was sited in the floor inboard of the entrance door of the main compartment. This gun was mounted in a ring that rotated 360 degrees, and could be tilted to allow a greater rearward angle of fire. The ring was mounted on steel trunnions that had a galvanic reaction with the aluminum floor, and after a while the corroded mount would tear out of the floor. The gun was supplied with 600 rounds of ammunition in six 100 round boxes located on the right hand side of the compartment, and just forward of the gun ring.

But the chief military development of the B-23 lay in the first glazed tail gunner's position installed on an operational American bomber. The B-17 was not equipped with its tail twin .50s until the E model in September 1941, and the Douglas XB-19 ordered in 1937 was designed with a tail turret, but this aircraft did not fly until mid-1941. The position contained a single .50 caliber M-2 machine gun with 99 rounds in three 33 round boxes, and a fixed telescopic sight mounted about 13 inches above the gun. The glazed portion of the tail cone aft of Station 713 opened into two clamshell like halves to permit machine gun movement. The left hand side window had a release mechanism, and could be used as an emergency exit by the tail gunner. The sight incorporated a movable reticle coordinated with the gun by cables, pulleys, and gears. The sight had a wide angle field of view, and the crosshair that appeared in the field of view automatically moved with the gun, so as to show at all times at what field the gun was aimed. The long bar of the crosshair pattern indicated a short range (200 yards), and the short bar corresponded to the trajectory drop at 600 yards. The width of the bars corresponded to the appearance of 37 feet of span at those distances. The objective lens of the sight was protected by a glass cover and shutter. This defensive position was primitive, as the gunner laid in a prone position with his feet forward, and before he could manually traverse the gun he had to open the transparent framework doors that enclosed the gun, which pointed directly to the rear during normal flight. The gun had a relatively narrow cone of fire at 40 degrees, but for a fast bomber it did not have to be very wide as, except for diving attacks, contemporary fighter attacks were from dead astern, where the field of fire did not have to be very wide. Also, the tail gun was the first .50 caliber machine gun used defensively on an American bomber, and at the time the B-23's .50 caliber gun far outranged the smaller bore, albeit more numerous guns of would-be foreign enemies. Type H-2 gun cameras were used on each of the four machine guns.

B-23 Acceptance and Unit Assignment

After several Douglas factory proving flights, the first two B-23s (39-27 and 39-28) were accepted by the ACC, and sent to the Air Materiel Command, which evaluated the aircraft at Wright Field, where they remained throughout their careers. Several individual aircraft were initially sent to various non-tactical units, such as the Service Test Center and the Mechanics School at Chanute Field, IL, and for photo-reconnaissance evaluation at Lowrey Field, CO, while Douglas retained several examples for continued development. The last B-23 rolled off the production line in September 1940.

The first B-23s entered operational service when four were sent to the 89RS at March Field, CA, but soon the squadron moved to McChord Field, Tacoma, WA. The next production B-23s were delivered between February and September 1940, replacing the Northrop A-17 attack bombers of the 34th, 37th, and 95th Bomb Squadrons of the 17BG(M), based at March Field, Riverside, CA. But the B-23s never completely equipped these squadrons, instead mixing with B-18s. The 95BS received its B-23s after the 17BG moved to McChord in June 1940. The 34BS received their B-23s at March Field before moving to McChord, where it remained until moving to Pendleton, OR, in June 1941, and then to Portland, OR, in December. This unit was then transferred to the East without their B-23s. The 37BS was stationed at Barksdale, LA, when it received its B-23s and relocated to Lowrey Field in July 1940, and then to Pendleton, OR, in June 1940. In late 1941 the 17BG's B-23s were replaced by B-25s, with the 34BS receiving the first of the type. The Dragons were transferred to the 12BG at McChord Field, WA, and then to the 13BG at Orlando. The operational B-23s displayed

After the delivery of the first four B-23s, the three squadrons of the 17BG received the next production B-23s delivered between February and September 1940, replacing the Northrop A-17. (USAF)

McChord AFB, Tacoma, WA, was a major base for B-23s. Members of the 12BG pose for a group photo in front of a B-23. (Pima)

such a notable improvement in performance and payload compared to the B-18A that 125 B-23As were ordered, but were then cancelled before production, as reality overcame enthusiasm. However, two B-23s (39-27 and 39-28) did become B-23As, as it was AAC practice to assign a series letter to production aircraft that were consigned to test programs. These test aircraft were exempt from Time Compliance Technical Orders or Change Orders, which could hold up production aircraft, and thus the test program would not be delayed. But the initial B-23A enthusiasm soon wore off, as it became apparent that any B-23 as a medium bombardment aircraft was, in truth, a transport modified to a bomber, like the B-18. It was plainly mediocre to the four engine B-17E, which was considered the first combat model of the Flying Fortress. As a medium bomber contender the Dragon was also slower, and was less heavily armed than the twin engine North American B-25 Mitchell and the Martin B-26 Marauder, which were entering production.

After Pearl Harbor the B-23s were transferred to the 12th, 17th, and 47th BGs, and the 89RS at McChord Field, WA. A few B-23s were used to patrol off the Pacific Coast to search for a possible Japanese carrier task force, and when that threat subsided they looked for Japanese submarines. After this anxious time the B-23 was left in operational limbo, and like the B-18 tried to find a role, being relegated mainly to transport, training, and a few were used for testing and experimental roles.

The B-23 Takes on the B-25 and B-26 in the Air Corps 1939 Proposal for a Medium Bomber

Toward the end of the 1930s, it became evident that America would eventually be at war, and that new designs were needed to match those of the *Luftwaffe*, as reported by Charles Lindbergh. On 11 March 1939 the Air Corps issued AC 39-640, calling for a new medium bomber with a crew of five that would have a maximum speed in excess of 300mph, with 350mph desired, carrying a bomb load of 3,000lbs over 2,000 miles, and was to be armed with four .30 caliber machine guns. It was to operate at altitudes between 8,000 and 14,000ft with a service ceiling of 20,000ft; primarily to support ground units, but also to be able to operate in a tactical role. The engine specifications called for the P&W R-2800, the Wright R-2600, or the new Wright R-3350. The bidding companies could submit any number of proposals, offering design options, mainly engine/propeller combinations, but also airframe, armament, and other redesigns. This procedure offered the Air Corps an a la carte selection of designs while saving the company the cost of building a prototype that was expensive and, more important, with war looming, time consuming, and usually required extensive modifications. The prize was an order for 385 aircraft. Seven companies submitted designs. Consolidated, Vought-Sikorsky, and Burnelli submitted bids, but quickly dropped out of the contest. Martin (Model 179), North American Aviation (NA-62), and the Stearman (XA-21) joined the Douglas B-23 as the only companies to remain in contention to meet the 10 September deadline.

The Contenders

Martin Model 179 (Future B-26)

The Martin Company was stunned when its Model 167 was not chosen by the Air Corps for production, but the French order for 115 aircraft in January 1939 allowed the company to enter the high speed medium bomber competition with a new design. To reach the desired high speeds the Martin engineering team, led by the 26 year old U.S. Naval Academy graduate Peyton Magruder, was encouraged to incorporate as many advanced design elements as possible, and to forfeit performance where necessary. Martin submitted their bid on 5 July 1939—the Model 179, which consisted of 15 different proposals varying mainly in engine type and supercharger arrangement. Five of the arrangements proposed the Wright R-2600, six the P&W R-2800, and four the Wright R-3350, which was only in the early development stage. Four of the engine proposals were based on a wing area of 650sqft. The winning proposal would be the Martin Proposal No. 6, which required the P&W R-2800 with a one-stage, two-speed supercharger and 600sqft wing. The P&W R-

2800-5 Double Wasp air-cooled radials fitted with two-speed mechanical superchargers and the 13.5ft four-bladed Curtis electric propeller fitted with spinners and root cuffs for engine cooling were introduced to U.S. service in this aircraft. These engines gave the aircraft a maximum speed of 315mph at 15,000ft, a cruising speed of 265mph, a service ceiling of 25,000ft, and its combat range was to be 1,000 miles at 265mph carrying a 3,000lb bomb load. However, the R-2800 were not fully developed, and were on the test stands at the time.

The bomber was very aesthetic, with a clean, low drag, cigar shaped monocoque fuselage that was circular in cross section, and supported by a hydraulic tricycle landing gear. The Model 179 measured 56ft long, 19.83ft high, and weighed 21,375lbs empty and 32,025 loaded. The original design mounted the fashionable twin tail, but wind tunnel tests demonstrated that the single tail would be more successful, both aerodynamically and defensively, as it gave the gunners a better field of fire. The horizontal tailplane was the first with a marked dihedral (8 degrees). The wings were unusual, in that they carried no fillets, and were only of 65ft span and 602sqft, which gave the aircraft a very high wing loading of 51lbs/sqft. Since speed was the major design concern, and the ACC made no stipulations concerning wing loading, Magruder designed the small wings with high loading that gave high speed, but also high landing speeds (130mph), and required long take off and landing distances. The aircraft was called the "Baltimore Whore," because its small wings gave it no visible means of support. The bomber was regarded as a "hot" aircraft, and gained a reputation as being dangerous to fly, so much so that aircrews did not wish to be assigned to the B-26 "widow maker." The bomber required more intensive training, and once mastered, the B-26 became an outstanding medium that had the lowest operational loss rate of all American bombers.

The small, shoulder-mounted wings left room in the center fuselage for a large bomb load; Magruder actually measured the bomb bay of a B-17 at Wright Field, and then duplicated it in the Model 179, while adding a second smaller tandem bomb bay aft. The forward bay doors folded in half when opening, while the aft doors were the standard hinged type. Two 2,000lb bombs could be carried in the forward bay, and a total bomb load of 5,800lbs could be accommodated. For a time the B-26 would be able to carry a larger bomb load than the B-17 model of the time.

Its armament was a .30 caliber ball-joint mounted in the nose operated manually by the bombardier, and another .50 caliber in the pointed tail cone, operated manually by a sitting gunner. There was another .30 caliber flexible gun installed in a manually operated tunnel opening in the floor of the rear fuselage. The original design called for four flexible .30 machine guns, but the Model 179 was the first aircraft to carry a power operated turret. Martin designed the electric 25CE power turret mounting twin .50 caliber machine guns located behind the bomb bay just forward of the tail. These turrets were also used later on the B-25, B-17, and B-24.

Magruder designed an innovative bomber that could be easily mass manufactured. To facilitate manufacture, innovations included: the use of high strength, lightweight plastic materials in place of metal; spot welding; large alloy castings; and large aluminum forgings instead of riveted parts. Magruder had Martin purchase stretch presses used in automobile manufacture to reduce the amount of drop hammer work on complex skin fairings. To enhance air flow, it was the first U.S. bomber to use butted seams for the skin covering, as opposed to the previous lapped seams. Martin utilized many sophisticated electrical and hydraulic systems that caused initial problems and complicated maintenance, but once resolved made the aircraft easy to fly. It was the first U.S. bomber to use rubber; self-sealing fuel tanks installed as regular equipment, which Martin had invented and developed as "Mueng Cells." Also, it was the first bomber to use an all Plexiglas nose (vs. glass) and emergency air brakes.

The bottom line of the Model 179 was that, while it appeared to be an excellent design on paper, it pushed contemporary engineering limits, its engine was unproven, and there were significant uncertainties that it could be delivered on time, as a number of major components required substantial time to be developed, both by Martin and its subcontractors. The projected cost was twice that of the NA-62, and equivalent to that of a four engine B-17.

NA-62 (Future B-25)

On the announcement of AC 39-640 Lee Atwood, NAA Chief Engineer, recently named Assistant General Manager, headed the design team for the medium bomber project that was given the company designation NA-62. Atwood had been on the design team for the successful BT-9, BC-1 (AT-6 precursor), and the O-47 projects, but the NA-21 and NA-40 bomber projects were not contracted for production, and Atwood was determined to produce a successful medium bomber. Over the years Kindelberger and Atwood had developed a number of precepts in developing a new design. Importantly, before proceeding with the design, the requirements of the Proposal needed to be accurately understood. Using complicated components with a long developmental lead time were to be avoided. With the world aviation scene constantly changing, the Air Corps deemed the use of new and more powerful engines necessary, but these new engines needed to be available when required, and their predicted performance met. Before America entered the war, government money was not thrown at manufacturers as it was later, and bidding was bidding. The proposed aircraft had to be able to be manufactured at a competitive price, be easy to maintain and repair, not only meet but exceed the Proposal's requirements, and most importantly, be easy to fly. NAA was in a good competitive position in the industry, and was able to respond quickly and efficiently to the Air Corp's Proposal. The Inglewood plant had been designed and built for military aircraft production, and as discussed earlier on the BT-9 project, Kindelberger and Atwood had pioneered the subassembly fabrication technique of splitting airframes.

The NA-40B design had proven viable in testing before its unfortunate accident, and Atwood maintained its basic characteristics by designing a larger version, with better performance and a greater bomb load. The new design gave the tubby NA-40B a more slender look by straightening the lower fuselage profile. The fuselage was constructed of riveted formed frames, stringers, longerons, and skin. It was composed of four separate and removable sections: nose, forward, center, and aft. Access doors were located in the navigator's compartment floor, and in the aft fuselage. Another escape hatch was located in the cockpit window directly over the pilots.

There was to be a crew of five: pilot, co-pilot, bombardier/gunner, navigator/radio operator, and gunner. The design was 54.2ft long (vs. 47.8ft for the NA-40B), 16.3ft high (vs. 15ft), and had a wingspan of 67.5ft (vs. 66ft) with a wing area of 610sqft (vs. 599sqft). The numerous "design add-ons" increased the weight of the NA-62 from an empty weight of 13,961lbs to 17,258lbs, and the loaded weight increased from 19,741lbs to 28,577lbs. To meet the increased bomb load requirements (3,000 vs. 2,100lbs), the raised tandem cockpit of the NA-40B was replaced by a side-by-side configuration, with its ceiling at the same level as the top of the main fuselage. The aircraft width was increased to 4.7ft, which was adequate to accommodate two side-by-side pilots and the glazed bombardier's position, and provided sufficient space for the proposed bomb load. The longer nose area provided the cockpit with outstanding visibility, particularly downward and to the sides, which would later become significant, when the B-25 took on its low level attack role. The bombardier had to reach his position in the nose via a crawl tunnel under the flight deck. The navigator/radio operator's station was located directly behind the flight deck. Since the bomber was not to be pressurized, a circular cross section was not necessary.

The wing was lowered from its high shoulder position on the NA-40B to a new mid-fuselage position, and passage to the rear fuselage was through a crawl space above the bomb bay. NAA designed a moderately tapered wing of continuous dihedral. NAA preferred the unproven 2,000hp P&W R-2800 because of its increased power and smaller frontal area. However, the Air Corps preferred the established 1,700hp Wright R-2600, but it was stipulated on the ACC's order. The nacelles covering the R-2600s were extended over the wing, and the main gear retracted backwards completely into them.

The armament consisted of four flexible .30 caliber machine guns. There was a .30 in the nose that could be readily mounted on any one of three ball-and-socket mounts: one pointed forward, and one to each side. Another flexible .30 was mounted behind a Plexiglas hatch located in the upper rear fuselage. The third flexible .30 could fire from mounts at each waist window, and through a hole in the floor. There was a .50 caliber gun located in the streamlined Plexiglas canopy in the extreme tail. The tail gunner had to lay prone behind clamshell doors that opened sideways to allow the traverse of the guns. The design had been pioneered on the Douglas B-23. The bomb load was designed to carry a varied bomb load to provide for its assorted missions. The bomb permutations varied from clusters of twelve 100lb bombs to varying quantities of 250, 300, 500, 600, and 1,100lb bombs, or combinations of these types, or just a single 2,000lb bomb. NAA worked 40 days, including weekends and overtime, to meet the 10 September 1939 deadline, and NAA offered the Air Corps 83 design options.

Boeing-Stearman XA-21

A fourth (non) contender was the Stearman XA-21, developed from the Boeing Stearman X-100, which was an unworthy challenger in the previous Air Corps Proposal (AC 98-102 and Circular Proposal 38-385) in the late December 1937 attack bomber competition, losing out to the Douglas DB-7.

The Decision

The AAC sent a bulletin entitled "The Method of Evaluation" to each bidder that described how the competing designs were to be evaluated and a winner selected. A number of weighted design characteristics were listed, with speed being the most important. Each aircraft design would receive a total numerical score, consisting of numerical values given to each design characteristics, with 1,000 being the highest theoretical score.

The Air Corps studied and evaluated each of the proposals, with the North American and Martin proposals being "off the drawing board," with no XB-25s or -26s, with the Douglas and Stearman proposals partially being garnered from "hard copies." The results were: Martin Model 179 scored 813.6 points; North American's NA-62 trailed by a sizeable 140.2 points at a distant 673.4 points; surprisingly, the Douglas B-23 was not too far behind the NA-62 at 610.3 points; while the Boeing-Stearman XA-21 was a non-contender at a lowly 442.7 points. Both Martin and NAA were requested to submit per unit cost bids. Martin quoted $7,868.15 for its Model 179, and NAA quoted $6,397.28 for the NA-62, which was a significant 19% under the Martin bid.

On 10 August the AAC ordered 201 Model 179s, designated as the B-26, from Martin at $15,815,000, with the contract approved on the 19th. The Martin contract was for 201 aircraft, instead of the 385 promised, as Martin was unable to provide that many B-26s, but did the $1,479.87 difference in per item bid also have an influence on the AAC decision? The company was too committed in building the Model 167s as the Maryland (309 total) for the French, and as the more powerful Model 187 Baltimore (1,575 total) for the British. In addition, the company was about to go into production of its PBM Mariner for the Navy. On 4 September 1939, in anticipation of receiving the contract, Martin had begun expansion of its Middle River Plant in Maryland to produce the B-26.

On 20 September 1939 the Secretary of War, Henry Stimson, contracted the remaining 184 medium bombers of AC 39-640 as the B-25 (plus one test airframe) to NAA under contract #AC-13258 for $11,771,000. North American began its initial shop work under North American General Order NA-62. The NAA engineering team worked on three versions of NA-62: 25 B-25s (NA-62); 40 B-25As (NA-62A); and 119 B-25Bs (NA-62B).

Although it faired well against the NA-62/B-25 in the competition, the only order it would receive was Change Order #2661 of late 1938, which specified that the last 38 B-18As ordered under contract AC9977 were to be built, and delivered as B-23s, and would retain the same B-18A serial numbers (s/n 39-27 through 39-64). The B-23 had encountered no production problems, as the way had been paved by the use of B-18A lines and stock DC-3 parts. Just eight months after the Change Order contract approval, the first B-23 flight took place at Clover Field, Santa Monica, on 27 July 1939, long before the 19 August 1940 maiden flight of the B-25, and the 25 November 1940 first flight of the B-26. But the "ready-to-go" B-23 was clearly a lackluster performer, and the Air Corps wisely decided to order and wait for the "off the drawing board" B-25s and B-26s to be developed. The 38 B-23s had to continue to try and find their place in the Air Corps.

The B-23 Tries to Find a Niche

Anti-submarine Role

Because of its new narrow fuselage, the B-23 could not be fitted with MAD and radar equipment, as had the 122 rotund B-18As that were converted to anti-submarine B-18Bs. The 520BS(H), formed on 13 October 1942, and equipped with North American O-47s, was one of the units that was quickly transformed into an anti-submarine unit. On 29 November 1942 it was reconstituted as the 15th Anti-submarine Squadron, and was equipped with an assortment of aircraft that could be spared for patrol duty: Mitchells, Hudsons, and a few Dragons. The unit operated along the Atlantic Coast, including Langley Field, VA, and Drew Field, FL, before transferring to Batista Field, Cuba, on 3 August 1943. It returned to the U.S. in October 1943.

Glider Snatch Tests

After the German airborne successes at Fort Eben Emael, in Holland, and the invasion of Crete, U.S. airborne tacticians saw a need to develop a method of retrieving flyable gliders from combat zones so that they could be used again. Also, the method could be used to snatch gliders from the factory to be transported to training bases. B-23 (39-28) was used in glider pick up testing at Wright Field in late 1942 and early 1943.

During the tests, the ground pick up station consisted of a socket plate that held two 20 foot poles in the socket, and whose ends were spread 20 feet apart in a "V." The poles had spring clips at their ends that held a long nylon loop tightly between the poles. The glider's nylon tow line was held off the ground, and was attached to the loop. The pick up method was made possible by the newly developed nylon rope tow line that was resilient, returning to its original length after being stretched on the pick up, and preventing the tow aircraft's tail from being yanked out of the fuselage. The newly developed Model 80 winch, rated at 8,000 pounds, was installed in the tail of the B-23, and was able to pick up a fully loaded XCG-4A glider. The winch used a hook at the end of a steel cable, and a 15 foot boom held the cable and hook, so that the open end of the hook could engage the nylon loop between the two poles. After the hook engaged the nylon line the resistance increased, and the steel cable was allowed to play out. As the glider began to fly within 200 feet and reached the speed of the tug, the brake was applied to the winch drum to gradually stop the cable play out. Then the winch motor began to reel in the steel cable onto the winch drum. The pick up applied great stress to the fuselage and engines, and the winch motor put a large load on the aircraft's generators.

Later the C-47, derived from the DC-3—the B-18/B-23 precursor—became the primary glider tow and snatch vehicle. They were used to develop "double" snatches: picking up the first glider and then returning to get the second. CG-4A gliders were used in eight major operations during World War II, and most were so badly damaged that they were not suitable for pick up, but a number of undamaged gliders were loaded with wounded soldiers and snatched for evacuation to hospitals in the rear. During *Varsity*, the crossing of the Rhine River, nearly 150 reusable gliders were snatched from the landing ground.

Miscellaneous Uses

B-23 (39-28) was delivered to Emerson Electric to test various remotely controlled turret arrangements, while B-23 (39-32) was delivered to Pratt & Whitney on 20 August 1940 to test the 2,000hp R-2800-5 engines as part of the B-26 and XB-28 programs. This aircraft later tested various propeller applications, including two and four configurations, and a six-blade counter-rotating mounting.

Five B-23s were transferred to Class 26 for use as non-flying instructional aircraft at mechanics school:

(39-34) Instructional airframe 20 June 1943
(39-40) Instructional airframe 17 September 1942
(39-45) Instructional airframe 8 May 1942
(39-49) Instructional airframe 26 November 1942
(39-54) Instructional airframe 31 August 1943

"Smart Bomb" Testing

In October 1941 Hap Arnold assigned a permanent detachment of scientists to test an assortment of unmanned controlled winged bombs and aerial torpedoes at Muroc Dry Lake Bombing and Gun-

B-23 (39-28) was used in glider pick up testing at Wright Field in late 1942 and early 1943, as U.S. airborne tacticians saw a need to develop a method of retrieving flyable gliders from combat zones so that they could be used again. The tow line can be seen under the fuselage. (USAF)

B-23 (39-28) tested various propeller applications, including two and four configurations, and a six-blade counter-rotating mounting. (USAF)

nery Range, CA. B-23 (39-53) was allocated as the mother aircraft to carry these first "smart bombs." The GB was essentially a 2,000 pound bomb fitted to a winged airframe that was controlled by flight servo equipment directed by radio. The initial airframe testing of the glide bomb (GB) took place on 8-9 November 1941, when the B-23 released a GB at 7,000 feet over the lakebed. The GB was a standard 2,000 pound general purpose bomb fitted with wooden wings and a simple stabilizing gyro. The concept was for the GB to be released at the beginning of the bombing run, so that the B-23 mother ship could stand off from enemy anti-aircraft fire while the GB glided, and was guided to the target. The test detachment evaluated several GB versions over two years, including simple gliders, to more complex versions that were radio-controlled from the mother ship.

The GB testing led to the development of a TV system known as the SCR-549-T-1, which was based on broadcast standards where possible, but had to be simplified and miniaturized into the 100-megacycle SCR-550-T-1. The camera-transmitting unit weighed about 60 pounds, the monitor 20 pounds, the 14 volt battery 37 pounds, and the receiver 40 pounds. Flight tests were conducted at Wright Field in early 1942, with the set installed in the Culver PQ-8 radio-controlled target drone, and the B-23 acting as a controller. The drone had a nacelle installed under the right wing to house the camera-transmitter, and the bomb was placed in the cockpit. The system encountered problems when the spinning propeller interfered with the TV camera reception, and while the 120 line picture resolution was satisfactory, it was not possible to identify a small structure at 9,000 feet. However, the tests demonstrated that airborne TV bomb control was practical, but work continued to reduce weight and size, resulting in the 300-megacycle SCR-550-T-2. In May 1943 tests on the improved set were conducted at Muroc Dry Lake using a GE "Bug" equipped with a ventral camera-transmitter housing out of the propeller arc. After several test flights being controlled through various maneuvers by the B-23 mother ship, the expensive Bug was put into a dive and guided by radio-control to a target on the ground by its TV camera. During August 1943 the TV equipment was deemed sufficiently developed to conduct an official demonstration at Muroc using a YPG-12A aircraft as the drone. The YRG was successfully put through its paces, and on the final bomb run a 500 pound was placed in the cockpit, and the drone was radio-controlled behind a second radio-controlled aircraft where it was exploded.

During June 1943 a small number of GB-4 air frames became available to test the performance of an improved TV system, the SCR-549-T3. The servo control mechanism, radio control equipment, and television transmitter were located in the airframe body, while the camera was housed in a faired nacelle under the bomb. The T2 and T3 versions were essentially similar, except the T3 operated for technical reasons at 300 megacycles instead of 100 MC and, in the T3 the camera and transmitter were separate units, which increased the unit weight to 90 lbs. Early tests were conducted to determine the satisfactory installation and function of the equipment. In August five bombs were guided by a B-23 to ground targets, and a number of problems were found, mostly relating to interference on the monitor, resulting in very poor TV picture resolution. Analysis of the tests showed that most of the problems could be traced to acoustic problems solved by sound proofing, and vibration difficulties solved by better bonding of the glued wooden airframe. To solve lens problems, a yellow filter was placed, and a heater was installed to reduce fogging. Tests and initial tactical hypothesis showed that range would have to be increased to make full combat use of the system. A directional antenna in the mother aircraft linked to a gyrostabilized antenna mount on the GB-4 increased the range.

B-23 (39-53) was allocated as the mother aircraft to carry these first "smart bombs" in early November 1941. The GB was a standard 2,000 pound general purpose bomb fitted with wooden wings and a simple stabilizing gyro. In the photo the B-23 is test dropping a conventional 2000 pound bomb. (USAF)

B-23 #39-23 with glide bomb tracking movie camera in doorway. (USAF)

Eventually GB-1 development progressed, freeing funds and personnel. Slowly GB-4 difficulties were eliminated, and satisfactory television pictures were finally transmitted, allowing a procurement order for 2000 GB-4s equipped with SCR-549-T3 transmitting units to be made. In January 1944 letters of intent to contractors were issued for airframes and controls. The Signal Corps was instructed to procure the television guidance system. In spring 1944 Wright Field stated that "while the bomb itself functioned with excellent reliability, the degree of skill required to hit a target with desired accuracy was evidently more than anticipated." Additional training of personnel was conducted, and by June 1944 development had progressed to a point to warrant limited operational testing. A glide bomb unit coded "*Batty*" was sent to the 8th Air Force in England for tactical employment of the GB-4 and GB-8 (controlled by flares and radio only) using B-17s. The TV units were also used to guide war-weary B-17 and B-24 drones in the *Castor* Program.

U-67 Transport

At least 14 (possibly 18) B-23s were modified to the UC-67 transport and VIP role by deleting armament and bomb bays, as well as adding seats and windows. The "U" prefix was adopted in 1942 for transports carrying fewer than eight passengers, while the "C" prefix was reserved for more than eight passengers. Even though the B-23 could and did carry more than eight passengers, the UC-67 designation remained. Later in the war several B-23s served as transports, especially in the VIP role, without the UC-67 designation. Known UC-67 serials were: 39-29/-31/-34/-35/-39/-41/-43/-44/-47/-54/-59/-61/-63, and -64.

A TV camera was experimentally placed in the nose compartment entrance during the GB tests. (USAF)

UC-67 (39-63) was assigned to 6AF HQ/HQ Squadron on 7 October 1944, where it served until 31 January 1945, when it was sent to the Caribbean Defense Command-Panama Canal Department (CDC/PCD). It was part of Gen. George Brett's fleet of VIP aircraft until 10 May 1945, when it was sent back to 6AF HQ/HQS. The aircraft was turned over to the RFC and sold to a private buyer on 26 December 1945.

"R" For Restricted

The B-23's military career was brief and inconspicuous, and it was not based overseas, and never fired a shot in anger. In October 1942 the B-18 and B-23 were placed on the "R," or "Restricted" classification, redesignated as the RB-18 and RB-23, which meant that the aircraft was considered obsolete, and no longer acceptable for its designated role of bombardment (in 1948 the "R" prefix denoted Reconnaissance).

Crashes

Five B-23s are recorded to have crashed during the war:

39-27: Damaged at Eglin Field, FL, 2 February 1942
39-42: Crashed near Beaumont, CA, in October 1941
39-43: Crashed as a UC-87 near Flomaton, AL, 26 June 1943, killing ten
39-55: Crashed near Washington, DC, on 22 January 1943
39-52: Crashed near Loon Lake, ID, on 29 January 1943, all eight crew survived

Post War B-23s

It was after the war as civilian aircraft that the B-23s became more well known, and finally found a niche. After the war at least 23 of the 38 B-23s and UC- 67s survived, and were sold as surplus for civilian use with further modifications made. Many well known airlines and corporations bought the B-23s and UC-67s from the RFC. In 1945 the government assessed the postwar civilian prospects of a number of military aircraft in order to issue licenses. The B-17 was allotted a "Limited" license, and was allowed to be used

As was to become standard in post war passenger aircraft, dual pilot controls were added, as seen in the Pima Air & Space Museum B-23. (Author/Pima)

as non-revenue transport for carrying a company's own personnel, or for non-passenger activities, such as aerial mapping or "weather modification," such as rain making. The B-18s and B-23s, because of their Douglas airliner legacy, received "Standard" (NC) licenses, as they qualified for "Memo" approval that originated in 1928. Memo approval meant that aircraft not completely meeting the technical requirements of a full Approved Type Certificate (ATC) could do business as a revenue passenger carrier. At the end of November 1945 a number of B-18s were allowed to operate as cargo carriers under Memo 2-577, and a few B-23/UC-67s did charter work under Memo 2-576. It seems that no B-18, B-23, or UC-67 ever operated as a scheduled passenger airliner.

The UC-67s were sold at $20,000 to $25,000 each, while the DC-3s sold at $30,000 to $50,000, depending on their condition. An average cargo C-47 sold for $20,000, a recently overhauled cargo C-47 at $27,500, and a new cargo C-47 at $36,000. An average passenger C-47 sold for $25,000, a recently overhauled passenger C-47 at $35,000, and a new passenger C-47 at $43,500. A B-18 or B-23 sold at only $3,000, as opposed to $32,500 for a B-29, and $13,500 for a B-17 or B-24, so the more easily certified UC-67 B-18s and B-23s were a bargain

The aircraft's twin Wright R-2600s gave it high power and high speed, and its DC-3 heritage gave it docile handling that made it an ideal commercial executive transport. Military B-23s were reconfigured to the preference of the customer. As was to become standard in post war passenger aircraft, dual pilot controls were added, and a longer new all metal nose compartment was added. The most cost effective seating was for 12 in two compartments in the fuselage, along with bathroom and washroom facilities. Windows were added in appropriate positions. As long as there were five or less passenger seats the existing escape hatches were sufficient for licensing, but with increased seating up to 14 additional hatches had to be furnished to gain a license. The military Wright R-2600-3s had two-stage superchargers that had to be locked irreversibly in the low speed position for civilian use. Some civilian adaptations installed the R-2600-29 engines, which had 1,700hp on take off, and others had two 114 gallon fuel tanks installed to the outboard ends of each wing panel.

In the fall of 1945 the powerful Reconstruction Finance Corporation (RFC) was the government department charged with the disposal of surplus military aircraft and aircraft equipment. B-23s disposed of by the RFC were:

RFC disposal on 19 February 1944: -50
Post war disposal:
RFC at McKellar Field: -28, -30, -31, -48, -51, -57, -59, -60, -61, -62, -64
RFC at Bush Field: -35, -36, -37, -38, -41, -44, -47
RFC at Patterson Field, GA: -32, -39, -56
RFC at Ontario, CA: -46
RFC at Grenler Field, NH: -58

Corporate Purchasers

Howard Hughes bought 39-33 (1945), 39-44 (for Hughes Tool 1945), and 39-60 (1948), while Pan Am bought 39-30 (1945), 39-31 (1947), 39-57 (1945), 39-59 (1945), 39-61 (1946), 39-62 (1945), and 39-64 (1945). Pan Am President Juan Trippe used one as his personal aircraft for a number of years. Other corporate buyers included: Union Oil (39-37), Henry J. Kaiser (39-38), General Motors (39-41), Garwood Industries (39-33 and 39-46), Standard Oil (39-63), Rexall Drug 39-47), Roscoe Turner (39-51), and Paul Mantz (39-36). Mantz's B-23 was called the "Honeymoon Express," as it was used to fly smitten Hollywood stars to Las Vegas for quickie marriages.

In January 1947 Howard Hughes took actor Cary Grant on a flight in his converted B-23 from Los Angeles to New York, and on the return flight the aircraft disappeared. After having lost radio contact with the Indianapolis tower the aircraft was presumed lost, and the media reported Hughes and Grant missing, and possibly dead. But Hughes and Grant had decided to change their plans and fly to Mexico. To keep their plans secret Hughes bribed airport officials along the way, but once in Mexico they read headlines in

Paul Mantz' B-23 (39-36) was called the "Honeymoon Express," as it was used to fly smitten Hollywood stars to Las Vegas for quickie marriages. (Pima)

The B-23 stored at the Pima Air & Space Museum was owned by Roscoe Turner in the mid-1950s. (U. of WY via Pima)

Spanish reporting their premature deaths, and Grant called his publicist to straighten matters out.

In 1974 the University of Washington bought a B-23 from a civilian source, probably Paul Mantz's 39-36, and it was equipped with antennas, sensors, and observation blisters for aerological research. Professor David "Rainman" Laskin, for the Air Force Cambridge Research Laboratory, used this modified B-23 in a study to sample cloud particles, and flew a number of missions over the erupting Mount St. Helens volcano in 1980.

By the early 1960s several ex-B-23s were operating in South America, and disappeared while smuggling drugs. There are four surviving B-23s in existence:

39-37: Air Force Museum, Wright-Patterson AFB, OH
39-45: Castle AFB, CA
39-36: McChord Air Museum, WA
39-51: Pima Air & Space Museum, AZ

B-23 Assessed

The B-23 proved to be ill-conceived, and its military career was brief and inconspicuous: it was not based overseas, and never fired a shot in anger. But the chief military development of the B-23 lay in the first glazed tail gunner's position installed on an operational American bomber. The aircraft's twin Wright R-2600s gave it high power and high speed, and its DC-3 heritage gave it docile handling characteristics, and ex-military B-23s were reconfigured to the preference of the customer to make it an ideal commercial executive transport.

The Pima Air & Space Museum B-23 in 2006 sits in a non-public area waiting for eventual restoration. (Author/Pima)

12

Obscurity and Disposition

As the B-24 Liberator took over the B-18's ASW role the Bolo's combat responsibility concluded, and as we have seen, the majority of the B-18 types served in a training or transport role until the end of the war, with a few grounded and shipped to maintenance schools to be used as instructional aircraft.

The B-18's stability in flight made it particularly practical as a crop sprayer and water bomber in fire fighting. (Pima)

Postwar Disposition

Once the war ended the Air Force could not utilize the obsolete B-18s, B-17s, and B-24s in the post war, so a small number of B-18s and a much larger number of B-17s and B-24s were sold by the War Assets Administration to be ingloriously salvaged for their aluminum and other metals. But the characteristics that made the B-18 unsuitable as a bomber made it attractive to civilian buyers, as it was a better aircraft than the DC-2 and DC-3, being more soundly constructed, and capable of carrying a larger load in its deep fuselage. Many marginal commercial operators saved money by purchasing the B-18s, which were less popular and cost less, while the popular C-47s continued to increase in price. Its stability in flight made it particularly practical as a crop sprayer and water bomber in fire fighting. Many B-18s were purchased for their parts, which were interchangeable with the DC-3/C-47. Also, the 42nd Troop Carrier Wing in Alaska requisitioned a number of B-18s to use their parts to keep their C-47s flying.

Tracing the postwar B-18s is difficult, as the FAA sporadically clears out records that it deems as unimportant, and many records have been lost. By 31 March 1947 the FAA issued U.S. Civil Register to at least 38 B-18s as transports under a Category Type 2 certificate A2-577. The last flight of a B-18 was on 26 October 1981 by s/n #37-29 (N52056) from Hawkins & Powers Aviation, Inc,. at Graybull, WY, to Castle Air Museum, CA.

Existing B-18s

Castle Air Museum, Atwater, CA, B-18 (37-29/N52056) It was used as a borate bomber for firefighting after WWII.

Wright-Patterson AFB, OH, B-18A (37-469/N56847), stationed at Wright Field 1939-42. Restored to 38RS in 1939 scheme.

The B-24 Liberator took over the B-18's ASW role, with the Bolo's combat responsibility was concluded. (USAF)

McChord AFB Museum, WA, B-18A (37-505/N67947) After delivery on 21 November 1938 the bomber served in the western U.S. with various units before being wrecked on 10 June 1942 at Wendover AFB, CA. Once repaired, it was sent to be converted to a B-18B in January 1943, but in July 1943 it was put into storage at Morrison Army Air Field, FL, until February 1945, when it was sold by the RFC to Plains Aero Service. After 20 years with PAS the aircraft was sold to a Mexican cooperative, and served as a fish transport until 1971, when it was donated to the Tucson Air Museum, and stored at Davis-Mothan until 1983, when it was loaned to McChord Air Museum for restoration.

Wings over the Rockies Aviation & Space Museum, Denver, CO, B-18A (39-25/N62477)
This aircraft was the next to last B-18A built by Douglas, and was delivered to Chanute Field, IL, on 20 February 1940; during the war it served as a bombardier trainer at several bases. It then became the property of the RFC, and was sold for civilian use successively as a cargo hauler, oil exploration surveyor, agricultural sprayer, Alaskan cargo hauler, and finally as a Miami to Havana cargo hauler, until it was seized by U.S. Customs Agents for illegal arms exportation. It was then stored at the Air Force Museum, Wright-Patterson Field, OH, and in 1988 it was disassembled and transferred to Lowrey Field, CO, in a Hercules C-130 transport for restoration by the Wings over the Rockies Museum.

Pima Air & Space Museum, Tucson, AZ, B-18B (38-593/N66267)
The Roberts Aircraft Company of Boise, ID, modified B-18B (38-593) for use as an aerial sprayer and fire bomber to operate in Department of Agriculture Area #4. Its civilian registration was N66267, and it was assigned the tanker code #D18. The photo shows the under wing spray bar and nozzles, and the dump doors of the fire retardant tank installed in the bomb bay. It operated until the early 1970s, spraying for the spruce budworm larvae. The almost derelict B-18 had been abandoned at Falcon Field, in the Phoenix area, when Pima Air & Space Museum (PASM) first became interested in acquiring it. Robert "Lefty" Lumpkin initiated and oversaw the B-18 project, with Robert Strand volunteering nearly 2,400 hours to its restoration.

Robert Strand. (Strand)

Restoration of the Pima Air & Space Museum Douglas B-18B
by Robert C. Strand, Restoration Team Leader (PASM)
After it was sold as surplus at the end of WWII it was fitted with surplus C-47 engines. This entailed removal of the original over-

The Roberts Aircraft Company of Boise, ID, modified B-18B (38-593) for use as an aerial sprayer and fire bomber. The photo shows the under wing spray bar and nozzles, and the dump doors of the fire retardant tank installed in the bomb bay. (Pima)

the-wing exhaust system, and installation of the C-47 engines, which had side cowling exhaust stacks. Since the original engines did not have cowl flaps, the replacement engines also required a cowl flap control system to be installed in the cockpit. We fortunately had the original over-the-wing exhaust stacks, and the original stainless steel trailing wing edge exhaust heat shields. When we reconverted the exhaust system we had to reconstruct the fairings that faired the exhaust stacks into the top surface of the wing. This was a big job for the Restoration crew, since the stacks are twelve feet long.

My early restoration work was primarily in the cockpit. I decided to remove everything I could take out for repair, cleaning, and repainting. The instrument panel was a major project. It had been cut up to accommodate a small radio control panel, as well as other flight instruments. The panel was patched, and original instruments were installed. Many of these instruments were created by making reproductions of the dial faces as shown in a 1942 War Department technical manual, since original examples were no longer in existence. Commercial "Press Type" saved the day.

Post war, the Pima B-18B was fitted with surplus C-47 engines that entailed removal of the original over-the-wing exhaust system, and installation of the C-47 engines, which had side cowling exhaust stacks (shown). (Strand)

The cockpit was also modified so that the pilot could start the engines. In the original military arrangement, the co-pilot controlled the primer valve and the hand operated fuel wobble pump. In the post war civilian arrangement, the pilot had electric primer switches and an electric auxiliary fuel pump, which required that the original 12 volt electric system had to be converted to the more modern 24 volt system.

At one point in its civilian life the airplane sustained a bad fire in the lower left side of the fuselage, in what was originally the forward nose gunner's position. This is where the hydraulic system fluid reservoir is mounted. When I started working on the airplane in 1994, I found that the interior insulation around this entire reservoir area was soaked with hydraulic fluid. I suspect that a careless smoker or an electrical spark was the source of this fire. Extensive repairs to this area had to be made to keep the airplane flying. Keep in mind that these were the days before "Skydrol," the non-flammable hydraulic fluid, came into use.

The removal of the rudder pedals produced some interesting observations. I was concerned with repairing the seals around the pedal mounting posts. To do this I had to remove the cockpit floor, and to my dismay, I discovered that the floor was secured by screws with nuts. These were apparently installed early in the construction of the airplane, because about half the securing nuts were hidden in blind spaces where the nuts were inaccessible. I ended up having to drill off most of the screw heads by crawling under the instrument panel. A secondary benefit of the floor removal project was the discovery of the original aircraft data plate. We had speculated that the data plate probably had been stolen when the airplane was abandoned, since we had never been able to find it. It showed up as I was removing the left hand rudder pedals. It was factory installed on the lower left side of the central throttle pedestal. It now resides in the storage vault at PASM. I incidentally reported this to Ray Burger, the B-18 restorer at Castle AFB Museum, who immediately rushed to his B-18 to make a similar discovery.

The 1935 technology that I observed was interesting, in light of how things would have been done in a modern airplane. The control wheels in the PASM B-18 are typical of what is found in a

Nose and engine restoration. The Plexiglas lower nose section has been installed while work is being done to fabricate the Plexiglas radar dome. The photo on the left shows that the engines still have the C-47 side exhaust stacks, while the over-the-wing stacks are installed on the right. (Strand)

The 8 December 1995 roll out ceremony of Pima's beautifully restored B-18B in the markings of the 4th Antisubmarine Squadron. (Strand)

C-47. They are a half circle around the bottom of the mounting hub, so that the pilots have an unobstructed view of the instrument panel. In the B-18, as it came from the factory, they were full circle wheels typical of 1930s airplanes, such as the Ford Tri-motor Transport. The airplane had a hydraulic system for operation of the landing gear and the bomb bay doors. The hydraulic lines, however, were bundled together in their runs with leather thongs—essentially leather shoe laces. This dampened any tendency of the lines to vibrate, and was certainly inexpensive. The landing gear was raised or lowered hydraulically. To insure that the landing gear did not drop precipitously, however, there was a system of seven very large rubber bungees attached to the top of the wheel strut to cushion the fall of the wheel from the retracted position. It was simple but effective for that time period. One could summarize that the aircraft technology incorporated shoe laces and rubber bands.

When the B-18As were converted to B models, the bombardier's position was replaced by a radar dome. Apparently, in 1942 there was no fiberglass technology available to manufacture the dome, and instead, the dome was made from Plexiglas. What we found at PASM was because Plexiglas is a thermoplastic, the dome would soften in the hot Arizona sun. Since the airplane was tied down facing the southeast, the morning sun shined on the black painted dome, causing it to absorb lots of heat. It then regularly began to develop a flat spot on its upper left hand side where the morning sun first hit, and this resulted in a slow deterioration of the entire dome. Ultimately we had to remove the original dome and manufacture one from fiberglass.

Robert C. Strand, Team Leader, Restoration
Pima Air and Space Museum

B-18 Legacy

The pug nosed B-18 simply was not pugnacious enough when it went into a war whose aerial tactics had evolved and the Bolo hadn't. The B-18 was ordered into production at a time when aerial tactics were determined by a General Staff that was dominated by infantry and artillery generals, who looked on the airplane as aerial artillery and reconnaissance to support their infantry, and as a transport for infantry supplies and equipment, or as an occasional VIP carrier.

The B-18 was built using contemporary but dead end design specifications, and it soon showed itself to be what it was, an "airliner-bomber" that allowed little latitude for future development to enable it to attack distant targets with 5,000 pound bomb loads or more at speeds and altitudes that would enable it to defend itself against more contemporary fighter aircraft. Thus, the bomber would not measure up to the matchless four engine B-17s and B-24s, nor the twin engine B-25 or B-26s, and its mass production delayed the development and production of these superior, more modern bombers that could have made a difference in the grim days after Pearl Harbor.

But at the time of Pearl Harbor, the B-18 was the most numerous American bomber to be based overseas, and then proved itself to be useful as a stopgap bomber and jack of all trades until the more modern bombers, whose development it interrupted, could become operational. However, as a stopgap the Bolo did leave an important legacy. At a time when the B-17 and B-24 could not be spared the B-18 was available in numbers, and played a significant role in America's early anti-submarine operations, first off the American coast, and later in the Caribbean, where the U-Boats were ravaging merchantmen carrying vital oil and bauxite cargoes.

Appendix A: B-18, B-18A, B-18B in Air Forces, Commands & Wings

B-18, B-18A, and B-18B Air Forces

Northeast Air District (1940) Based at Mitchel Field, NY, 18 Dec 40-42.
Became the First Air Force in 1941,
I Bomber Command (1941-42)
Southeast Air District (1940) Based at MacDill Field, FL, 18 Dec 40; Tampa, FL, Jan 41-25 Feb 42.
Became the Third Air Force in 1941.
III Bomber Command (1941-42)
Philippine Department Air Force (1941) Based at Nichols Field, Luzon, 20 Sept 41; Darwin, Australia, Dec 41; Java, Jan-Feb 42.
Became the Far East Air Force in Oct 41 and then the Fifth Air Force in Feb 1942.
V Bomber Command (1941-42)
Panama Canal Air Force (1940) Based at Albrook Field. CZ, 20 Nov 40-43,
Became the Caribbean Air Force in Aug 41 and then the Sixth Air Force in Feb 1942.
VI Bomber Command (1941-43)
Hawaiian Air Force (1940) Based at Ft. Shatter, TH, 1 Nov 40; Hickam Field, TH, c. 12 Jul 41-42
Became the Seventh Air Force in Feb 1942.
VII Bomber Command (1942)
58th Bombardment Squadron (1 Jan 41) (1941-42) Based at Wheeler Field, TH, 1 Jan 41; Bellows Field, TH, 18 Mar 41; Hickam Field, TH, 29 Apr 41; Bellows Field, TH, 11 Dec 41; Wheeler Field, TH, 19 Dec 41-1942.
Became the 531st Fighter-Bomber Squadron 14 Aug 43,
86th Observation Squadron (26 Feb 42) (1940-43) Based at Bellows Field, TH, 15 Mar 41; Hilo Airport, TH, Jun 42; Wheeler, Field, TH, 17 Aug 42-c. Jun 43. Became the 43rd Reconnaissance Squadron 16 Jun 45.
Alaskan Air Force (1942) Based at Elmendorf Field, AK, 15 Jan 42-c. Dec 42.
Became the Eleventh Air Force in Feb 1942.
28th Composite Group (1940-42)

B-18, B-18A, and B-18B Commands

Army Air Force Anti-submarine Command (1942) Based at New York, NY, 15 Oct 2-Aug 43. Was assigned directly to Army Air Forces
25th Anti-submarine Wing (1942-43)
26th Anti-submarine Wing (1942-43)
I Bomber Command (1941) Based at Langley Field, 5 Sept 41; New York, NY, c. 12 Dec 41-15 Oct 42.
Engaged primarily in anti-submarine operations along the east coast. Units??
III Bomber Command (1941) Based at Drew Field FL, 5 Sept 41; MacDill Field, FL, c. Sept 41; Savannah, GA, c. 10 Dec 41; MacDill Field, FL, c. 15 Dec-25 Feb 42.
13th Bombardment Group (1941-42)
V Bomber Command (1941) Based at dark Field, Luzon, 14 Nov 41; Darwin, Australia, Dec 41; Java, Jan-Mar 42.
3rd Bombardment Group (1941)
7th Bombardment Group (1941)
19th Bombardment Group (1941-42)
27th Bombardment Group (1941-42)
VI Bomber Command (1941) Based at Albrook Field, CZ, 25 Oct 41-1 Nov 46
6th Bombardment Group (1941-42)
9th bombardment Group (1941-42)
25th Bombardment Group (1941-43)
40th Bombardment Group (1941-42)
VII Bomber Command (1942) Based at Hickam Field, TH, 29 Jan 42-c.Oct 43.
5th Bombardment Group (1942)
11th Bombardment Group (1942)
Antilles Air Task Force (1943) Based at San Juan, PR, 1 Mar 43-c. Mar 44.
Became the Antilles Air Command in Jun 1943.
25th Bombardment Group (1942-44)
72nd Observation Group (1943-44)
Newfoundland Base Command
21st Reconnaissance Squadron (1941)
41st Reconnaissance Squadron (1941-42)
Air Transport Command
1103rd AAF Base Unit Based at Morrison AAF, FL, Sep 44 *(36-593)*
1110th AAF Base Unit Based at Zandery Field, Surinam (Dutch Guiana) Feb 43. *(38-593)*
4007th AAF Base Unit Based at Brookley AAF, AL, Aug 44 *(37-29)*
4119th AAF Base Unit Based at Greenville AAF, SC, Oct 44 (37-23)

B-18, B-18A, B-18B WINGS

1st Bombardment Wing (1931) Based at March Field, CA, 1 Apr 31; Tucson, AZ, 27 May 41-Jul 42.
Became one of the original wings of GHQAF in 1935, in the western part of the US until 1941. Was known as the 1st Wing in 1935.
7th Bombardment Group (1935-41)
17th Bombardment Group (1931-41)
19th Bombardment Group (1935-41)
41st Bombardment Group (1941)
2nd Bombardment Wing (1929) Based at Langley Field, VA, 8 Aug 22-5 Sept 41; Detrick Field, MD, Oct 41- 15 Aug 42.
Became one of the original wings of GHQAF in 1935, in the eastern part of the US until 1941. Was known as the 2nd Wing in 1935.
2d Bombardment Group (1929-41)
9th Bombardment Group (1935-40)
44th Bombardment Group (1942)
3rd Bombardment Wing (1935) Based at Barksdale Field, LA, 27 Feb 35-Sept 40; MacDill Field, FL, 2 Oct 40-5 Sept 41.
Became one of the original Wings of GHQAF in 1935, until 1941. Was known as the 3rd Wing in 1935.
3rd Bombardment Group (1935-40)
13th Bombardment Group (1941)
29th Bombardment Group (1940-41)
44th Bombardment Group (1941)
4th Bombardment Wing (1940) Based at Mitchel Field, NY, 18 Dec 40-Feb 41; Westover Field, MA, 20 Mar 41-1 Oct 41.

34th Bombardment Group (1941
43rd Bombardment Group (1941)
5th Bombardment Wing (1940) Based at McChord Field, WA, 18Dec 40; Ft George Wright, WA, 9 Jan 41-Jul 42.
12th Bombardment Group (1941)
17th Bombardment Group (1941)
47th Bombardment Group (1942)
13th Composite Wing (1940) Based at Langley Field, VA, 10-26 Oct 40; Borinquen Field, PR, 1 Nov 40; San Juan, PR, c. 6 Jan 41; Borinquen Field, PR, c. 1 May 41-25 Oct 41.
25th Bombardment Group (1940-41)

40th Bombardment Group (1941)
15th Bombardment Wing (1940) Based at March Field, CA, 18 Dec 40; Fresno, CA, c. 2 Aug 41-3 Sept 41.
47th Bombardment Group (1941)
48th Bombardment Group (1941)
16th Bombardment Wing (1940) Based at Langley Field, VA, 18 Dec 40; Bowman Field, KY, Mar 41-1 Sept 41.
5th Bombardment Group (1941)
17th Bombardment Wing (1940) Based at Savannah, GA, 18 Dec 40-1 Sept 41.
3rd Bombardment Group (1940-41)
27th Bombardment Group (1940-41)
18th Bombardment Wing (1937) Based at Hickam Field, TH, 30 Oct 37-29 Jan 42.
Was known as the 18th Wing 1937.
5th Bombardment Group (1937-42)
11th Bombardment Group (1940-42)
19th Bombardment Wing (1940) Based at Albrook Field, CZ, 25 Jan 33-25 Oct 41.
Was known as the 19th Wing in 1937.
6th Bombardment Group (1933-41)
9th Bombardment Group (1940-41)
20th Bombardment Wing (1940) Based at Ft Douglas, UT, 18 Dec 40-1 Sept 41.
7th Bombardment Group (1940-41)
42nd Bombardment Group (1941)
24th Composite Wing (1942) Based at Iceland, 25 Dec 42-15 Jun 44.
342nd Composite Group (1942-44)
Trinidad Wing, Antilles Air Command (1943) Based at Waller Field, Trinidad, 15 May 43-15 Mar 44. Units??

The B-18, B-18A, B-18B Groups and Squadrons 1936 to 1945

2nd Bombardment Group (1 Sep 36) Aircraft coded "BB" and "R"
20th Bombardment Squadron (25 Jan 23) Based at Langley Field, VA. 1936-Nov 41; Mitchel Field, NY, 8 Dec 41-24 Jan 42; Ephrata, WA, 29 Oct 42.
49th Bombardment Squadron (25 Jan 23) Based at Langley Field, VA, 1936-Nov 41; Newfoundland AB, Newfoundland, 13 Dec 41; Agentia,, Newfoundland, 16 Jan-Jun 42?
Langley Field, VA, 24 Jun-1 Oct 42.
96th Bombardment Squadron (25 Jan 23) Based at Langley Field, VA, 1936-1 Oct 42; Ephrata, WA, 29 Oct 42.
21st Reconnaissance Squadron (1 Sep 36) Based at Langley Field, VA, 1 Sep 36-c. 1 Sep 39 (with B-18s) Aircraft numbered between 16 and 30. Became the 411th Bombardment Squadron 22 Apr 42.
41st Reconnaissance Squadron (1 Feb 40) Based at Langley Field. VA. 1 Feb 40-23 Aug 41; Ephrata, WA, 29 Oct 42. Became the 429th Bombardment Squadron 22 Apr 42.

3rd Bombardment Group Aircraft coded "BC"
8th Bombardment Squadron (15 Sep 39) Based at Barksdale Field. LA, 15 Sep 39; Savannah, GA, 8 Oct 40-19 Jan 41.
13th Bombardment Squadron (15 Sep 39) Based at Barksdale Field. LA, 15 Sep 39; Savannah GA, 8 Oct 40-19 Jan 41.
89th Bombardment Squadron (14 Aug 41) Based at Savannah, GA. 15 Jan 41-20 Jan 42. Was the 10th Reconnaissance Squadron 15 Jan 41.
90th Bombardment Squadron (15 Sep 39) Based at Barksdale Field. LA, 15 Sep 39; Savannah, GA, 10 Oct 40-19 Jan 42. Aircraft numbered between 70 and 90.

5th Bombardment Group (12 Oct 36) Aircraft coded "BE" and "R"
23rd Bombardment Squadron (25 Jan 23) Based at Hickam Field. TH, 1 Jan 39; Mokuieia, TH, 24 Mar-3 Nov 42.
31st Bombardment Squadron (1 Apr 31) Based at Hamilton Field. CA, 1937; Hickam Field, TH, 8 Feb 38; Kipapa, TH, 23 May 42; Kualoa, TH, 9 Sep-9 Nov 42.
72nd Bombardment Squadron (1 May 23) Based at Hickam Field. TH, 4 Jan 39; Bellows Field, TH, 11 Dec 41-18 Sep 42.
4th Reconnaissance Squadron (25 Jan 38) Based at Hickam Field. TH, 1 Jan 39-1 May 42. Became the 394th Bombardment Squadron 22 Apr 42.
Note: The entire Group was stationed in the Territory of Hawaii during the attack of December 7, 1941.

6th Bombardment Group (1 Feb 40) Aircraft coded "BF" and "R"
3rd Bombardment Squadron (1 Jan 38) Based at France Field. CZ. 1 Feb 40; Rio Hato, Panama c. 8 Dec 41; Galapagos Islands, 4 May 42.
25th Bombardment Squadron (25 Jan 23) Based at France Field, CZ, 30 Apr 22-7 Dec 41; Rio Hato, Panama, 8 Dec 41; Salinas, Ecuador, c. 21 Jan 42
74th Bombardment Squadron (1 Nov 39) Based at Albrook Field. CZ, 1 Oct 33; Howard Field, CZ, 14 Jul 41; Aguadulce, Panama, 8 Nov 41; Rio Hato, Panama, c.11 Dec 41; Guatemala City, Guatemala, 9 Jan 42.
7th Reconnaissance Squadron (1 Sep 37) Based at France Field, CZ, 1 Sep 37; Howard Field, CZ, 26 Nov 41; David, Panama, 11 Dec 41; Talara, Peru, 18 Aug 42. Became the 397th Bombardment Squadron 22 Apr 42.

7th Bombardment Group (1 Apr 31) Aircraft coded "BG" and "R"
9th Bombardment Squadron (1 Apr 31) Based at Hamilton Field, CA, 1937; Ft Douglas, UT, 7 Sep 40; Salt Lake City, UT. 13 Jan-13 Nov 41.
11th Bombardment Squadron (1 Jun 28) Based at Hamilton Field, CA, 1937; Ft Douglas, UT, c. 18 Jan-13 Nov 41.
22nd Bombardment Squadron (20 Oct 39) Based at Hamilton Field, CA, 20 Oct 30; Ft Douglas, UT, 7 Sep 40; Salt Lake City, UT, c, 21 Jun-13 Nov 41.
88th Reconnaissance Squadron (1 Sep 36) Based at Hamilton Field, CA, 1937; Ft Douglas, UT, 7 Sep 40 Salt Lake City, UT c. 15 Jan-11 Nov 41. Became the 436th Bombardment Squadron 22 Apr 42.

9th Bombardment Group (1 Mar 35) Aircraft coded "BI" and "R"
1st Bombardment Squadron (1 Mar 35) Based at Mitchel Field, NY, 1938-6 Nov 40; Rio Hato, Panama, 13 Nov 40; Piarco Airport, Trinidad, 24 Apr 41; Waller Field, Trinidad, 29 Oct 41; Edinburgh Field, Trinidad, 23 Aug 42; Orlando AB, FL, 31 Oct 42-1 Dec 42.
5th Bombardment Squadron (1 Mar 35) Based at Mitchel Field, NY, 1938-6 Nov 40; Rio Hato. Panama, 13 Nov 40; Beane Field, St. Lucia, c. 28 Sep 42; Orlando AB, FL, 31 Oct 42. Aircraft numbered between 40 and 69
99th Bombardment Squadron (1 Mar 35) Based at Mitchel Field, NY, 1938-6 Nov 40; Rio Hato, Panama, 13 Nov 40; Zandery Field, Surinam, 3 Dec 41; Orlando AB, FL, 31 Oct 42.
18th Reconnaissance Squadron (1 Sep 36) Based at Mitchel Field. NY, 1 Sep 36-14 Nov 40. Became the 408th Bombardment Squadron 22 Apr 42.
44th Reconnaissance Squadron (1 Sep 36) Based at Albrook Field, CZ, 31 Jan 40-7 Jul 41; Howard Field, CZ, 8 Jul-27 Oct 41; Atkinson Field, British Guiana 4 Nov 41; Orlando AB, FL, 31 Oct 42 Became the 430th Bombardment Squadron 22 Apr 42.

11th Bombardment Group (1 Feb 40)
26th Bombardment Squadron (1 Feb 40) Based at Hickam Field, TH, 1 Feb 40; Wheeler Field, TH, 20 Dec 40-19 Jul 42.
42nd Bombardment Squadron (1 Feb 40) Based at Hickam Field. TH, 1 Feb 40; Kualoa Point, TH, 5 Jun 42.
58th Bombardment Squadron (1 Jan 41) Based at Wheeler Field, TH, 1 Jan 41; Bellows Field, TH, 18 Mar 41; Hickam Field, TH, 29 Apr 41; Bellows Field, TH, 11 Dec 41; Wheeler Field, TH, 19 Dec 41-1942. Became the 531st Fighter-bomber Squadron 14 Aug 43.
50th Reconnaissance Squadron (25 Jan 38) Based at Luke Field, TH, 1938; Hickam Field, TH, 9 Oct 39-1941. Became the 431st Bombardment Squadron 22 Apr 42. All of the Squadrons were at Hickam Field, Oahu, HT, during the Pearl Harbor attack.

12th Bombardment Group (15 Jan 41)
81st Bombardment Squadron (15 Jan 41) Based at McChord Field, WA, 15 Jan 41; Ester Field, LA, 27 Feb-3 Jul 42; Stockton Field, CA, 24 May-24 Jun 42).
82nd Bombardment Squadron (15 Jan 41) Based at McChord Field, WA. 15 Jan 41; Ester Field, LA, 27 Feb-3 Jul 42; (Stockton Field, CA, 24 May-24 Jun 42).
83rd Bombardment Squadron (15 Jan 41) Based at McChord Field. WA, 15 Jan 41; Ester Field, LA, **27** Feb-3 Jul 42; (Stockton Field, CA, 24 May-24 Jun 42).
19th Reconnaissance Squadron (15 Jan 41) Based at McChord Field. WA, 15 Jan 41; Ester Field, LA, 27 Feb-3 Jul 42; (Stockton Field, CA, 24 May-24 Jun 42). Became the 94th Bombardment Squadron 14 Aug 41 and then became the 434th Bombardment Squadron 22 Apr 42.

13th Bombardment Group (15 Jan 41)
39th Bombardment Squadron (15 Jan 41) Based at Langley Field, VA, 15 Jan 41; Orlando, FL, 6 Jun 41; Savannah, GA, 8-14 Dec 41. Became the 3rd Anti-submarine Squadron 29 Nov 42, then the 819th Bombardment Squadron 22 Apr 43.
40th Bombardment Squadron (15 Jan 41) Based at Langley Field. VA, 15 Jan 41; Orlando, FL, 7 Jun-31 Dec 41. Became the 4th Anti-submarine Squadron 29 Nov 42,
41st Bombardment Squadron (15 Jan 41) Based at Langley Field, VA, 15 Jan 41; Orlando, FL, 7 Jun-31 Dec 41. Became the 5th Anti-submarine Squadron 29 Nov 42, then the 827th Bombardment Squadron 1 Oct 43.
3rd Reconnaissance Squadron (15 Jan 41) Based at Langley Field, VA, 15 Jan 41; Orlando, FL, 7 Jun-31 Dec 41. Became the 393rd Bombardment Squadron 22 Apr 42, then the 6th Anti-submarine Squadron 29 Nov 42.

16th Pursuit Group (1 Dec 32)
44th Reconnaissance Squadron (1 Sep 36) Based at Albrook Field, CZ, 1938- 31 Jan 40. Became the 430th Bombardment Squadron 22 Apr 42.

17th Bombardment Group (17 Oct 39)
34th Bombardment Squadron (17 Oct 39) Based at March Field, CA, 17 Oct 39; McChord Field, WA, 24 Jun-31 Dec 40.
73rd Bombardment Squadron (17 Oct 39) Based at March Field, CA, 17 Oct 39; McChord Field, WA, 26 Jun 40-10 Mar 41.
95th Bombardment Squadron (17 Oct 39) Based at March Field, CA, 17 Oct 39; McChord Field, WA, 26 Jun 40.
89th Reconnaissance Squadron (1 Feb 40) Based at March Field, CA, 1 Feb 40; McChord Field, WA, 26 Jun 40. Became the 432nd Bombardment Squadron 22 Apr 42.

19th Bombardment Group (24 Jun 32) Aircraft coded "BS", "AQ" and "R".
28th Bombardment Squadron (25 Jan 23) Based at Nichols Field, Philippines 1941; Batchelor, Australia, 21 Dec 41; Singosari, Java, 30 Dec 41-1 Mar 42.
32nd Bombardment Squadron (24 Jun 32) Based at March Field. CA, 25 Oct 35; Albuquerque, NM, c. 4 Jun-22 Nov 41; Bakersfield, CA, 17 Dec 41-31 Dec 41. Aircraft
numbered between 10 and 39.
93rd Bombardment Squadron (20 Oct 39) Based at March Field, CA, 20 Oct 39; Albuquerque, NM, Jun-27 Sep 41.
38th Reconnaissance Squadron (1 Sep 36) Based at March Field, CA, 1 Sep 36; Albuquerque, NM, 5 Jun-22 Nov 41; Bakersfield, CA, 17 Dec 41-31 Dec 41. Aircraft numbered between 31 and 45. Became the 427th Bombardment Squadron 22 Apr 42.

21st Bombardment Group (1 Feb 42)
313th Bombardment Squadron (1 Feb 42) Based at Bowman Field, KY, 1 Feb 42; Jackson AAB, SC, 24 Apr 42; Key Field. MS, 26 May 42; Hattiesburg Army Air Field, MS, 8 Jun 42-11 Jun 42.

22nd Bombardment Group (1 Feb 40)
2nd Bombardment Squadron (1 Feb 40) Based at Bolling Field, DC, 1 Feb 40; Langley Field, VA, 14 Nov-1 Dec 41.
19th Bombardment Squadron (1 Feb 40) Based at Patterson Field, OH, 1 Feb 40; Langley Field, VA, 16 Nov-1 Dec 41.
33rd Bombardment Squadron (1 Feb 40) Based at Patterson Field. OH, 1 Feb 40? Langley Field, VA, 16 Nov 40-30 Nov 41.
18th Reconnaissance Squadron (1 Sep 36) Based at Langley Field, VA, 15 Nov 40; Muroc, CA, 9 Dec-31 Dec 41. Became the 408th Bombardment Squadron 22 Apr 42.

23rd Composite Group (1 Dec 39)
24th Bombardment Squadron (1 Dec 39) Based at Maxwell Field, LA, 1 Dec 39; Orlando, FL, 2 Sep 40; Eglin Field, FL, 29 Jun-1 May'42.
54th Bombardment Squadron (6 Dec 39) Based at Maxwell Field, LA, 6 Dec 39; Orlando, FL, 2 Sep 40. Then became the Air Corps Proving Detachment; Air Force Proving Ground Group and tested equipment 1940-42.

25th Bombardment Group (1 Feb 40)
10th bombardment Squadron (1 Feb 40) Based at Langley Field, VA, 1 Feb-16 Oct 40; Borinquen Field, PR, 1 Nov 40; Edinburgh Field, Trinidad, c. 1 Nov 42; Port of Spain, Trinidad, 27 Aug 12 Oct 43; Waller Field, Trinidad, 1 Oct-1 Dec 43.
12th Bombardment Squadron (1 Feb 40) Based at Langley Field, VA, 1 Feb-26 Oct 40; Borinquen Field, PR, 1 Nov 40; Benedict Field, St Croix, c. 8 Nov 42; Dakota Field, Aruba NWI, c. 10 Oct 42; Coolidge Field, Antigua, 23 Nov 43-24 Mar 44.
35th Bombardment Squadron (1 Feb 40) Based at Borinquen Field, PR, 31 Oct 41; Coolidge Field, Antigua, 11 Nov 41; Zandery Field, Surinam, 1 Nov 42; Atkinson field, British Guiana, 1 Nov 42-c. 7 Oct 43 and Port of Spain, Trinidad, 27 Aug-12Oct 43) Vernam Field, Jamaica, c. 7 Oct-31 Dec 43.
59th Bombardment Squadron (1 Jan 41) Based at Edinburgh Field, Trinidad, 12 Jul 43; Beane Field, St Lucia, 20 Oct 43-24 Mar 44.
27th Reconnaissance Squadron (16 Sep 39) Based at Camaguey, Cuba, 13 Apr 42-Aug 43) Vernam Field, Jamaica. 24 Sep 42; Losey Field, PR, 29 May 43-24 Mar 44. Became the 417th Bombardment Squadron 22 Apr 42.

27th Bombardment Group (1 Feb 40)
15th Bombardment Squadron (1 Feb 40) Based at Barksdale Field, LA, 1 Feb 40; Lawson Field, GA, 7 Oct 40-31 Dec 40. The Squadron was used for the First Mass Airborne jump held in the U.S.
16th Bombardment Squadron (1 Feb 40) Based at Barksdale Field, LA, 1 Feb 40; Hunter Field, GA, 7 Oct 40-19 Oct 41. Became the 522d Fighter-Bomber Squadron 23 Aug 43 and finished as the 522d Fighter Squadron 30 May 44.
17th Bombardment Squadron (1 Feb 40) Based at Barksdale Field. LA, 1 Feb 40; Hunter Field, GA, 7 Oct 40-19 Oct 41. Became the 523d Fighter-

Bomber Squadron 23 Aug 43 and finished as the 523d Fighter Squadron 30 May 44.
11th Reconnaissance Squadron (15 Jan 41) Based at Hunter Field, GA, 15 Jan-19 Oct 41 Became the 91st Bombardment Squadron 14 Aug 41, then the 524th Fighter-Bomber Squadron 23 Aug 43 and at last the 524th Fighter Squadron 30 May 44.

28th Composite Group (1 Feb 40)
36th Bombardment Squadron (1 Feb 40) Based at March Field. CA, 1 Feb 40; Lowry Field, CO, 9 Aug 40-23 Mar 41; Elmendorf Field, AK, 31 Mar 41 -c. Feb 42.
37th Bombardment Squadron (1 Feb 40) Based at Barksdale Field, LA, 1 Feb 40; Lowry Field, CO, 10 Jul 40-22 Apr 41.
73rd Bombardment Squadron (17 Oct 39) Based at Elmendorf Field, AK, 14 Mar 41.- Jan 42. 406th Bombardment Squadron (22 Apr 42) - based at Elmendorf Field, AK, c. Jun 42-c. Oct 43. Was the 16th Reconnaissance Squadron.

29th Bombardment Group (1 Feb 40)
6th Bombardment Squadron (1 Feb 40) Based at Langley Field, VA, 1 Feb 40; MacDill Field, FL, 21 May 40; Pope Field, NC, c. 7 Dec-31 Dec 41.
43rd Bombardment Squadron (13 Mar 40) Based at Langley Field, VA,1 Feb 40; MacDill Field, FL, 21 May 40; Pope Field, NC, c. 7 Dec 41; MacDill Field, FL ,c. 1 Jan 42. 52nd Bombardment Squadron (1 Feb 40) Based at Langley Field, VA, 1 Feb 40; MacDill Field, Fl, 21 May 40-31 Jan 41.
21st Reconnaissance Squadron (1 Sep 39) - Based at MacDill Field. FL, 5 Sep 41-25 Jun 42.

30th Bombardment Group (15 Jan 41)
21st Bombardment Squadron (15 Jan 41) Based at March Field, CA, 15 Jan 41; New Orleans, LA, 22 May 41 Savannah, GA, 8-14 Dec 41.
27th Bombardment Squadron (15 Jan 41) Based at March Field, CA, 15 Jan 41; New Orleans, LA, 25 May-1 Dec 41.
38th Bombardment Squadron (15 Jan 41) Based at March Field, CA, 15 Jan 41; New Orleans, LA, 3 Jun 41; Savannah, GA, 8-14 Dec 41; Muroc, CA, 24 Dec 41; March Field, CA, 9 Feb 42.
2nd Reconnaissance Squadron (15 Jan 41) Based at March Field. CA, 15 Jan 41; New Orleans, LA, 5 May-1 Dec 41. Became the 392d Bombardment Squadron 22 Apr 42.

34th Bombardment Group (15 Jan 41)
4th Bombardment Squadron (15 Jan 41) Based at Langley Field, VA, 15 Jan 41; Westover Field, MA, 29 May-31 Dec 41.
7th Bombardment Squadron (15 Jan 41) Based at Langley Field, VA, 15 Jan 41; Westover Field, MA, 29 May-31 Dec 41.
18th Bombardment Squadron (15 Jan 41) Based at Langley Field, VA, 15 Jan 41; Westover Field, MA, 29 May-31 Dec 41.
1st Reconnaissance Squadron (15 Jan 41) Based at Langley Field, VA, 15 Jan 41; Westover Field, MA, c. 29 May-31 Dec 41. Became the 391st Bombardment Squadron 22 Apr 42.

38th Bombardment Group (15 Jan 41)
69th Bombardment Squadron (15 Jan 41) Based at Langley Field, VA, 15 Jan 41; Jackson AAB, MSc. 5 Jun-31 Dec 41 .
70th Bombardment Squadron (15 Jan 41) Based at Langley Field, VA, 15 Jan 41; Jackson AAB, MS, c. 5 Jun-31 Dec 41.
71st Bombardment Squadron (15 Jan 41) Based at Langley Field, VA, 15 Jan-c. 1 Jun 41.
15th Reconnaissance Squadron (15 Jan 41) Based at Langley Field, VA, 15 Jan-5 Jun 41. Became the 405th Bombardment Squadron 22 Apr 42.

40th Bombardment Group (1 Apr 41)
29th Bombardment Squadron (1 Apr 41) Based at Borinquen Field, PR, 1 Apr 41; Aguadulce, Panama, 19 Jun 42.
44th Bombardment Squadron (1 Apr 41) Based at Borinquen Field, PR, 1 Apr 41; Howard Field, CZ, 16 Jun 42: Guatemala City, Guatemala, 6 Jul-31 Jul 42.
45th Bombardment Squadron (1 Apr 41) Based at Borinquen Field, PR, 1 Apr 41; France Field, CZ, 17 Jun 42; David, Panama, 13 Nov-30 Nov 42.
5th Reconnaissance Squadron (1 Apr 41) Based at Borinquen, PR, 1 Apr 41; Rio Hato, Panama, 17 Jun 42-16 Jun 43. Became the 395th Bombardment Squadron 22 Apr 42.

41st Bombardment Group (15 Jan 41)
46th Bombardment Squadron (15 Jan 41) Based at March Field, CA, 15 Jan 41; Davis-Monthan, AZ, 18 May 41; Muroc bombing range, 10 Dec 41; Hammer Field, CA, 26 Feb 42; Alameda NAS, CA, 9 May-1 Jul 42. Became the 22nd Anti-submarine Squadron 3 Mar 43.
47th Bombardment Squadron (15 Jan 41) Based at March Field, CA, 15 Jan 41; Tucson, AZ, 16 May 41; Muroc, CA, 10 Dec-31 Dec 41.
48th Bombardment Squadron (15 Jan 41) Based at March Field, CA, 15 Jan 41; Tucson, AZ, 16 May 41; Muroc, CA, 10 Dec 41; Bakersfield, CA, 9 Jan 42; Alameda NAS, CA, 2 Jul-30 Nov 42.
6th Reconnaissance Squadron (15 Jan 41) Based at Match Field, CA, 15 Jan 41; Tucson, AZ, 16 May 41; Muroc, CA, 8 Dec 41; Sacramento, CA, c. 11 Jan 42; Alameda, CA, Apr-10 May 42; Cherry Point, NC, Jun-Aug 42. Became the 396th Bombardment Squadron 22 Apr 42.
16th Reconnaissance Squadron (15 Jan 41) Based at Paine Field, WA, 21 Jan-c. Jun 42. Became the 406th Bombardment Squadron 22 Apr 42.

42nd Bombardment Group (15 Jan 41)
75th Bombardment Squadron (15 Jan 41) Based at Ft. Douglas. UT, 15 Jan 41; Boise, ID, 6 Jun 41-1 Jan 42.
77th Bombardment Squadron (15 Jan 41) Based at Salt Lake City, UT, 15 Jan 41? Boise, ID, 4 Jun-14 Dec 41; Elmendorf, AK, 29Dec 41-Jan 42.
390th Bombardment Squadron (20 Mar 42) Based at Gowen Field, ID, 20 Mar 42; McChord Field, WA, 21 Mar-1 Apr 42.
16th Reconnaissance Squadron (15 Jan 41) Based at Ft Douglas, UT, 15 Jan 41; Boise, ID, 4 Jun 41- Jan 42. Became the 406th Bombardment Squadron 22 Apr 42.

43rd Bombardment Group (15 Jan 41)
63rd Bombardment Squadron (15 Jan 41) Based at Langley Field, VA, 15 Jan 41; Bangor, ME. 28 Aug 41-17 Feb 42.
64th Bombardment Squadron (15 Jan 41) Based at Langley Field, VA, 15 Jan 41; Bangor, ME. 29 Aug 41-17 Feb 42.
13th Reconnaissance Squadron (15 Jan 41) Based at Langley Field, VA, 15 Jan 41; Bangor, ME, 29 Aug-18 Jan 42. Became the 403rd Bombardment Squadron 22 Apr 42.

44th Bombardment Group (15 Jan 41)
66th Bombardment Squadron (15 Jan 41) - Based at MacDill Field, FL, 15 Jan 41; Barksdale Field, LA, 9 Feb-30 Jun 42.

45th Bombardment Group (15 Jan 41)
78th Bombardment Squadron (15 Jan 41) Based at Savannah, GA, 15 Jan 41; Manchester, NH, 20 Jun 41; Langley Field, VA, 9 Apr 42; Jacksonville, FL, 16 May 42. Became the 7th Anti-submarine Squadron 29 Nov 42 and then the 851st Bombardment Squadron 27 Sep 43.
79th Bombardment Squadron (15 Jan 41) Based at Savannah, GA, 15 Jan 41; Manchester. NH, 19 Jun 41; Cherry Point, NC, 12 May 42? Miami, FL, 11 Sep 42. Became the 8th Anti-submarine Squadron 29 Nov 42 and then the 839th Bombardment Squadron 12 Oct 43.

48th Bombardment Group (15 Jan 41)
9th Reconnaissance Squadron (15 Jan 41) Based at Savannah, GA, 15 Jan-25 May 41. Became the 88th Bombardment Squadron 14 Aug 41, then the 88th Bombardment Squadron (Dive) 28 Aug 41 and at last the 495th Fighter-Bomber Squadron 10 Aug 43.

65th Observation Group (1 Sep 41)
121st Observation Squadron (10 Apr 41) Based at Washington, DC, 10 Apr 41; Owens field, SC, 23 Sep 41; Lexington County Apr, 8 Dec 41? Langley Field, VA, 26 Dec 41; Birmingham, AL, 18 Oct-30 Oct 42.

72nd Observation Group (26 Sep 41)
1st Observation Squadron (13 Jan 42) Based at Rio Hato, Panama, 14 Jan 42; Howard Field, CZ, 19 Jan 42; David, Panama, 17 Apr 42; Rio Hato, Panama, 10 May 42; Howard
Field, CZ, 20 Jun-7 May 44. Became the 1st Reconnaissance Squadron (Special) 25 Jun 43, Then the 41st Photographic Reconnaissance Squadron 25 Nov 44.
39th Reconnaissance Squadron (25 Jun 43) Based at Waller Field, Trinidad, 6 Aug 42-c. 26 Feb 44. Became the 101st Photographic Reconnaissance Squadron 12 Jun 44, then the 101st Bombardment Photographic Squadron 7 Feb 45.
108th Reconnaissance Squadron (25 Jun 43) Based at Howard Field, CZ, 25 Jun 43-1 Nov 43.

308th Bombardment Group (15 Apr 42)
373rd Bombardment Squadron (15 Apr 42) Based at Gowen Field, ID, 15 Apr 42; Davis-Monthan Field, AZ, 20 Jun 42; Alamogordo, NM, 23 Jul-31 Jul 42.
374th Bombardment Squadron (15 Apr 42) Based at Gowen Field. ID, 15 Apr 42; Davis-Monthan Field, AZ, 18 Jun 42, Alamogordo, NM, 24 Jul-31 Jul 42.
375th Bombardment Squadron (15 Apr 42) Based at Gowen Field, ID, 15 Apr 42; Davis-Monthan Field, AZ, 18 Jun 42; Alamogordo, NM, 24 Jul-31 Jul 42.
36th Reconnaissance Squadron (15 Apr 42) Based at Gowen Field, ID, 15 Apr 42; Davis-Monthan Field, AZ, 18 Jun 42; Alamogordo, NM, 24 Jul 42. Became the 425th
Bombardment Squadron 22 Apr 42.

342nd Composite Group (11 Sep 42)
There was one B-18 on Iceland used for anti-submarine patrols (c.11 Sep 42-18 Mar 44).

7th Naval District
21st Reconnaissance Squadron (1 Sep 36) Based at Miami Municipal Airport, FL, 9 Sep 39-22 Apr 42 {with B-16s). Became the 411th Bombardment Squadron 22 Apr 42.

Puerto Rican Department
27th Reconnaissance Squadron (16 Sep 39) Based at Langley Field, VA, 16 Sep-17 Nov 39; Borinquen Field, PR, 20 Nov 40.

Note: There were four B-18s evacuated from the Philippines during Dec 41. There were given Australian Civil Registration codes and used as Transports. One of the B-18s was shot down over Java (no date). The following are the units that the planes were assigned:
Air Transport Command US Army Forces in Australia (3 Apr 42); Became the 374th Troop Carrier Group (12 Nov 42).
21st Transport Squadron (3 Apr 42) - Based at Archerfield, Australia, 3 Apr 42. Became the 21st Troop Carrier Squadron 5 Jul 42.
22nd Transport Squadron (3 Apr 42) - Based at Essendon Airdrome, Australia, 3 Apr 42; Garbutt Field, Australia. 11 Oct 42. Became the 22nd Troop Carrier Squadron 5 July 42.

The B-18, B-18A, and B-18B Anti-submarine Wings, Groups, Squadrons

1st Sea-Search Attack Group
2nd Sea-Search Attack Squadron (17 June 42) Based at Langley Field. VA. 17 June 42-23 June 43. First designated squadron formed for Anti-submarine Warfare (ASW) with 10 B-18s (One "A" and Nine "Bs").
3rd Sea-Search Attack Squadron (10 Dec 42) Based at Langley Field, VA, 10 Dec 42-21 Nov 43. The two Squadrons tested electronic equipment and trained crews for its use in ASW.
Note: After World War II started, B-18s were used mostly for Anti-submarine duty and the Army Air Force formed the Anti-submarine Wing. Anti-submarine Commands were not assigned to Numbered Air Forces and Squadrons were not assigned to Groups. The following is the Squadrons that were assigned under the two Wings.
25th Anti-submarine Wing
2nd Anti-submarine Squadron (23 Nov 42) Based at Langley Field. VA. 18 Oct-26 Dec 42. Was the 523d Bombardment Squadron 18 Oct 42.
3rd Anti-submarine Squadron (29 Nov 42) Based at Westover Field, MA. 20 Jan 42; Dover, DE, 19 July 42-31 July 42. Was the 39th Bombardment Squadron 15 Jan 41; Became the 819th Bombardment Squadron 22 Sept 43.
4th Anti-submarine Squadron (29 Nov 42) Based at Westover Field. MA, 22 Jan 42; Langley Field, VA, 16 June 42; Mitchel Field, NY, 3 Aug 42-c. 6 June 43 (Operated from Guantanamo, Cuba, 30 Aug-Sept 42). Was the 40th Bombardment Squadron 15 Jan 41.
6th Anti-submarine Squadron (29 Nov 42) Based at Mitchel Field, NY, 22 Jan 42; Westover Field, MA, 3 Aug 42-1 Apr 43. Was the 3rd Reconnaissance Squadron 20 Nov 42 and redesignated the 393rd Bombardment Squadron on 22 Apr 42.
12th Anti-submarine Squadron (29 Nov 42) Based at Langley Field, VA, 12 Jan-18 Sept 43. Was the 517th Bombardment Squadron 18 Oct 42, became the 492d Bombardment Squadron 24 Sept. 43.
17th Anti-submarine Squadron (29 Nov 42) Based at Lantana, FL. 18 Oct 42; Boca Chica, FL, 9 Jan 43; Miami, FL, 3 July 43-15 Aug 43. Was the 522nd Bombardment Squadron 18 Oct 42.
46th Bombardment Squadron (15 Jan 41) Based at Cherry Point NAS, NC, 28 Aug 42-2 Mar 43. Became the 22nd Anti-submarine Squadron 3 Mar 43.
362nd Bombardment Squadron (15 July 42) Based at Salt Lake City AAB. UT, 15 July 42; Geiger Field, WA, 15 Sept 42; Ephrata, WA, 1 Oct 42; Langley Field, VA. 29 Oct 42-1 Oct 43. Became the 18th Anti-submarine Squadron 29 Nov 42, then became the 4th Sea-Search Attack Squadron 23 Oct 42. This was the Operational Training unit for the 25th ASW

26th Anti-submarine Wing
7th Anti-submarine Squadron (29 Nov 42) Based at Jacksonville, FL, 29 Nov 42; Edinburgh Field Trinidad, 20 Apr 43-20 July 43. Was the 78th Bombardment Squadron 15 Jan 41, Became the 851st Bombardment Squadron 27 Sept 43.
8th Anti-submarine Squadron (29 Nov 42) Based at Miami, FL, 11 Sept 42; operated from Trinidad and other bases in the area, Jul-28 Aug 43. Was the 79th Bombardment Squadron 15 Jan 41. Became the 839th Bombardment Squadron 14 Oct 43.
9th Anti-submarine Squadron (29 Nov 42) Based at Edinburg Field, Trinidad, Nov 42-Mar 43. Was the 80th Bombardment Squadron 15 Jan 41. Became the 835th Bombardment Squadron 23 Sept 43.
(**Note**: Much of the preceding work on the disposition of the B-18s was from unpublished material by Robert Lumpkin at the PASM and the Maurer & Maurer books on Combat Squadrons and Combat Units of the AAF).

Appendix B: B-18 Crashes and Crash Sites in the U.S.

10 September 1941, Mt. Constance, WA
B-18 (37-518) Crashed into Mt. Constance, WA, killing all six crewmen.

24 October 1941, Twin Sisters Peak, CA
Capt. F.C. Nelson and his crew of four left Hamilton Field, CA in 37-498 and climbed into partly cloudy Northern California skies headed to McClellan Field, near Sacramento enroute to Fort Douglas, UT. Nelson and his co-pilot 2Lt. E.W. Sell were instructed to fly around the low-laying clouds over the Vaca Mountains that were directly in the flight path on the short 30 minute flight. Nelson appears to have been on instruments, unaware of the Twin Sister peaks in his path and the B-18 ploughed directly into the south peak, killing all aboard.

12 December 1941, Sierra Nevada Mountains, CA
Probably the most significant crash of a B-18 occurred when Maj.Gen. Herbert Dargue, CO of the First Air Force at Mitchel Field, NY, was killed while flying a B-18 that crashed enroute to Hamilton Field, CA. Dargue and a large number of air staff and vitally needed crew chiefs aboard was headed for post Pearl Harbor Hawaii to aid in the investigation of the lack of preparedness at Hawaii. At the same time Gen. Ira Eaker, Dargue's old friend from the 1926 Pan American Goodwill Flight, was flying in from San Diego and the further north he flew the worse weather became and he radioed Dargue, whose aircraft was heading into California from Arizona, warning him of the deteriorating weather, high winds, and heavy fog. Eaker continued to fly north but finally had to land at Fresno and take a bus to San Francisco. When he arrived Eaker received a message informing him that Dargue's bomber was overdue. The wreckage of the aircraft was finally found in the Sierras near Bishop, CA in March 1942. Dargue received a posthumous Distinguished Service Medal for his lifelong military service.

14 January 1942, White Mountains, NH
The doomed B-18 piloted by 1Lt. Anthony Benevenuto, left Westover, AFB at Chicopee, MA, and after its anti-submarine patrol off Newfoundland it returned in poor weather toward the coast. The navigator misidentified Concord, NH, for Providence, RI, and then plotted a course that led the bomber toward New Hampshire's White Mountains As the bomber neared Woodstock, NH, icing and poor visibility caused the pilot to crash into the side of Mt. Wateromee at 2,700 feet. Benevenuto managed to prevent a head-on crash and the bomber skidded along the side of the mountain before the bomb load exploded. Five of the seven crewmembers, including the pilot, survived as a rescue party slogged through heavy winter snow to save them.

3 April 1942, Albuquerque, NM
A B-18A crashed after both engines quit 25 miles south of Albuquerque during a bombardier training flight from Kirtland Field.

2 October 1942, Flagstaff, AZ
1Lt. Arnold King and his crew were flying a B-18A (37-515) from the Sacramento Air Depot to transport equipment to Albuquerque, NM, via Bakersfield and Daggett, CA. At 9,000 feet the aircraft encountered unforecasted rain, snow, and high winds that caused turbulence and icing and cut visibility to nil. King was off course and hit the tops of trees on the slopes of the San Francisco Peaks which rise to over 12,000 feet north of Flagstaff.

22 August 1944
A B-18 crashed near Watkins, CO in poor weather.

Bibliography

Books

Andrade, John, *US Military Aircraft Designations and Serials since 1909*, Midland, UK, 1979.

Archer, Robert & Victor, *USAAF Aircraft Markings & Camouflage: 1941-47*, Schiffer, PA, 1997.

Auphan, Paul & Jacques, Mordal, *French Navy in World War II*, Naval Institute Press, MD, 1959.

Avery, Norm, *North American Aircraft 1934-1998, Vol. 1*, Narkiewicz/Thompson, CA, 1998.

Arakaki, Leatrice, & Kuborn, John, *7 December 1941: The Air Force Story*, Pacific Air Forces, Hawaii, 1991.

Bartsch, William, *Doomed at the Start*, Texas A&M Press, TX, 1992.

Baxter, James, *Scientists Against Time*, MIT Press, MA, 1946.

Bell, Dana, *Air Force Colors: Vol.1 1926-42*, Squadron/Signal, TX, 1979.

Bell, Dana, *Air Force Colors: Vol.2 ETO & MTO 1942-45*, Squadron/Signal, TX, 1980.

Blair, Clay, *U-Boat War: The Hunters, 1939-42*, Random House, NY, 1996.

Bodie, Warren, *American War Eagles 1937-42*, Widewing, NC, 2001.

Bowman, Martin, *B-24 Liberator 1939-45*, Wensum, UK, 1979.

Bowman, Martin, *USAAF 1926-1942*, Sutton, UK, 2000.

Bowers, Peter, *Fortress in the Sky*, Sentry, CA, 1976.

Breuer, William, *Geronimo!*, St. Matins, NY, 1982.

Campbell, John, *Naval Weapons of World War Two*, Naval Institute Press, MD, 1985.

Carey, Alan, *Galloping Ghosts of the Brazilian Coast*, iUniversal, NE, 2004.

Chesneau, Roger, *Aircraft Carriers of the World, 1914 to Present*, Arms & Armor Press, UK, 1984.

Christman, Albert, *Sailors, Scientists, and Rockets*, Naval History Division, Wash. DC, 1971.

Cohen, Stan, *Forgotten War Vol.1*, Pictorial Histories, MT, 1981.

Cohen, Stan, *Forgotten War Vol.2*, Pictorial Histories, MT, 1983.

Cloe, John, *Aleutian Warriors: Part 1*, Pictorial Histories, MT, 1990.

Cloe, John, *Top Cover for America*, Pictorial Histories, MT, 1986.

Conn, Stetson et al, *Guarding the United States and its Outposts*, Center of Military History, Wash. DC, 1964.

Copp, DeWitt, *A Few Great Captains: Men and Events that Shaped the Development of US Air Power*, Doubleday, NY, 1980.

Copp, DeWitt,
Forged in Fire: Strategy and Decisions in the Air War Over Europe 1940-45, Doubleday, NY, 1982.

Couhat, Jean, *French Warships of World War II*, Ian Allen, UK, 1971.

Cox, Gary, *Beyond the Battle Line: US Air Attack Theory and Doctrine 1919-1941*, Air University Press, AL, 1996.

Craven, W.F. & Cate, J.L., *Army Air Forces in World War II, Vol.1: Planes and Early Operations*, U. of Chicago Press, IL, 1948.

Davis, Larry, *C-47 Skytrain in Action*, Squadron/Signal, TX. 1995.

Day, Charles, *Silent Ones*, Private Printing, MI, 2004.

Devlin, Gerard, *Paratrooper!*, St. Martins, NY, 1979.

Dolan, Col. W.C., *History of the First Sea-Search Attack Group*, Langley Field, VA, 1943.

Dorr, Robert, *7th Bombardment Group/Wing: 1918-1995*, Turner, KY, 1996.

Dorr, Robert, *US Bombers of World War II*, Arms & Armour, UK, 1989.

40th Bomb Group Association, *40th Bombardment Group (VH) History*, Turner, KY, 1989.

11th Bomb Group, *Grey Geese Calling: Pacific Air War History of the 11th BG (H)*, American Publishing, MN, 1981.

Flanagan, E.M., *Combat History of the American Airborne Forces*, Ballantine, NY, 2002.

Franks, Norman, *Conflict over the Bay*, Kimber, UK, 1986.

Freeman, Roger & Osbourne, David, *B-17 Flying Fortress Story*, Arms & Armour, UK, 1998.

Gannon, Michael, *Operation Drumbeat*, Harper & Row, NY, 1990.

Garfield, Brian, *The Thousand Mile War*, Ballantine, NY, 1969.

Gregory, Barry, *US Airborne Forces*, Gullen Books, NY, 1990.

Gilman, William, *Our Hidden Front*, Reynal & Hitchcock, NY, 1945.

Glines, Carroll, *The Doolittle Raid*, Orion, NY, 1990.

Hagedorn, Dan, *Alae Supra Canalem: Wings Over the Canal*, Turner, KY, 1995.

Hartcup, Guy, *The Challenge of War*, David & Charles, UK, 1970.

Henry, John, *A History of the 38th Bombardment Group (M)*, Aerospace Historian, KS, 1978.

Hobbs, Peter, "Rain Man, David Laskin," *Smithsonian Magazine*, October, 2002.

Holly, Irving, *Buying Aircraft: Materiel Procurement for the Army Air Forces*, Office of the Chief of Military History, Wash., DC, 1964.

Hoyt, Edwin, *Airborne: History of the American Parachute Forces*, Stein & Day, 1979.

Ingells, Douglas, *The Plane that Changed the World*, Aero, CA, 1966.

Jane, Fred, *Jane's All The World's Aircraft 1938*, Arco, NY, reprint.

Jones, Lloyd, *US Bombers*, Aero, CA, 1974.

Kelsey, Benjamin, *Dragon's Teeth: Creation of US Air Power for World War II*, Smithsonian, Wash. DC, 1982.

Kemp, Paul, *U-Boats Destroyed*, Arms & Armour, UK, 1999.

Koberger, Charles, *The Cyrano Fleet: France and Its Navy 1940-42*, Praeger, NY, 1989.

Kostenik, Samuel & Griffin, John, *RCAF Squadrons and Aircraft*, Hakkert, Canada, 1977.

Larkins, William, *Surplus WW-II US Aircraft*, BAC, CA, 2006.

LeMay, Curtis & Kantor, McKinley, *Mission with LeMay: My Story*, Doubleday, NY, 1965.

Lent, Henry, *Bombardier*, MacMillan, NY, 1944.

Lloyd, Alwyn & Moore, Terry, *B-17 Flying Fortress in Scale & Detail: Part 1*, Aero, UK, 1981.

Manvell, Roger, *Films of the Second World War*, Dent, UK, 1974.

Maurer, Maurer, *Aviation in the US Army: 1918-1939*, Office of Air Force History, Wash. D.C., 1987.

Maurer, Maurer, *Combat Squadrons of the Air Force*, USAF Historical Division, Wash. DC, 1969.

Maurer, Maurer, *Air Force Combat Units of World War II*, US Government Printing Office, Wash. DC, 1961.

McFarland, Stephen, *America's Pursuit of Precision Bombing, 1910-1945*, Smithsonian, Wash, DC, 1995.

Milberry, Larry & Hugh, *Royal Canadian Air Force at War 1939-45*, CANAV, Canada, 1990.

Mitchell, John, *On Wings We Conquer*, Pvt. Printing, MO., 1990.

Mizrahi, Joseph, *Air Corps*, Sentry Books, CA, 1970.

Morrison, Wilbur, *Donald W. Douglas: A Heart with Wings*, Iowa State University Press, IA, 1991.

Morton, Louis, *US Army in World War 2: War in the Pacific*, Office of the Chief of Military History, Wash. DC, 1953.

Munson, Kenneth, *Airliners Between the Wars: 1919-39*, MacMillan, NY, 1972.

Nelson, Craig, *First Heroes*, Viking, NY, 2002.
Orriss, Bruce, *When Hollywood Ruled the Skies*, Aero Associates, CA, 1985.
Pardini, Albert, *Legendary Secret Norden Bombsight*, Schiffer, PA, 1999.
Parrant, Erik (ed.), *Defenders of Liberty: 2nd Bombardment Group/Wing: 1918-1993*, Turner, KY, 1996.
Paterson, Lawrence, *First U-Boat Flotilla*, Leo Cooper, UK. 2002.
Paterson, Lawrence, *Second U-Boat Flotilla*, Leo Cooper, UK. 2003.
Pearcy, Arthur, *Dakota at War*, Ian Allan, UKL, 1982.
Pearcy, Arthur, *DC-3*, Ballantine, NY, 1975.
Price, Alfred, *Aggressors: Patrol Aircraft vs. Submarine*, Howell, VA, ND.
Price, Alfred, *Aircraft vs. Submarines*, William Kimber, UK, 1973.
Rathbone, A.D., *He's in the Paratroops Now*, McBride, NY, 1943.
Rossler, Eberhard, *The U-Boat*, Cassell & Son, UK, 1998.
Savas, Theodore, *Silent Hunters*, Naval Institute Press, MD, 1997.
Schoenfeld, Max, *Stalking the U-Boat*, Smithsonian, Wash. DC, 1995.
Seventh Air Force, *Operational History of the Seventh Air Force, 7 December 1941 to 6 December 1943*, Maxwell AFB, AL.
Schiller, Bill (ed.), *19th Bomb Group*, Turner, KY, 2000.
Sharpe, Peter, *U-Boat Fact File*, Midland, UK, 1998.
Shore, Louis, *Highways in the Sky: The Story of the AASC*, Barnes and Noble, NY, 1947.
Smith, Myron, *Passenger Airliners of the United States: 1926-1986*, A Pictorial History, Pictorial Histories, MT, 1986.
Steinbeck, John, *Bombs Away!*, Viking, NY, 1942.
Syrett, David, *The Defeat of the German U-Boats*, U. of South Carolina, SC, 1994.
Terrett, Dulany, *Signal Corps: The Emergency*, Office of the Chief of Military History, Wash. DC, 1956.
Thompson, George, et al, *Signal Corps: The Test*, Office of the Chief of Military History, Wash. DC, 1957.
USAAF, *History of the US Air Force and Air Corps in the Caribbean 1940-45*.
USAAF, *History of the US Air Force and Air Corps in the Trinidad Sector 1940-45*
VI Bomber Command, *VI Bomber Command in Defense of the Panama Canal, 1941-45*, VI Bomber Command, ND.
Vincent, Carl, *Canada's Wings 2*, Canada's Wings, Toronto, 1975.
Warnock, A. Timothy, *Air Power versus U-Boats*, Air Force History and Museums Program, Wash., DC, 1999.
Wilkinson, Paul, *Aircraft Engines of the World, 1945*, Wilkinson, NY, 1945.
Wolf, William, *America's First Smart Bombs: Air-to-Surface Guided Missiles in WW-II*, Unpublished manuscript.
Wolf, William, *Boeing B-29 Superfortress: The Ultimate Look from Drawing Board to VJ-Day*, Schiffer, PA, 2005.
Wolf, William, *Consolidated B-32 Dominator: The Ultimate Look from Drawing Board to Scrap Yard*, Schiffer, PA, 2006.
Wolf, William, *Victory Roll: The American Fighter Pilot and Aircraft in World War II*, Schiffer, PA, 2003.
Wynn, Kenneth, *U-Boat Operations of the Second World War, Vol.2*, NIP, MD., 1998.
Yenne, Bill, *McDonnell Douglas: A Tale of Two Giants*, Crescent, NY, 1985.

Magazines, Booklets, Studies, and Official Correspondence

AAFAC: Anti-submarine Manual SOP of 29 July 1943.
AAFAC: *Antisubmarine Monthly Intelligence Report,* August 1943.
Aeroplane, "Masterly Martin," *Aeroplane*, Vol. 5, No. 2.
Air Corps Newsletter, various articles February 1937-May 1939.
Air Classics, "Bolo for the Airborne," *Air Classics*, July 1985.
Annis, A/M Clare, "Eastern Air Command Recalled," *Canadian Aviation Historical Society Journal,* Canada, 1979.
Aviation, "Send It to the Depot," *Aviation*. December 1940.
The Bomber, "Propellers Whirling Today," 23 January 1940.
Bowers, Peter, "Reluctant Dragon," *Airpower*, Vol. 10, No. 6.
Boyne, Walt, "B-10 Baltimore's Best," *Airpower*, Vol. 1, No.3.
Biddle, Wayne, "How America Eagerly Built Her Arsenal," *APF Reporter*, Vol. 11, No. 3.
Cronenberg, Allen, "U-Boats in the Gulf: The Undersea War in 1942," *Gulf Coast Historical Review*, Vol. 5, No. 2.
Close, Winton, "B-18-A Reminiscence," *Aerospace Historian*, Summer 1982.
Frazer, Charles, "Night Mission over the Caribbean," *Air Force Magazine*, April, 1943.
Greenway, Sean, "Donald Douglas & the Gooney Bird," *Air Classics Review IV*, 1993.
Dean, Jack, "The Depression Bomber," *Wings*, Vol. 22, No. 4.
Dorr, Robert, "A Bolo, Up Close and Personal," *Air Force Times*, September 18, 2000.
Enyedt, "Douglas' Reluctant Warriors," *Air Classics*, Vol.3, No.4.
First Sea Search Attack Group, "History of the First Sea Search Attack Group," October 1943.
Goldsworthy, Harry, "Whale Bomber's Respite," *Aerospace Historian*, summer, 1984.
Goodall, Geoff, "Bolos Down Under," *Air Classics,* Vol.17, No. 10.
Guess, Joseph, "6th Air Force," *Air Force Magazine*, October, 1945.
Hagedorn, Daniel, "Alae Supra Canalem," *American Aviation Historical Society Journal,* Spring, 1974.
Hagedorn, Daniel, "Sentimental Journey: The Airfields of the Sixth Air Force: Albrook Field," *American Aviation Historical Society Journal*, Spring, 2005.
Hagedorn, Daniel, "Sentimental Journey: The Airfields of the Sixth Air Force: France Field," *American Aviation Historical Society Journal*, Summer, 2003.
Jesse, William, "Short-Lived Dragon," *Air Enthusiast*, May/June 1999.
Johanson, Robert, "Night Bombardment," *Flying & Popular Aviation*, February 1941.
Johnson, Frederick, "The Great Bolo Boondoggle," *Wings*, Vol. 3, No. 4.
Joubert, AM J.B., Letter to Lord Halifax, 16 August 1942.
King, Adm. Ernest, Letter to Gen. George C. Marshall, 21 June 1942.
Larson, George, "The Douglas B-18," *American Aviation Historical Society Journal*, Spring 2001.
Larson, George, "Douglas B-23 Dragon," *AAHS Journal*, Winter, 2004.
Leary, Michael, "Those Mexican Bolos," *Warbirds International,* May/June 1991.
Ludkiewicz, William, "Airlines: Letters to the Editor," *Air Classics*, February 1981.
Lumpkin, Robert ("Lumpy"), "B-18, B-18A, and B18B Groups and Squadrons between 1936-1945," unpublished study, Pima Air and Space Museum, 1995.
Lumpkin, Robert ("Lumpy"), "Anti-submarine Wings, Groups, and Squadrons using the B-18, B-18A, and B18B," unpublished study. Pima Air and Space Museum, 1995.
Lumpkin, Robert ("Lumpy"), "B-18, B-18A, and B-18B Groups and Squadrons between 1936-1945," unpublished study, Pima Air and Space Museum, 1995.
Lumpkin, Robert ("Lumpy"), "B-18, B-18A, and B18B Wings, Commands, and Air Forces between 1936-1945," unpublished study, Pima Air and Space Museum, 1995.
Marshall, Charles, "Television Equipment for Guided Missiles," *InterAvia*, January, 1947.
Marshall, Gen. George, Letter to Adm. Ernest King, 14 September 1942.
Meilinger, Phillip, "When the Fortress Went Down," *Air Force: Journal of the Air Force Association,* October, 2004.
Merriam, Ray, "Douglas B-18 Bolo," *World War II Journal #15*.
Mitchell, John, "Flying Fortresses of the 19th Bomb Group," AAHS Journal, Winter, 1984.
Mitchell, William, "Panama is Defenseless," *Popular Aviation*, Sept 1929.
Neto, Roberto, "Brazil-1, U-199-0," *Air Enthusiast*, Winter, 1994.
O'Leary, Michael, Dragon's Last Flight," *Air Classics*, April 1986.
O'Leary, Michael, "A Dragon Bows Out," *Aeroplane Monthly*, April 1988.

Parker, King, "Canal Zone-World War II," *Aerospace Historian*, Winter, 1986.
Perna, Lee, "People & Planes," *Aviation History*, November 2005.
Puffer, Raymond, "Bombs with Wings," Air Force Flight Center, Wash. D.C., No Date. Roberts, Lt.Col. Jack, "Radar Performance Report," March 1943.
Salisbury, Winfield, "SCR-584 Jamming, *SMEC Vintage Electrics*, Vol. 2, 1990.
Skywriting, "B-18 Update," *Skywriting*, Spring 1994.
Slater, Richard, "And Then There Were None!: The American Army Air Corps' Last Stand in the Philippines," *Air Power*, November 1987.
Stimson, Henry, Memorandum of 7 July 1942 to Secretary of the Navy Frank Knox.
Thompson, Jim, "The Air Corps' Forgotten Bomber," *International Air Review*, Fall 1991.
Trimble, Robert, "Bolo: America's Forgotten Warrior," *Air Classics*, Vol. 16, No. 6.
Unknown author, "Borinquen Field, Early History: 1936-41," Research Study, ND.
Vincent, Carl, "North Atlantic Squadron," *Air Extra*, No. 8.
Vincent, Carl, "Distended Douglas," *High Flight*, Vol. 2, No. 5 & 6, Vol. 3, No. 1 & 2.
Wright, Roger, "Birth of a Legend: History of the DC-3," *Air Extra*, June 1985.

Manuals and Technical Reports

USAAC, *Handbook of Service Instructions for the Model B-18 Bombardment Airplanes; TO No. 01-40EA-1*, Air Materiel Command, March 1938.
USAAC, *Handbook of Operation and Flight Instructions for the Model B-18 Bombardment Airplanes, TO No. 01-40E-3*, Air Materiel Command, March 1938
USAAC, *Handbook of Service Instructions for the Model B-18A Bombardment Airplanes; TO No. 01-40E-2*, Air Materiel Command, February 1939.
USAAC, *Handbook of Operation and Flight Instructions for the Model B-18 Bombardment Airplanes; TO No. 01-40EB-1*, Air Materiel Command, July 1940.
USAAC, *Handbook of Operation and Flight Instructions for the Model B-18A Bombardment Airplanes; TO No. 01-40EB-1*, Air Materiel Command, November 1940.
USAAC, *Handbook of Operation and Flight Instructions for the Model B-23 Bombardment Airplanes*, Air Materiel Division, OH, 20 July, 1940.
USAAF, *Antisubmarine Command*, AAFRH No. 7, April, 1945.
USAAF, *History of the 75mm Gun M5A1 with Mount M9*, 1945.
USAAF, *History of the 75mm Gun M4 with Mount M6*, 1944.
USAAF, *Preliminary Handbook of Service Instructions for the Model B-23 Bombardment Airplane*, Air Service Command, OH, 15 October 1940.

Websites

www.aerofiles.com
www.aerothentic.com , "B-18 *Damfino*"
www.boeing.com/history, "Donald Wills Douglas Sr.,"
www.fas.org, "Magnetic Anomaly Detection (MAD)"
www.filmsite.org, Various movie postings
www.40thbombgroup.org, Matthews, Ira, Eighty-one War Stories: "France Field Patrol"
www.home.att.net/~jbaugher, Baugher, Joe, Various postings
www.website.lineone.net, Norman Groom, "Air to Surface Vessel radar ASV MkII"
www.mat.ufrgs.br "History of the Brazilian Air Force"
www.navsource.org "Location of USAAF December 7, 1941"
www.secesh.net/b23, B-23 Bomber Crash Site and Wrecks
www.tighar.org, *Lady of the Lake*
www.turnerclassicmovies.com, 49th Parallel
www.uboat.net, "Fighting the U-Boats"
www.uboat.net, Gustin, Emmanuel, "British ASV Radar"
www.uboat.net, Gustin, Emmanuel, "The Leigh Light"
www.uboat.net, Gustin, Emmanuel, "U-Boat Radar Detectors"
www.USAAF.net/chron "Combat Chronology of the US Army Air Forces in WW-2"
www.warbirdsresourcegroup.org
www.uswarplanes.com
www.zoe73.net, B-23 after the war

Index

Note: Pages in BOLD indicate photo of subject
Separate Digby and B-23 indices follow this main index

B-23 Index

Other Books by William Wolf

13th Fighter Command in World War II
Death Rattlers: VMF-323 over Okinawa
American Fighter-Bombers in World War II
Boeing B-29 Superfortress: The Ultimate Look
Consolidated B-32 Dominator: The Ultimate Look